# Lecture Notes in Networks and Systems

## Volume 595

The series "Lecture Notes in Networks and Systems" publishes the latest developments in Networks and Systems—quickly, informally and with high quality. Original research reported in proceedings and post-proceedings represents the core of LNNS.

Volumes published in LNNS embrace all aspects and subfields of, as well as new challenges in, Networks and Systems.

The series contains proceedings and edited volumes in systems and networks, spanning the areas of Cyber-Physical Systems, Autonomous Systems, Sensor Networks, Control Systems, Energy Systems, Automotive Systems, Biological Systems, Vehicular Networking and Connected Vehicles, Aerospace Systems, Automation, Manufacturing, Smart Grids, Nonlinear Systems, Power Systems, Robotics, Social Systems, Economic Systems and other. Of particular value to both the contributors and the readership are the short publication timeframe and the world-wide distribution and exposure which enable both a wide and rapid dissemination of research output.

The series covers the theory, applications, and perspectives on the state of the art and future developments relevant to systems and networks, decision making, control, complex processes and related areas, as embedded in the fields of interdisciplinary and applied sciences, engineering, computer science, physics, economics, social, and life sciences, as well as the paradigms and methodologies behind them.

Indexed by SCOPUS, INSPEC, WTI Frankfurt eG, zbMATH, SCImago.

All books published in the series are submitted for consideration in Web of Science.

For proposals from Asia please contact Aninda Bose (aninda.bose@springer.com).

Javier Prieto · Francisco Luis Benítez Martínez ·
Stefano Ferretti · David Arroyo Guardeño ·
Pedro Tomás Nevado-Batalla
Editors

# Blockchain and Applications, 4th International Congress

Springer

*Editors*
Javier Prieto
Departamento de Informática y Automática
University of Salamanca
Salamanca, Spain

Francisco Luis Benítez Martínez
Parque Tecnológico de la Salud (PTS)
Granada, Spain

David Arroyo Guardeño
Spanish National Research Council (CSIC)
Madrid, Spain

Stefano Ferretti
University of Urbino Carlo Bo
Urbino, Italy

Pedro Tomás Nevado-Batalla
Facultad de Derecho
University of Salamanca
Salamanca, Spain

ISSN 2367-3370          ISSN 2367-3389  (electronic)
Lecture Notes in Networks and Systems
ISBN 978-3-031-21228-4        ISBN 978-3-031-21229-1  (eBook)
https://doi.org/10.1007/978-3-031-21229-1

This Springer imprint is published by the registered company Springer Nature Switzerland AG
The registered company address is: Gewerbestrasse 11, 6330 Cham, Switzerland

# Organization

## General Chair

Javier Prieto Tejedor   University of Salamanca, (Spain) and AIR Institute, Spain

## Advisory Board

Abdelhakim Hafid    Université de Montréal, Canada
Ashok Kumar Das    IIIT Hyderabad, India
António Pinto        Instituto Politécnico do Porto, Portugal
Paulo Leitao         Technical Institute of Bragança, Portugal

## Program Committee Chair

Francisco Luis Benitez Martínez   Fidesol, Spain

## Local Chair

Stefano Ferretti   University of Bologna, Italy

## Organizing Committee Chair

Pedro Nevado   University of Salamanca, Spain

## Local Organizing Committee

| | |
|---|---|
| Pierpaolo Vittorini (Co-chair) | University of L'Aquila, Italy |
| Tania Di Mascio (Co-chair) | University of L'Aquila, Italy |
| Federica Caruso | University of L'Aquila, Italy |
| Anna Maria Angelone | University of L'Aquila, Italy |

## Organizing Committee

| | |
|---|---|
| Juan M. Corchado Rodríguez | University of Salamanca, Spain, AIR Institute, Spain |
| Fernando De la Prieta | University of Salamanca, Spain |
| Sara Rodríguez González | University of Salamanca, Spain |
| Javier Prieto Tejedor | University of Salamanca, Spain, AIR Institute, Spain |
| Pablo Chamoso Santos | University of Salamanca, Spain |
| Liliana Durón | University of Salamanca, Spain |
| Belén Pérez Lancho | University of Salamanca, Spain |
| Ana Belén Gil González | University of Salamanca, Spain |
| Ana De Luis Reboredo | University of Salamanca, Spain |
| Angélica González Arrieta | University of Salamanca, Spain |
| Emilio S. Corchado Rodríguez | University of Salamanca, Spain |
| Alfonso González Briones | University of Salamanca, Spain |
| Yeray Mezquita Martín | University of Salamanca, Spain |
| Beatriz Bellido | University of Salamanca, Spain |
| María Alonso | University of Salamanca, Spain |
| Sergio Marquez | University of Salamanca, Spain |
| Marta Plaza Hernández | University of Salamanca, Spain |
| Guillermo Hernández González | AIR Institute, Spain |
| Ricardo S. Alonso Rincón | University of Salamanca, Spain |
| Raúl López | University of Salamanca, Spain |
| Sergio Alonso | University of Salamanca, Spain |
| Andrea Gil | University of Salamanca, Spain |
| Javier Parra | University of Salamanca, Spain |

## Program Committee

| | |
|---|---|
| Mahmoud Abbasi | University of Salamanca, Salamanca |
| Ermyas Abebe | ConsenSys R&D, USA |
| Iván Abellán | Outpost24, Sweden |

Mo Adda                         University of Portsmouth, UK
Imtiaz Ahmad Akhtar             XLENT Link, Sweden
Sami Albouq                     Islamic University of Madenah, Saudi Arabia
Ricardo S. Alonso               AIR Institute, Valladolid
Onur Ascigil                    University College London, UK
Anusha Avyukt                   University of Southern California, USA
Syed Badruddoja                 University of North Texas, USA
Richard Banach                  The University of Manchester, UK
Aritra Banerjee                 Trinity College Dublin, Ireland
Badr Bellaj                     Mchain, Morocco
Yahya Benkaouz                  Mohammed V University in Rabat, Morocco
José Vicente Berná Martínez     University of Alicante—Computer Science
                                Department, Spain
Cyrille Bertelle                Le Havre University, France
Andrea Bondavalli               University of Florence, Italy
Carlos Bordons                  University of Seville, Spain
William J. Buchanan             Napier University, UK
Arnaud Castelltort              Université de Montpellier, France
Bishakh Chandra Ghosh           Indian Institute of Technology Kharagpur, India
Sang-Yoon Chang                 University of Colorado Colorado Springs, USA
Mohammad Jabed Morshed          La Trobe University, Australia
Chowdhury
Giovanni Ciatto                 University of Bologna, Italy
Victor Cook                     University of Central Florida, USA
Manuel E. Correia               CRACS/INESC TEC; DCC/FCUP, Portugal
Gaby G. Dagher                  Boise State University, USA
Sankarshan Damle                IIIT Hyderabad, India
Ashok Kumar Das                 International Institute of Information Technology,
                                India
Giovanni De Gasperis            DISIM, Università degli Studi dell'Aquila, Italy
Josep Lluis De La Rosa          TECNIO Centre EASY Innovation, UdG, Spain
Volkan Dedeoglu                 CSIRO, Australia
Roberto Di Pietro               Hamad Bin Khalifa University—College of
                                Science and Engineering, Qatar
Ba-Lam Do                       Hanoi University of Science and Technology,
                                Vietnam
Katerina Doka                   National Technical University of Athens, Greece
Claude Duvallet                 LITIS—Université Le Havre Normandie, France
Joshua Ellul                    University of Malta, Malta
Sante Dino Facchini             Università degli Studi dell'Aquila, Italy
Tooba Faisal                    King's College London, UK
Wenjun Fan                      Xi'an Jiaotong-Liverpool University, China
Xinxin Fan                      IoTeX, USA
Manuel J. Fernandez             University of Seville, Spain

| | |
|---|---|
| Christof Ferreira Torres | University of Luxembourg, Luxembourg |
| Ernestas Filatovas | Vilnius University, Lithuania |
| Nikos Fotiou | AUEB, Greece |
| Miguel Frade | Instituto Politécnico de Leiria, Portugal |
| Paula Fraga-Lamas | University of A Coruña, CITIC, Spain |
| Muriel Franco | University of Zurich, Switzerland |
| Felix Freitag | Universitat Politècnica de Catalunya, Spain |
| Shahin Gheitanchi | IEEE, UK |
| Radu Godina | NOVA University Lisbon, Portugal |
| Seep Goel | IBM Indian Research Labs, India |
| Mongetro Goint | Université Le Havre Normandie, France |
| Hélder Gomes | Escola Superior de Tecnologia e Gestão de Águeda, Universidade de Aveiro, Portugal |
| Dhrubajyoti Goswami | Concordia University, Canada |
| Volker Gruhn | Universität Duisburg-Essen, Germany |
| Susana Gutiérrez | Fundación CARTIF, Spain |
| Abdelatif Hafid | University of Montreal, Canada |
| Christopher G. Harris | University of Northern Colorado, USA |
| Yahya Hassanzadeh-Nazarabadi | Ferdowsi University of Mashhad, Turkey |
| Pham Hoai Luan | NAIST, Japan |
| Tim Hoiss | Universität der Bundeswehr München, Germany |
| Hsiang-Jen Hong | University of Colorado Colorado Springs, USA |
| Qin Hu | George Washington University, USA |
| Yining Hu | IBM Australia, Australia |
| Maria Visitación Hurtado | University of Granada, Spain |
| Shahid Hussain | National University of Ireland, Ireland |
| Fredrick Ishengoma | The University of Dodoma, Tanzania |
| Hans-Arno Jacobsen | University of Toronto, Canada |
| Marc Jansen | University of Applied Sciences Ruhr West, Germany |
| Zakwan Jaroucheh | Edinburgh Napier University, UK |
| Eder John Scheid | University of Zurich, Switzerland |
| Christina Joseph | National Institute of Technology Karnataka, India |
| Taeho Jung | University of Notre Dame, USA |
| Raja Jurdak | QUT, Australia |
| K. Chandrasekaran | NITK, India |
| Ayten Betul Kahya | University of Southern California, USA |
| Dimitris Karakostas | University of Edinburgh, UK |
| Christos Karapapas | Athens University of Economics and Business, Greece |
| Samuel Karumba | UNSW, Australia |
| Shoji Kasahara | Nara Institute of Science and Technology, Japan |
| Latifur Khan | UTD, USA |
| Nectarios Koziris | National Technical University of Athens, Greece |

Benfano Soewito               Bina Nusantara University, Indonesia
Mark Staples                  CSIRO, Australia
Denis Stefanescu              Ikerlan, Spain
Marko Suvajdzic               University of Florida, USA
Stefan Tai                    TU Berlin, Germany

Chamseddine Talhi             École de Technologie Supérieure, Canada
Teik Guan Tan                 Singapore University of Technology and Design,
                              Singapore
Keisuke Tanaka                Tokyo Institute of Technology, Japan
Subhasis Thakur               National University of Ireland, Galway, Ireland
Llanos Tobarra                UNED, Spain
Natkamon Tovanich             IRT SystemX, France
Tuan Tran                     CMPS232, USA
Florian Tschorsch             TU Berlin, Germany
Kritagya Upadhyay             University of North Texas, USA
Aitor Urbieta                 IK4-Ikerlan Technology Research Centre, Spain
Julita Vassileva              University of Saskatchewan, Canada
Massimo Vecchio               FBK, Spain
Andreas Veneris               University of Toronto, Canada
Paulo Vieira                  Insituto Politécnico da Guarda, Portugal
Luigi Vigneri                 IOTA Foundation, Germany
Marco Vitale                  Foodchain Spa, Italy
Chenggang Wang                University of Cincinnati, USA
Alexander Weinert             German Aerospace Center (DLR), Germany
Tatjana Welzer                University of Maribor, Faculty of Electrical Engi-
                              neering and Computer Science, Slovenia
Lei Xu                        University of Texas Rio Grande Valley, USA
Yury Yanovich                 Skoltech, Russia
Kosala Yapa Bandara           NUI Galway, Ireland
Amr Youssef                   Concordia University, Canada
Jiangshan Yu                  Monash University, Australia
Uwe Zdun                      University of Vienna, Austria

Kaiwen Zhang                  École de Technologie Supérieure de Montréal,
                              Canada
Chen Zhao                     UTDallas, USA
Haofan Zheng                  UC Santa Cruz, USA
Mirko Zichichi                Universidad Politécnica de Madrid, Spain
Avelino F. Zorzo              PUCRS, Brazil
André Zúquete                 University of Aveiro, Portugal

## Workshop on Beyond the Promises of Web3.0: Foundations and Challenges of Trust Decentralization (WEB3-TRUST)

Web 3.0 has arisen as the next step into the configuration of a more secure and trustworthy Internet. This new stage in the deployment of web technologies is based on the implementation of new architectures for trust decentralization. Distributed ledger technologies and, in specific, blockchain are used as the main components to manage trust without any central authority. Nonetheless, the deployment of such technologies in real and practical scenarios is of problematic nature, and in many occasions this leads to the re-centralization of decision taking. This being the case, governance and the supposed equality provided by blockchain and DLT are hindered, which eventually determine cybersecurity risks of major impact than those in web 2.0. This workshop is devoted to discuss these shortcomings and the associated cyber-risks, taking into account the current state of maturity of blockchain, smart contracts, and the new governance schemes in the blockchain era.

## Organizing Committee Chairs

David Arroyo      CSIC, Spain
Jesús Díaz Vico   IOHK, EE.UU.

## Program Committee

Luca Nizzardo          Protocol Labs, Spain
Antonio Nappa          Universidad Carlos III de Madrid, Spain
Andrés Marín López     Universidad Carlos III de Madrid, Spain
Andrea Vesco           Head of cybersecurity at LINKS Foundation, Italy
Pedro López            CSIC, Spain
Mayank Dhiman          Dropbox, USA

**BLOCKCHAIN'22 Sponsors**

# Preface

The 4th International Congress on Blockchain and Applications 2022 was held in L'Aquila from 13th to 15th of July. This annual congress reunites blockchain and artificial intelligence (AI) researchers who share ideas, projects, lectures, and advances associated with those technologies and their application domains.

Among the scientific community, blockchain and AI are seen as a promising combination that will transform the production and manufacturing industry, media, finance, insurance, e-government, etc. Nevertheless, there is no consensus with schemes or best practices that would specify how blockchain and AI should be used together. Combining blockchain mechanisms and artificial intelligence is still a particularly challenging task.

The BLOCKCHAIN'22 congress has been devoted to promoting the investigation of cutting-edge blockchain technology, to exploring the latest ideas, innovations, guidelines, theories, models, technologies, applications, and tools of blockchain and AI for the industry, and to identifying critical issues and challenges those researchers and practitioner must deal with in future research. BLOCKCHAIN'22 wants to offer researchers and practitioners the opportunity to work on promising lines of research and to publish their developments in this area.

In this 4th edition of the congress, the technical program has been diverse and of high quality, focused on contributions to both well-established and evolving areas of research. The congress has had an acceptance rate close to 50%. More than 75 papers were submitted, and 37 full papers were accepted. The volume also includes three papers from the WEB3-TRUST workshop and two papers from the Doctoral Consortium.

Shipments came from more than 35 different countries (Spain, USA, Sweden, UK, Saudi Arabia, Ireland, Morocco, France, Italy, India, Australia, Portugal, Qatar, Greece, Vietnam, Malta, China, Luxembourg, Lithuania, Switzerland, Canada, Germany, Turkey, Japan, Tanzania, Croatia, Brazil, Israel, Romania, Indonesia, Bangladesh, Singapore, Slovenia, Russia, and Austria).

We would like to thank all the contributing authors, the members of the program committee, the sponsors (IBM, Indra, DISIM, MESVA, Armundia, Reply White Hall, Technologies and Communication, LCL, MDPI, AEPIA, APPIA, and AIR Institute), and the Organizing Committee for their hard and highly valuable work. Their work contributed to the success of the BLOCKCHAIN'22 event. And finally, special thanks to the local organization members and the program committee members for their hard work, which was essential for the success of BLOCKCHAIN'22.

Javier Prieto
Francisco Luis Benítez Martínez
Stefano Ferretti
David Arroyo Guardeño
Pedro Tomás Nevado-Batalla

# Contents

Workshop on Beyond the Promises of Web3.0: Foundations and
Challenges of Trust Decentralization (WEB3-TRUST)

Doctoral Consortium

# BLOCKCHAIN 2022 Main Track

# Modelling of the Internet Computer Protocol Architecture: The Next Generation Blockchain

AoXuan Li[1(✉)], Luca Serena[2], Mirko Zichichi[3], Su-Kit Tang[1],
Gabriele D'Angelo[2], and Stefano Ferretti[4]

[1] Faculty of Applied Sciences, Macao Polytechnic University, Macao SAR, China
aoxuan.li@mpu.edu.mo, sktang@ipm.edu.mo
[2] Department of Computer Science and Engineering,
University of Bologna, Bologna, Italy
luca.serena2@unibo.it, g.dangelo@unibo.it
[3] Ontology Engineering Group, Universidad Politécnica de Madrid, Madrid, Spain
mirko.zichichi@upm.es
[4] Department of Pure and Applied Sciences, University of Urbino "Carlo Bo",
Urbino, Italy
stefano.ferretti@uniurb.it

**Abstract.** The Internet Computer Protocol is described as a third-generation blockchain system that aims to provide secure and scalable distributed systems through blockchains and smart contracts. In this position paper, this innovative architecture is introduced and then discussed in view of its modeling and simulation aspects. In fact, a properly defined digital twin of the Internet Computer Protocol could help its design, development, and evaluation in terms of performance and resilience to specific security attacks. To this extent, we propose a multi-level simulation model that follows an agent-based paradigm. The main issues of the modeling and simulation, and the main expected outcomes, are described and discussed.

**Keywords:** Internet Computer · Distributed Ledger Technology · Modelling and simulation · Blockchain

## 1  Introduction

Cloud computing has undoubtedly been the fastest growing and most successful in delivering technical and economic benefits for application and system development in recent years [26,30]. Starting from startups up to large companies, everyone is adopting cloud computing to get rid of the risk of capital

---

This work has received funding from the EU H2020 research and innovation programme under the MSCA ITN grant agreement No 814177 LAST-JD-RIoE; and the research grant (No.: RP/ESCA-04/2020) offered by Macao Polytechnic University.

© The Author(s): under exclusive license to Springer Nature Switzerland AG 2023
J. Prieto et al. (Eds.): BLOCKCHAIN 2022, LNNS 595, pp. 3–12, 2023.
https://doi.org/10.1007/978-3-031-21229-1_1

investment, cutting the cost of hardware and software infrastructure, and availing themselves of services according to their demand. This is why paradigms such as 'Infrastructure-as-a-Service (IaaS)', 'Platform-as-a-Service (PaaS)', and 'Software-as-a-Service (SaaS)' have emerged. In general, however, cloud service providers maintain their customers with an opaque knowledge about the location and storage of data, the privacy offered to users, and the type of hardware infrastructure used. This leads firstly to a problem of trust by users [19]. Secondly, security and privacy are undermined by the centrality of these solutions, which more easily attracts cyber-attacks, i.e. single points of failure [30]. In addition, it should not be forgotten that centralized solutions will not be able to support the huge amount of data generated globally by users and Internet-of-Things devices for much longer [26]. Finally, it is commonly difficult to assess if Quality of Service (QoS) guarantees are met and Service Level Agreements (SLA) negotiated between users and the cloud provider are satisfied, due to the absence of trusted logs [7]. All this motivates the transition towards a completely decentralized approach. The benefits of this solution are many. In fact, the decentralization of the system removes the presence of a single point of failure, allows for inherently increasing scalability, curbs illicit activities of malicious nodes, and can also provide for accountability guarantees. Clearly, in order to realize a similar kind of system, it becomes necessary to encourage node participation that can be somehow rewarded through incentive mechanisms [33].

The Internet Computer Protocol (ICP) architecture[1] aims to establish a network of networks by defining a protocol for combining the resources of several decentralized computers into the reading, replication, modification, and procurement of an application state. A network of nodes runs the protocol through independently-operated data centers to provide general-purpose (largely) transparent computations for end-users. On the other hand, the development of applications on top of the ICP is facilitated by reliable message delivery, transparent accountability, and resilience. The typical use-case would involve users interacting with a decentralized application as is on a public or private cloud. This is enabled by the use of Canisters, i.e. tamper-proof and autonomous smart contracts hosted on-chain, that can be run concurrently and interact with each other. With respect to other smart contract implementations, such as Ethereum's ones, the Canisters enable applications, systems, and services to be created and accessed by users without incorporating websites running on centralized cloud hosting, e.g. a canister can directly serve HTTP requests created by end-users through their browser. All of this paves the way for the creation of decentralized services where the user is constantly at the center of the process.

However, the design of the ICP requires a complete understanding of the technologies involved and the interactions among these building blocks. There is a need for viable modeling and simulation strategies that allow for what-if analyses and manageable evaluation studies. In this paper, we describe the rationale behind the design of an ICP digital twin that could serve this purpose.

---

[1] Authors are not sponsored or affiliated in any way with the DFINITY Foundation which is the not-for-profit organization that develops the Internet Computer.

Due to the complexity of the system, high levels of detail should be kept only when needed, while coarse simulations should be exploited when dealing with a high number of involved nodes. This leads to a multi-level simulator design [6].

This paper is structured as follows. Section 2 provides the necessary background about the technologies used in the ICP and a discussion of the related work. Section 3 presents a specific introduction of the ICP; while in Sect. 4, the main modelling and simulation issues of the ICP are discussed. Finally, Sect. 5 provides the concluding remarks.

## 2   Background and Related Work

### 2.1   Related Technologies

In this section, we briefly describe the background technologies and methodologies that are necessary for understanding the ICP architecture and evaluating the main problems that are related to its modelling and simulation.

**Blockchains** Informally, a blockchain is a public ledger that may hold any data, e.g., transactions between different parties, email records, or even daily grocery records. The ledger is distributed among all network participants, and it is immutable once written down. As the name suggests, a blockchain is a chain of blocks while each block contains a set of records. Moreover, a block also contains a timestamp and the hash value of the previous block. If any adversary user tries to change intermediate blocks, he/she has to change all following blocks. However, this is impossible since the ledger is decentralized. For any new block, it will not automatically join the chain until the majority of parties agree so. Blockchains have made impacts on various areas [20].

**Consensus Algorithms** Consensus algorithms allow (the majority of) nodes to agree on the status of the ledger. That is, they agree on the validity of transactions in a block, the validity of the block itself, and if there is more than one proposed block, on which block is appended to the chain. There are different types of consensus algorithms. Among them, two are worthy of mention here, i.e. proof-based algorithm and vote-based algorithm [24]. In a proof-based algorithm, parties need to solve a cryptography puzzle, and the first successful one gets the right to append the block. In a vote-based algorithm, if a party wants to append a block, there must be more than $T$ parties appending the same block where $T$ is a threshold number.

**Smart Contracts** Smart contracts are a set of instructions (or the source code from which such instructions were compiled from) stored in the blockchain and automatically triggered once the default condition is met [2]. This execution is triggered via a transaction and will produce a change in the blockchain state. Each node executing the instructions receives the same inputs and produces the same outputs, thanks to a shared protocol. Smart contracts enable the execution of a service without a trusted human third party validator to check the terms of an agreement, however the smart contract issuer must be sure that the behaviour implemented is correct [2]. For instance, the creation of smart contract-based services may enable users to interact with devices/vehicles or favor interoperability in smart cities [14,32].

## 2.2   Related Work

While there are no specific simulations of the ICP architecture, some simulators have been proposed for modelling blockchains and distributed ledgers technologies. The main difference between the simulators is the level of detail (and the corresponding simulation methodology) that has been chosen for modelling the system to be represented. For example, in [27], some of the authors of this paper have proposed an agent-based simulation to investigate some well-known network attacks on blockchains and distributed ledgers. In [1], the authors present Block-Sim which is a discrete-event simulator of blockchain systems (implemented in Python) that specifically considers the modeling and simulation of block creation through the proof-of-work consensus algorithm. A totally different approach is introduced in [22], in which the queuing theory is used for modelling blockchain systems. In [29], the authors propose VIBES which is another blockchain simulator but specifically designed and implemented for large-scale peer-to-peer networks and able to simulate blockchain systems beyond Bitcoin and support large-scale simulations with thousands of nodes. Finally, in [25], the authors propose an approach that is based on stochastic blockchain models (i.e. Monte Carlo simulations).

## 3   The Internet Computer Protocol (ICP)

The ICP is defined as the third generation of the blockchain systems [31], where the first generation is Bitcoin [23], and the second generation is Ethereum [4]. The ICP provides an infinite blockchain where we may hold everything. Unlike previous blockchain systems, it aims to be scalable and to run at web speed. The main technical components of the Internet Computer are the Canister [8] and the Network Nervous System (NNS) [9]. The canister is a special type of smart contract. Users may interact with a canister directly as long as they know the identity of the canister. In the ICP, communication between the different nodes is demanded to the Network Nervous System (NNS, see Sect. 3.1).

The ICP has a four-layer structure. From bottom to top, there are data centers, nodes, subnets, and canisters. Data centers are hardware devices for holding nodes, and each node is a physical computer providing computational power. Each data center may have many nodes, and nodes from different data centers could build up a subnet. Each subnet hosts many canisters, which is the application program on the ICP. Figure 1 reports a high-level representation of this design structure. Each subnet handles the trust and immutability of the

Canister with a blockchain. The blockchain grows in rounds, and, in each round, a randomly selected node proposes a block containing the canister inputs and the hash of the previous block. If the majority of nodes agree on the subnet's state and the validity of the new block, this new block is appended to the blockchain.

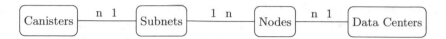

**Fig. 1.** The ICP high-level design architecture.

The ICP design guarantees the availability of canisters in subnets. In fact, by implementing a replication approach, the canisters do not suddenly stop running in case of localised failures. As long as more than two-thirds of replicas are online, the canister is available. A critical requirement for this approach is that all replicas must catch up with the latest state. In previous blockchains, like Bitcoin and Ethereum, this would require downloading the whole blockchain. The ICP provides a CatchUp Package (CUP) [10] so that a node only needs to download a limited amount of data to catch up with the current state of the blockchain. The CUP contains an intermediate replica state and a subsequent of the blockchain. With the replica state, the node can compute the next state on itself, and with the sub-blockchain, it can verify the replica states.

The ICP community claims that their blockchain could scale out to billions of users [11]. Since each canister can only support up to 4 GB of memory (i.e. due to the limitations of WebAssembly) then the Internet Computer uses a multi-canister architecture. For example, for a video-sharing application, it would be possible to split the user-uploaded content into multiple chunks and store them into multiple canisters. When a user wants to retrieve a video, the user makes a query call to the front-end canister, which in turn will make cross-canisters requests to multiple storage canisters. It is worth noticing that all these operations are transparent to users. Table 1 summarizes the main terms used to describe the ICP architecture.

**Table 1.** Main terms used in the ICP architecture.

| Term | Definition |
|---|---|
| Canister | A special type of smart contract |
| Catch Up Packages (CUP) | A schema for state synchronization |
| Data center | The decentralized hardware of the ICP architecture |
| Network Nervous System (NNS) | A special canister serving as the governance body |
| Node | The peer computer in data centers |
| Subnet | The blockchain for providing computing resources |

### 3.1   Network Nervous System

To obtain a scalable and highly efficient system, the ICP must be able to host any number of canisters and to run them concurrently. The ICP introduces a novel Decentralized Autonomous Organization (DAO) that is called Network Nervous System (NNS). The NNS is designed for managing all the base nodes of the system through a Proof-of-Stake consensus protocol.

More specifically, NNS is a set of initial canister programs that oversee the whole network. For example, a data center may apply to the NNS to join the network. NNS also manages how the subnets are formed and how the replication of the nodes is managed. Moreover, the NNS is in charge of upgrading the ICP. For example, the users are enabled to submit proposals for changing the ICP design and implementation. NNS will host the proposal and then allow users to vote on the proposal. Finally, the NNS will implement and deploy the proposal, if the majority of the users have approved it.

### 3.2   Chain Key Cryptography

Most likely, the main scientific breakthrough provided by the ICP is the Chain Key Cryptography [12,16]. In the ICP, a canister is replicated through a subnet, and those nodes in the subnet have to agree on the computation results. The high-level process is described below:

1. each node holds a secret key share;
2. if enough nodes agree on the result then they can jointly sign the message with their respective key share;
3. the user may verify the received message with a single public-key.

If some nodes have failed or crashed then the NNS will add new nodes to the subnet, and the remaining active nodes will reshare the secret key while keeping the same public key. In the ICP, all subnets have a public key and corresponding secret key shares, and all those public keys could be verified with a single 48-byte public key. Even if the Internet Computer had millions of nodes, the network would only need one public key to verify all messages. This technology is called Chain Key Cryptography [12]. The used protocol builds on Shamir secret sharing [28] and BLS signature [3], and moreover, it facilitates the secret sharing keys creation and refreshing.

## 4   Modelling and Simulation of the ICP Architecture

The aim of the Modelling and Simulation (M&S) techniques is to reproduce the behaviour of the system under investigation, in order to (i) study the dynamics of interaction among the various components, (ii) evaluate the resilience of the system under specific conditions (e.g. cyberattacks or failures) and (iii) assess the impact of possible future extensions or features to the system before its implementation or even to support their design. Specifically for the ICP architecture, different aspects are of interest under a M&S viewpoint. First of all, it is important to model the consensus protocol, in order to analyze how the block creation flow is working.

While state-of-the art works regarding the simulation of Distributed Ledger Technology (DLT) focus on Bitcoin-like protocols, with only one blockchain collecting the incoming data [27], we are interested in modeling a DLT where multiple partial blockchains (i.e. subnets) work asynchronously in parallel, exchanging information when necessary. Furthermore, we think that, what is needed to model, are also the aspects specifically related to the DAO (i.e. the Decentralized Autonomous Organization [18]), that is in charge of managing the policies and the future developments of the DLT. This is because the level of security, scalability, decentralization and also the economical sustainability of the whole architecture strongly depends on the DAO's decisions.

A common problem in M&S is the level of detail to be chosen when abstracting the real system to a simulation model. In fact, a fine grained model that considers the very detailed aspects (such as the transmission of every network packet in the distributed system of interest) would permit a high-level model accuracy at the cost of a relevant complexity in the management of the obtained model and in the execution overhead of the simulator. On the other hand, a high-level model in which, for example, the communication aspects in the distributed system are neglected, would permit the building of a very simple (and fast) simulator with poor accuracy. Choosing the most appropriate level of detail to be used in the M&S is not easy and it is strongly linked to the desired outcomes of the simulator. For example, the modelling of some specific cyber-security attacks in a distributed system often requires a very specific level of abstraction in the communications. For these reasons, we think that in a scenario such as the ICP architecture in which we are interested in many different aspects of different abstraction layers, a solution based on a fixed level of abstraction would not be optimal. In fact, we plan to employ an approach that is based on multi-level modelling and simulation [17]. This approach is not new but it is still not very common in the simulation of complex systems. More in detail, we plan to build an ICP model in which the different components are represented by two (or more) simulated models that will be alternatively used depending on the specific analysis that we are interested in. For example, when the specific aspects related to the DAO will be investigated, some low-level details of the model will not be required and therefore the "high-level" (i.e. coarse-grained) version of some components will be used. On the other hand, when the security of the consensus protocol will be investigated then the "fine-grained" models will be required.

## 4.1   Design and Implementation of an ICP Simulator

In order to model and simulate the ICP architecture, we decided to employ an agent-based approach. Agent-Based Simulation (ABS) is a widely diffused technique that in the years gained a lot of popularity in many different fields such as engineering, economics, and computational social sciences [21]. In ABS, the most relevant system components and modules are represented by means of agents. Every agent is then characterized by a specific behavior and interacts with other agents using interactions (that are often implemented as messages). In other words, the system evolution is represented through changes in the local state of the agents (and of the environment) in which they are located.

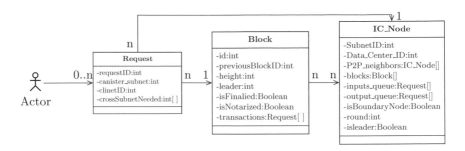

**Fig. 2.** Class diagram of the main components of the ICP architecture.

Referring more specifically to the modeling of the ICP architecture, two types of agents populate our simulation scenario: firstly, there are the clients of the system, which can carry out transactions and requests to the system. Secondly, there are the nodes of the ICP, each one localized in a specific data center, and operating in a specific subnet. Figure 2 shows a possible modelling of the ICP nodes. All the nodes are located in a certain datacenter, belong to a specific subnet and maintain a set of blocks as well as a set of transactions still to validate.

The current implementation of the ICP architecture relies on a very low number of nodes, since 32 subnets exist, each one with 13 nodes contributing to store the transactions (except the NNS, which is dealt with as a special subnet, composed of 40 nodes) [13]. Thus, for the modelling and simulation of the current setup of the ICP architecture, the simulator's scalability is not a big concern. However, it is expected and already planned that the future developments of the ICP will lead to a considerable growth of the network size, with many more nodes and subnets involved in the validation of transactions. We plan to use the developed simulation tools to be able to investigate and properly assess how such a network growth should be managed. For example, right now it is easy for the nodes to be directly in contact with all the other peers belonging to the same subnet, but with many more nodes managing a single subnet, a gossip algorithm might be adopted to efficiently disseminate blocks and transactions inside each subnet [5]. Moreover, from a simulation point of view, more simulated entities entail a larger amount of computing resources employed and a greater execution time. Thus, Parallel And Distributed Simulation (PADS) [15] approaches might be necessary to efficiently carry out the tests.

# 5   Conclusions

The Internet Computer Protocol (ICP) architecture is a third generation blockchain system that is being designed, implemented, and deployed to provide a secure and a scalable way for creating very large-scale distributed systems. In this position paper, we have introduced the ICP architecture and its main problems in terms of modelling and simulation. In fact, the usage of proper simulation techniques would permit us to investigate some very relevant aspects of the ICP architecture and support its design. The main issues related to the modelling and simulation of the ICP concern the specific level of detail used for abstracting the system in a model that can be then evaluated using a simulation. In the following of this paper, we described our current effort in the creation of an agent-based simulator of the ICP that is able to both provide the desired level of detail and the needed scalability. The creation of the ICP simulator is an ongoing activity that requires a relevant effort in many different phases (e.g., design, implementation, and validation) that will likely permit us to release a preliminary version of the simulator in the next months.

# References

1. Alharby, M., van Moorsel, A.: Blocksim: a simulation framework for blockchain systems. SIGMETRICS Perform. Eval. Rev. **46**(3), 135–138 (2019)
2. Becker, M., Bodó, B.: Trust in blockchain-based systems. Internet Policy Rev. **10**(2) (2021)
3. Boneh, D., Lynn, B., Shacham, H.: Short signatures from the Weil pairing. In: International Conference on the Theory and Application of Cryptology and Information Security, pp. 514–532. Springer (2001)
4. Buterin, V., et al.: A next-generation smart contract and decentralized application platform. White Paper **3**(37) (2014)
5. D'Angelo, G., Ferretti, S.: Highly intensive data dissemination in complex networks. J. Parallel Distrib. Comput. **99**, 28–50 (2017)
6. D'Angelo, G., Ferretti, S., Ghini, V.: Multi-level simulation of internet of things on smart territories. Simul. Model. Pract. Theory **73**, 3–21 (2017). Smart Cities and Internet of Things
7. D'Angelo, G., Ferretti, S., Marzolla, M.: A blockchain-based flight data recorder for cloud accountability. In: Proceedings of the 1st Workshop on Cryptocurrencies and Blockchains for Distributed Systems (CryBlock) (2018)
8. Dfinity: a closer look at software canisters, an evolution of smart contracts (2021). https://medium.com/dfinity/software-canisters-an-evolution-of-smart-contracts-internet-computer-f1f92f1bfffb
9. Dfinity: The network nervous system: governing the internet computer (2021). https://medium.com/dfinity/the-network-nervous-system-governing-the-internet-computer-1d176605d66a
10. Dfinity: Resumption: how internet computer nodes quickly catch up to the blockchain's latest state (2021). https://medium.com/dfinity/resumption-how-internet-computer-nodes-quickly-catch-up-to-the-blockchains-latest-state-5af6e53e2a7

11. Dfinity: A technical overview of the internet computer (2021). https://medium.com/dfinity/a-technical-overview-of-the-internet-computer-f57c62abc20f
12. Dfinity: Chain key cryptography: the scientific breakthrough behind the internet computer (2022). https://medium.com/dfinity/chain-key-technology-one-public-key-for-the-internet-computer-6a3644901e28
13. Dfinity: Internet computer network status (2022). https://dashboard.internetcomputer.org
14. Ferraro, P., King, C., Shorten, R.: Distributed ledger technology for smart cities, the sharing economy, and social compliance. IEEE Access **6**, 62728–62746 (2018)
15. Fujimoto, R.M.: Parallel and Distribution Simulation Systems, 1st edn. Wiley, USA (1999)
16. Groth, J.: Non-interactive distributed key generation and key resharing. Cryptology ePrint Archive, Report 2021/339 (2021). https://ia.cr/2021/339
17. Huraux, T., Sabouret, N., Haradji, Y.: A multi-level model for multi-agent based simulation. In: ICAART (2), pp. 139–146 (2014)
18. Jentzsch, C.: Decentralized autonomous organization to automate governance. White paper, November (2016)
19. Khan, K.M., Malluhi, Q.: Establishing trust in cloud computing. IT Prof. **12**(5), 20–27 (2010)
20. Lei, I.S., Tang, S.K., Tse, R.: Integrating consortium blockchain into edge server to defense against ransomware attack. Procedia Comput. Sci. **177**, 120–127 (2020)
21. Macal, C., North, M.: Introductory tutorial: agent-based modeling and simulation. In: Proceedings of the Winter Simulation Conference 2014, pp. 6–20 (2014)
22. Memon, R.A., Li, J.P., Ahmed, J.: Simulation model for blockchain systems using queuing theory. Electronics **8**(2) (2019)
23. Nakamoto, S.: Bitcoin: a peer-to-peer electronic cash system. Decent. Bus. Rev. 21260 (2008)
24. Nguyen, G.T., Kim, K.: A survey about consensus algorithms used in blockchain. J. Inf. Process. Syst. **14**(1), 101–128 (2018)
25. Piriou, P.Y., Dumas, J.F.: Simulation of stochastic blockchain models. In: 2018 14th European Dependable Computing Conference (EDCC), pp. 150–157 (2018)
26. Sadeeq, M.M., Abdulkareem, N.M., Zeebaree, S.R., Ahmed, D.M., Sami, A.S., Zebari, R.R.: IoT and cloud computing issues, challenges and opportunities: a review. Qubahan Acad. J. **1**(2), 1–7 (2021)
27. Serena, L., D'Angelo, G., Ferretti, S.: Security analysis of distributed ledgers and blockchains through agent-based simulation. Simul. Model. Pract. Theory **114**, 102413 (2022)
28. Shamir, A.: How to share a secret. Commun. ACM **22**(11), 612–613 (1979)
29. Stoykov, L., Zhang, K., Jacobsen, H.A.: Vibes: fast blockchain simulations for large-scale peer-to-peer networks: demo. In: Proceedings of the 18th Middleware Conference: Posters and Demos, pp. 19–20. Association for Computing Machinery (2017)
30. Subramanian, N., Jeyaraj, A.: Recent security challenges in cloud computing. Comput. Electr. Eng. **71**, 28–42 (2018)
31. Team, D., et al.: The internet computer for geeks. Cryptology ePrint Archive (2022)
32. Zichichi, M., Ferretti, S., D'Angelo, G.: A framework based on distributed ledger technologies for data management and services in intelligent transportation systems. IEEE Access 100384–100402 (2020)
33. Zichichi, M., Ferretti, S., D'angelo, G.: On the efficiency of decentralized file storage for personal information management systems. In: 2020 IEEE Symposium on Computers and Communications (ISCC), pp. 1–6 (2020)

# Assessing Blockchain Challenges
# in the Maritime Sector

Rim Abdallah[1,2(✉)], Jérôme Besancenot[2], Cyrille Bertelle[1], Claude Duvallet[1],
and Frédéric Gilletta[2]

[1] University of Le Havre, Le Havre 76600, France
[2] HAROPA PORT, 71 Quai Colbert, 76600 Le Havre, France
{rim.abdallah}@etu.univ-lehavre.fr
{rim.abdallah,jerome.besancenot,frederic.gilletta}@haropaport.com
{cyrille.bertelle,claude.duvallet}@univ-lehavre.fr
https://www.haropaport.com

**Abstract.** The maritime sector is the main contributor to global trade but has generally been weakened by its composite anatomy, miscellaneous information sources and systems, and dispersed processes. In this regard, the innovative blockchain technology, characterized by its distinctive features: distribution, immutability, and transparency, promotes great transformative possibilities for the sector. And blockchain applications, use cases, and pilot projects are increasingly emerging in the industry. However, not all applications were successful. And mass adoption of the technology that allows unlocking its full potential in the sector is still in its infancy. This article highlight blockchain's contribution to the maritime sector and list its applications anchoring literary recommendations for technological advancements in the maritime sector. Second, we unravel the reason that halted Blockchain advancement in the industry and its widespread adoption. Furthermore, we categorize those reasons into challenges, assess them and provide possible solutions.

**Keywords:** Blockchain · Smart port · Maritime sector · Shipping

## 1 Introduction

There has been a growing interest in blockchain technology across the maritime sector to streamline operational processing and reduce delays and costs. Blockchain has completed the single-use and localization phases of recent technology adoption compared to the adoption cycle of network technologies in business. The first phase can be represented by the emergence of bitcoin as a reputable payment alternative. And the second phase can be represented by the different blockchain solutions in a limited number of enterprises, especially with the emergence of consortium and permissioned blockchain [16]. These solutions are mostly restricted to one or multiple organizations that are connected together and consist of a high novelty level. Currently, we can consider blockchain based

J. Prieto et al. (Eds.): BLOCKCHAIN 2022, LNNS 595, pp. 13–22, 2023.
https://doi.org/10.1007/978-3-031-21229-1_2

on its use-cases and applications to be alternating between the last two phases of its global adoption. The third can be represented by public use-cases that remove restrictions between actors in the ecosystem and it is often characterized by low novelty and high coordination. And finally, the fourth phase will be a completely transformative phase that could revolutionize the ecosystem as we know it. These two last phases consist of introducing new ways of doing things in the ecosystem that normally face many challenges. And despite the emerging blockchain applications, the technology's challenges have been very little addressed in the literature particularly for the maritime sector to unleash its full potential and fully enable interoperability and traceability across the sector. This document investigates barriers that blockchain needs to overcome in the maritime sector and their solutions to facilitate its transition through the final phases and avoid it being substandard.

## 1.1 Technological Literary Recommendations for the Maritime Sector

The literature identified some main recommendations for any digitally transformative technology adoption in the maritime environment. We have grouped them into four main categories.

1. **Specify governance policies and accessibility**: The maritime sector is heterogeneous and consists of public and private organizations collaborating across the entire supply chain. Most of its data digitization and governance programs are fragmented and operate independently causing data centralisation, replication and operational inefficiencies. This brings forward data governance concerns with the hesitancy to trust a single centralized entity across the whole supply chain.
2. **Measure visibility**: The term visibility needs clarification to ensure the digital platform's reliability. Greater visibility should not revoke data confidentiality and endanger competitiveness. This raises questions and concerns around achievability. Particularly, in a global environment as the maritime sector where not only inter-organizational visibility should be addressed but also intra-organizational. This means contract re-definition for organizations' services to also include the correspondent inter-organizational level of visibility.
3. **Define transparency**: Despite that, actors in the maritime sector agree that the cost of data cut-offs and obscurity have been too high and are keen on transparency. The transparency that allows further traceability should adapt to new expectations. This poses the question of the depth of the traceability, maritime actors are willing to expose, for customers able to track their shipments in real-time. For example, will a customer shipping a car be able to track its parts manufacturers?
4. **Data and analytics**: The terms visibility, transparency, and traceability all align with data analysis possibilities for improved performance. With machine learning/AI and automation, better traceability and transparency allow more

reliable results and overall operational management. Data should be disposable for unleashing those possibilities while maintaining trust.

One seemingly promising technology that aligns with these recommendations is blockchain. An emerging technology that complies with the sector's effort to maintain its competitiveness and revolutionize itself. Blockchain can alleviate digitization burdens, reduce paperwork, eliminate processing redundancy and allow better visibility, transparency, and traceability.

## 2   Blockchain Overview

### 2.1   Blockchain Features

Blockchain and its features align coherently with the technological needs and constraints of the maritime sector highlighted in the aforementioned technological maritime literary recommendations. The distributed append-only ledger technology allows maritime actors to collectively maintain a shared ledger with collective governance over the network without establishing complete trust in one of the actors or introducing a trusted third party. It introduces a shift from centralized technologies to peer-to-peer networks that allows notarised real-time direct data transfer between maritime actors. These data exchanges are time-stamped, encrypted. They are called transactions and are grouped into blocks to be appended to the ledger using an agreed-upon collective consensus. Each block of transactions is linked to its previous using hashes forming an immutable chained ledger. This allows information to be tracked, traced, and certified [7]. Its decentralized and replicated nature alleviates single point of failure concerns, and its cryptographic algorithms and encryption help ensure data protection while maintaining visibility to authorized parties using permissioned blockchain networks. Moreover, smart contracts unlock further automation opportunities.

### 2.2   The Need to Identify Blockchain Challenges in the Maritime Sector

Maritime actors have acknowledged the potentials of this revolutionizing technologies and started testing use cases in different domains in the sector. Example use cases and projects are displayed in Table 1. However not all these projects were successful, some failed to understand blockchain challenges in the sector and how to properly address them. In the following section, we identify the challenges facing blockchain in the maritime sector and we propose possible solutions based on literature studies and applications. Therefore, a challenge can be represented from both a scientific and technological point of view in the sense that not only does it requires scientific research to be assessed but also research can be presented in the form of a technological application aiming to address the underlying problem.

**Table 1.** Blockchain use cases and applications in the maritime sector.

| Use case | Project | Owner/proposer | Blockchain used | References |
|---|---|---|---|---|
| Fuel quality and traceability | BunkerTrace | Blockchain labs for open collaborative (BLOC), main blockchain labs (MBL) | Ethereum | BunkerTrace website |
| Shipment tracking | TradeLens | Maersk and IBM | Hyperledger fabric | TradeLens website |
| | GSBN | Oracle, Microsoft, AntChain and Alibaba cloud | AntChain | GSBN website |
| | Silsal | Abu Dhabi Port | Hyperledger fabric | Additional information |
| | Calista | PSA International, and Global eTrade Services (GeTS) | Not a blockchain but an intensive API delivering end-to-end data (based on blockchain concepts) | Calista website |
| Track and Trace hazardous goods | (Pilot project) | BLOC, LLoyd's register foundation and rain-making consortium project | | Additional information |
| Smart bill of lading | CargoX | CargoX | Ethereum | CargoX website |
| | TradeLens | Maersk and IBM | Hyperledger fabric | TradeLens website |
| | Bolero's digital trade platform | Bolero | Volton Corda based | Bolero website |
| | Easy trading connect | Blue water shipping Louis Dreyfus | Ethereum Quorum | Additional information |
| | (Proof of concept) | Pacific international lines (PIL), PSA International and IBM Singapore | Hyperledger fabric | Additional information |
| | WAVEBL | https://wavebl.com/ | | WAVEBL website |
| | Tokio Marine | Tokio Marine Holdings | Corda | Additional information |
| Digitization | (storage on blockchain) | DNV GL | Vechain Thor | Additional information |
| Smart contracts | Blockconnect | 300Cubits | Ethereum | 300Cubits website |
| | ShipChain | ShipChain | Ethereum | Additional information |
| Insurance and finances | (proof of concept: maritime insurance platform) | A.P.Møller-Maersk, Willis Towers Waston, MS Amilin, and XL Catlin | Corda | Additional information |
| | (Proof of concept: Blockchain Insurance Industry Initiative) | B3i Services AG | Corda | b3i website |
| | (Enterprise-level blockchain consortium) | RiskStream Collaborativ | Corda | Additional information |
| | Shipowners | Shipowner.io | Ethereum | ShipOwner website |
| | Skuchain | Founded by Srinivasan Sriram | Hyperledger Fabric | SkuChain website |
| | Provenance | Owned by Morgan McKenney | Ethereum | Provenance website |
| | Tallysticks | Co-founded by Kush Patel | | Additional information |
| VGM portal | SOLASVGM | Kuehne + Nagel Group | Hyperledger fabric | Additional information |

# 3 Addressing Blockchain Challenges in the Maritime Sector

## 3.1 A Heterogeneous Ecosystem

Despite increasing interest in blockchain, a common international supply chain platform is still in its infancy, and we have yet to witness such a platform on the technological horizon in the maritime sector. The supply chain does not only consist of private actors and enterprises that have been considering the adoption of blockchain, but also of public entities (government, customs, ... ) whose roles are vital for the supply chain. And until now, blockchain use in the public sector is mainly restricted to theory and use case studies(identity management, healthcare, finance and insurance, land title registry, supply chains, energy plants, copyrights, voting systems, corruption prevention and identification) with no tangible widespread application. Although, the recognition of blockchain potentials by the public sector and their participation in pilot projects sound promising but absence of substantial applications in one of the two poles of the supply chain halts the development of a fully interlinked communication platform via a blockchain infrastructure. The deterrence of this pole can be linked to other blockchain challenges in the sector such as security threats, network limitations, lack of regulations, and others. Therefore, to address this particular challenge we need to overcome other underlying obstacles.

## 3.2 Transparency: A Double Edged Sword

Blockchain transparency, visibility, and traceability features can be a double-edged sword. From an economically unbiased point of view, those features result in increased efficiency and more accountability and are of excellent value to the maritime sector where margins of error are extremely narrow, and accountability remains fragmented across the whole supply chain. But if incompetently managed they might discourage entities from adopting distributed ledger technologies such as blockchain. Additional scrutiny and information availability (to consumers, auditors, and competitors) do not seem welcomed to businesses in the shipping industry in particular and to global business in general. And they affect competitiveness factors which oppose with blockchain adoption goals. Moreover, when we discuss transparency, we need to also address the lengths that maritime actors are willing to go and this is represented in food contamination traceability where production information remained concealed [12]. Even though data encapsulation in smart contracts, sounds innovative to solving data accessibility issues, the spread of contracts across the network expose data through several contract executing nodes. Therefore, a unanimous technological adoption remains lacking due to transparency consensus which is why literature coverage of total traceability using blockchain-based solutions remains conceptual.

### 3.3   Technology Predates Regulations

To regulate the problematical border-crossing shipping, the international maritime organization (IMO) was founded in 1948. Currently, regulations are locally enforced in ports using technological advancements such as port community systems and others across shipments' journeys. These checkpoints do not usually communicate with each other, incentivising blockchain solutions to globally abide laws. But, the lack of current blockchain regulatory oversight introduces unwanted ambiguity.

First, solutions based on pre-existing crypto blockchains clashes with the European Commission's 2018 warnings about price volatility. Second, any amended regulations considered only exchanges between cryptos and fiat currencies [17]. Third, clear global regulatory intentions are lacking, and crypto money remain controversial. Some countries such as El Salvador adopted bitcoin as a legal tender while France has begun regulatory process, the European Union announced their regulatory intentions, as opposed to China where miners were eradicated. And even without the financial regulatory conundrum, blockchain features contradict pre-existing data regulations such as the EU general data protection regulation (GDPR) [23]. First, Blockchain immutability conflict with the right to rectify and erase data. Second, its decentralization retains from establishing clear accountability. And finally, lawfulness can only be ensured by node basis which seems excessive. The literature proposes three approaches to address this challenge. The first is the use of a central authority to enforce GDPR regulations and assume responsibility, as for data rectifying, it can be through redactable blockchain or that each node removes the data and recalculate all consequent blocks [28]. The second is through distributed responsibility. This approach is similar to the first one, responsibility is assumed through the collective efforts of multiple central authorities [23]. And the third one is pseudonymization. This can be achieved through data encryption, hashing, and pseudo-identifiers. The third approach was used in a successful pilot project from the German federal office for migration and refugees [21].

## 3.4 Distinct Technical Capabilities

The maritime ecosystem consists of diverse actors that are not necessarily familiar with high technological advancements as pointed out in the following study [3]. This will eventually lead to increased technological deployment costs that will halt the widespread adoption of Blockchain which is key for a viable global platform.

## 3.5 Data Storage

Blockchain is not initially conceived for storing data bulks. However, the maritime sector is data abundant, especially when using IoT technologies. Therefore, the IoT-Blockchain data storage challenge should be addressed prior to deployment. In literature, several studies addressed the storage challenge of which we mention five main approaches.

**Distributive Storage approach** With the diverse numerous actors in the maritime sector this method is susceptible to information inconsistency and has limited efficiency [5].

**Simplified payment verification method and data erasure** This method targeted only specific use cases and is not adequate for the maritime sector. Data erasure compromises transparency and historical traceability [8,18].

**IPFS data storage** Despite being innovative, some challenges still remained such as data duplication, content piracy, and information availability issues. To prevent duplication a study suggested the use of a SPROV system architecture [9] but this approach lacked data querying. Another study suggested the use of an intermediary application between the IPFS and the blockchain network where encrypted files stored on the IPFS network can only be decrypted in the application using Blockchain keys and smart contracts [13]. But this method only addressed the data piracy problem. Data availability and duplication issues remained. Data availability can also be addressed using incentives [24,27] or erasure coding [4].

**Cloud storage** Some data provenance architecture including [14] had no policies or prohibitions on piracy threatening data integrity.

**Data Compression** The use of the Compression private data sharing (Cpds) framework proved effective but was still susceptible to fault tolerance and time constraints vulnerabilities. Fault tolerance in theory could be addressed using key sharding [22] or fully homomorphic encryption [10]. Additionally, supplementary efficiency could be reached using multi-center architecture to enhance privacy and security [25] and recycling smart contracts [19].

## 3.6 Smart Contracts

Businesses are often interested in one particular blockchain feature which is "smart contracts". A self-executing code that runs on the blockchain once its predetermined conditions are met and can be invoked by sending transactions to its address on the blockchain. It allows conditional self-execution and automation that priorly were not possible without third parties interventions and ensures more operational efficiency in terms of speed costs and transparency. But smart contracts should be carefully addressed in the maritime sector due to the following reasons.

**Payment Clause** Payments in the maritime sector use fiat currencies, and up till now, smart contracts have no way of enforcing fiat-based currencies without external intervention. Experiments such as 300Cubits TEU tokens have been made to test the sector's willingness to adopt cryptocurrencies.

**Coding Clause** Smart contracts are codes, and codes remain susceptible to bugs or hacks which can be damaging upon self-execution. This was demonstrated by the decentralised autonomous organization hack that resulted in forking the Ethereum network, where the theft was annulled by developers tampering with fundamental blockchain concepts. Afterward, studies in the literature addressed this challenge using two methodologies. The first focused on correct coding logic [6]. And the second acknowledged errors and bugs inevitability, and focused on the development and effectiveness of bug detection tools such as Oyente, ReGuard [15], NPChecker [26], Zeus and others. But since not all smart contracts' source codes are available, results can only be assessed based on those verified by their source codes. Some studies that proposed a new tool focused on effectiveness comparison between existing ones where Zeus was proved more effective than Oyente [11]. But in general, tools' results cannot be compared since their testing attributes were not the same. Another paper suggested invariant formalization for the creation of a more effective cross-platform detection tool since. It consisted of first detecting bug invariants and then feeding them to a Certora prover which is a tool used along with existing smart contract compilers to detect bugs. Results countered the unquantifiable clause fault that suggested automation is not feasible and the study highlighted the use of common rules between smart contracts and proved promising and cost-effective [2].

## 3.7 Blockchain Is Isolated from the Real-World

Blockchain interaction with external sources is not possible since it is a protected isolated decentralized environment, thus smart contracts involving real-world data cannot be enforced [20]. Communication with the real world is sometimes crucial in the maritime sector. An example is the use of IoT and smart containers which can cause hesitancy in the adoption of an isolated technology. An outside communication can be established through various design patterns(publish, subscribe, and immediate read) using oracles. However, the use of a centralized entity as a trusted information source conflicts with blockchain decentralization principles and can add security risks such as a Single point of failure and non-determinism. The literature has addressed this challenge by suggesting the implementation of decentralized oracle systems such as Provable, ChainLink, Astraea, Dia, and others.

## 3.8  Energetic Concerns

The most reputable Blockchains such as Ethereum and Bitcoin uses the proof of work (PoW) consensus to establish trust in the decentralized environment resulting in additional environmental worries. A single bitcoin transaction today has a power consumption around 1770.67 kWh equivalent to the consumption over 4 months of a typical French household with a carbon footprint of 841.07 kgCO$_2$. A single Ethereum transaction is comparable to a typical French household's consumption over 13 days, with a carbon footprint of 84.71 kgCO$_2$. And despite Ethereum's announcement of switching to a less exhaustive consensus, the current use of PoW delays the technology's adoption, particularly in the maritime sector where environmental awareness is rapidly rising and regulations are becoming more strict. To overcome this challenge, alternating between distinct types of blockchains (public, permissioned, and consortium) can alleviate computational consumption. It's important to note that the maritime sector differs from the financial sector. Hence, a difference between the notion of trust in currencies, payments, and the maritime sector. Financial decentralization and distribution aim to dissociate financial means from countries' economical situations without centralized trusted entities such as banks. Whereas, in the maritime sector the reign of governmental authorities over the supply chain is inevitable. Decentralization aims to remove barriers between governments to establish a collective trusted platform. Subsequently the increased use of permissioned Blockchain as opposed to public ones is recommended.

## 4  Conclusion

Recent reports show that improved digitized collaboration between actors in the maritime sector responsible for 90% of global commerce will largely improve operational efficiency, and increase security and durability across supply chains [1]. All of this promotes the diligent development of digitized maritime solutions with an ultimate goal of a digital international maritime platform for inclusive economical benefits. And currently, the innovative blockchain technology is oscillating between its last two adoption phases despite having features that align with the maritime industry's exigencies to establish a barrier removing border-crossing platform. The existing blockchain applications show that the benefits of blockchain have been very well perceived by the maritime sector. However, challenges remain and need to be addressed for successfully implementing the technology across the sector to cover its different use cases and allow more widespread adoption for optimal results. Our study sheds light on these challenges and refers to literary and applicative projects to overcome them. We analyze the challenges of blockchain adoption in the maritime sector. Despite having a literary analysis of Blockchain and its features, there are still very few concerning the maritime sector and fewer addressing the technology's challenges and how to overcome them. The lack of full academic literature about the disruptive technology challenges in the maritime sector was the key motive behind this research. The main research question driving this study is to shift the concept of blockchain application in the maritime industry from theory to demonstration by highlighting that the technology's challenges have not passed unnoticed and have been addressed through literature. By conducting such a study, we hope to lay the foundation stones for a proper Blockchain application that takes into consideration not only the technology's potential but also its limitations. For future research, we recommend an analysis of the challenges' solutions to identify the most adapted solutions for the maritime sector. Allowing to ultimately implement an unobstructed digital solution and attain optimal economical gains.

# References

1. Bank, W.: Accelerating Digitalization: Critical Actions to Strengthen the Resilience of the Maritime Supply Chain. World Bank (2020)
2. Bernardi, T., Dor, N., Fedotov, A., Grossman, S., Immerman, N., Jackson, D., Nutz, A., Oppenheim, L., Pistiner, O., Rinetzky, N., et al.: WIP: finding bugs automatically in smart contracts with parameterized invariants. Retrieved 14 July 2020
3. Botton, N.: Blockchain and trade: not a fix for Brexit, but could revolutionise global value chains (if governments let it). Tech. rcp., ECIPE Policy Brief (2018)
4. Chen, Y., Li, H., Li, K., Zhang, J.: An improved P2P file system scheme based on IPFS and blockchain. In: 2017 IEEE International Conference on Big Data, pp. 2652–2657. IEEE (2017)
5. Dai, M., Zhang, S., Wang, H., Jin, S.: A low storage room requirement framework for distributed ledger in blockchain. IEEE Access **6**, 22970–22975 (2018)
6. Delmolino, K., Arnett, M., Kosba, A., Miller, A., Shi, E.: Step by step towards creating a safe smart contract: lessons and insights from a cryptocurrency lab. In: International Conference on Financial Cryptography and Data Security, pp. 79–94. Springer (2016)
7. Dib, O., Brousmiche, K.L., Durand, A., Thea, E., Hamida, E.B.: Consortium blockchains: overview, applications and challenges. Int. J. Adv. Telecommun. **11**(1&2), 51–64 (2018)
8. Gao, J., Li, B., Li, Z.: Blockchain storage analysis and optimization of bitcoin miner node. In: International Conference in Communications, Signal Processing, and Systems, pp. 922–932. Springer (2018)
9. Hasan, R., Sion, R., Winslett, M.: SPROV 2.0: a highly-configurable platform-independent library for secure provenance. In: ACM Conference on Computer and Communications Security (CCS), p. 122 (2009)
10. Huang, H.L., Zhao, Y.W., Li, T., Li, F.G., Du, Y.T., Fu, X.Q., Zhang, S., Wang, X., Bao, W.S.: Homomorphic encryption experiments on IBM's cloud quantum computing platform. Front. Phys. **12**(1), 1–6 (2017)
11. Kalra, S., Goel, S., Dhawan, M., Sharma, S.: Zeus: analyzing safety of smart contracts. In: NDSS, pp. 1–12 (2018)
12. Kamath, R.: Food traceability on blockchain: Walmart's pork and mango pilots with IBM. J. Br. Blockchain Assoc. **1**(1), 3712 (2018)

13. Khatal, S., Rane, J., Patel, D., Patel, P., Busnel, Y.: Fileshare: a blockchain and IPFS framework for secure file sharing and data provenance. In: Advances in Machine Learning and Computational Intelligence, pp. 825–833. Springer (2021)
14. Liang, X., Shetty, S., Tosh, D., Kamhoua, C., Kwiat, K., Njilla, L.: Provchain: a blockchain-based data provenance architecture in cloud environment with enhanced privacy and availability. In: 2017 17th IEEE/ACM International Symposium on Cluster, Cloud and Grid Computing, pp. 468–477. IEEE (2017)
15. Liu, C., Liu, H., Cao, Z., Chen, Z., Chen, B., Roscoe, B.: Reguard: finding reentrancy bugs in smart contracts. In: 2018 IEEE/ACM 40th International Conference on Software Engineering: Companion, pp. 65–68. IEEE (2018)
16. Maple, C., Jackson, J.: Selecting effective blockchain solutions. In: Mencagli, G., B. Heras, D., Cardellini, V., Casalicchio, E., Jeannot, E., Wolf, F., Salis, A., Schifanella, C., Manumachu, R.R., Ricci, L., Beccuti, M., Antonelli, L., Garcia Sanchez, J.D., Scott, S.L. (eds.) Euro-Par 2018. LNCS, vol. 11339, pp. 392–403. Springer, Cham (2019). https://doi.org/10.1007/978-3-030-10549-5_31
17. Miseviciute, J.: Blockchain and virtual currency regulation in the EU. J. Invest. Compliance (2018)
18. Palai, A., Vora, M., Shah, A.: Empowering light nodes in blockchains with block summarization. In: 2018 9th IFIP International Conference on New Technologies, Mobility and Security, pp. 1–5. IEEE (2018)
19. Pontiveros, B.B.F., Norvill, R., State, R.: Recycling smart contracts: compression of the Ethereum blockchain. In: 2018 9th IFIP International Conference on New Technologies, Mobility and Security, pp. 1–5. IEEE (2018)
20. Reyna, A., Martín, C., Chen, J., Soler, E., Díaz, M.: On blockchain and its integration with IoT. Challenges and opportunities. Future Gener. Comput. Syst. **88**, 173–190 (2018)
21. Rieger, A., Lockl, J., Urbach, N., Guggenmos, F., Fridgen, G.: Building a blockchain application that complies with the EU general data protection regulation. MIS Q. Exec. **18**(4) (2019)
22. Shamir, A.: How to share a secret. Commun. ACM **22**(11), 612–613 (1979)
23. Tatar, U., Gokce, Y., Nussbaum, B.: Law versus technology: blockchain, GDPR, and tough tradeoffs. Comput. Law Secur. Rev. **38**, 105454 (2020)
24. Vorick, D., Champine, L.: SIA: simple decentralized storage (2014). Retrieved 18 May 2018
25. Wan, J., Li, J., Imran, M., Li, D., et al.: A blockchain-based solution for enhancing security and privacy in smart factory. IEEE Trans. Ind. Inform. **15**(6), 3652–3660 (2019)
26. Wang, S., Zhang, C., Su, Z.: Detecting nondeterministic payment bugs in Ethereum smart contracts. Proc. ACM Program. Lang. **3**, 1–29 (2019)
27. Wilkinson, S., Lowry, J., Boshevski, T.: Metadisk a blockchain-based decentralized file storage application. Storj Labs Inc., Technical Report, hal, pp. 1–11 (2014)
28. Zemler, F.: Concepts for GDPR-compliant processing of personal data on blockchain: a literature review. Anwendungen und Konzepte der Wirtschaftsinformatik **9**, 96–107 (2019)

# A Model of Decentralised Distribution Line Using Layer 2 Blockchains

Subhasis Thakur$^{(\boxtimes)}$ and John Breslin

National University of Ireland, Galway, Ireland
{subhasis.thakur,john.breslin}@nuigalway.ie,
subhasis.thakur@insight-centre.org

**Abstract.** At least one-third of the food produced for human consumption is lost and wasted due to inefficient perishable product distribution plans. Currently used distribution plans are static and centralised as a single entity decides on a distribution plan where products may move along a predefined path. In this paper, we developed a dynamic distribution plan generation algorithm where distribution paths are dynamic. We used blockchain offline channels to develop decentralised, secure, and privacy-preserving dynamic distribution plans in a multi-party supply chain. Our algorithm generates distribution plans using real-time quality information about perishable goods so that products may be sold before their shelf life expires. Our algorithm ensures the privacy of all actors in a supply chain as inventory and order information of all actors remains privacy-preserved.

**Keywords:** Perishable supply chains · Distribution lines · Food loss and waste · Blockchains · Layer 2 blockchains

## 1 Introduction

At least one-third of all food produced is wasted and lost due to not consuming it before its expiry date. Reducing food loss and waste is an important goal to develop sustainable supply chains. In this paper, we design sustainable supply chains for perishable food products using blockchains. In this paper, we study a distribution line of one manufacturer with multiple manufacturing units, multiple distribution units, and multiple retail units. The manufacturer does not own and controls operations of the distribution units and the retail units. The distribution and retail units are competitors and may not reveal their order and demand information. Additionally, we consider the uncertainty of the shelf life of perishable food products. All distribution units are assumed to have IoT devices monitoring the perishable food items and can detect the remaining shelf life of perishable items. In these settings, in a static distribution line, retailers estimate the expected demand of the item and place orders of items using such demand estimation with a lead time (estimated time to acquire the product from the distributors and remaining shelf life). A distributor can estimate demand by

© The Author(s): under exclusive license to Springer Nature Switzerland AG 2023
J. Prieto et al. (Eds.): BLOCKCHAIN 2022, LNNS 595, pp. 23–36, 2023.
https://doi.org/10.1007/978-3-031-21229-1_3

aggregating the demands from the retailers or other distributors who placed an order of the items to it, and it will place its order to the manufacturer or other retailers accordingly. In a static distribution line, the party that fulfils the order is fixed. If there is a disruption (products are expired due to environmental factors) then the order is not fulfilled. In this paper, we proposed to use dynamic distribution lines where the party who fulfils the order is not fixed.

The dynamic distribution line requires collaboration among various of a supply chain as order and inventory stock level information is needed to produce a dynamic distribution plan. However, there are security and privacy problems in developing a dynamic distribution line. Distributors and retailers are competitors and they may not wish to reveal their order information. It may be possible that a malicious distributor may not fulfil the order of a specific retailer or distributor. Thus the dynamic distribution lines route the products in a privacy-preserving fashion, where order information is not revealed to parties outside a group of trusted distributors. Additionally, the objective of dynamic distribution lines should reduce overall food loss and waste and all parties can verify a solution to such an optimisation problem without all order information. Our main contributions are as follows:

(1) We developed a blockchain-based dynamic distribution line management solution. We used public proof of work-based blockchains. We used blockchain offline channels to develop a high scale distribution line management algorithm.
(2) We developed a privacy-preserving distribution line management algorithm where order information of a distributor or retailer will not be known by any distributor.
(3) We developed a dynamic distribution line management algorithm that matches supply and demand information based on real-time product decay information.
(4) We prove the security and privacy-preserving property of the proposed distribution line management solution.
(5) Using experimental evaluation with simulations of distribution lines we show that the proposed distribution line management solution can significantly reduce food waste.

The paper is organised as follows: in Sect. 2 we discuss distribution line management literature, in Sect. 3 we discuss the dynamic distribution line management problem, in Sect. 4 we discuss the blockchain-based distribution line management algorithm, in Sect. 5 we analyse the security and privacy-preserving properties of the proposed algorithm, in Sect. 6 we present an experimental evaluation of the proposed solution, and we conclude the paper in Sect. 7.

## 2   Related Literature

In [4] a survey of supply chain optimisation methods was presented where optimisation conditions include waste reduction and sustainability parameters. In [5] the author developed methods to optimise perishable distribution lines to reduce food waste. In [3] the authors showed how to use IoT devices to monitor meat quality in a supply chain. In [8] the authors used Monte Carlo simulation to optimise distribution in a perishable supply chain. In [2] the authors developed an algorithm for designing an optimal transportation network for perishable goods. In [13] the authors developed a multi-agent systems-based simulation of perishable supply chains. In [1] the authors developed pricing strategies for perishable goods. In [7] the authors optimised the transportation network to reduce food loss. In [14] authors used IoT devices to monitor product quality in perishable supply chains. In [12] authors presented optimisation algorithm for perishable supply chains. In [11], authors used Bitcoin offline channel network to develop a decentralised product serialisation method. In this paper, we used proof of work-based blockchains. Proof of work-based blockchains was proposed in [9]. There are several variations of blockchains in terms of consensus protocols. Offline channels for Bitcoin, i.e., Bitcoin Lightning network was proposed in [10], which allows peers to create and transfer funds among them without frequently updating the blockchain. We advance the state of the art as follows: (1) We use public blockchain and offline channel of public blockchains to execute dynamic distribution line generation algorithm. Public blockchain makes the participation of supply chain actors in blockchain-based solution and offline channel improves the scalability of the solution. (2) We used public blockchains to execute the distribution line generation algorithm. It improves interoperability among actors of a supply chain as each actor should only be interoperable with the public blockchain transaction and smart contract data structure. (3) The proposed solution preserves the privacy of actors of a supply chain as inventory and order information of all actors.

## 3   A Model of Perishable Supply Chain

In this paper, we will investigate distribution lines for the following supply chain: (a) The supply chain has multiple manufacturing units producing identical perishable items. (b) Demand of a distributor or retailer can be satisfied by any item in this supply chain, i.e., they can not differentiate among the items from different manufacturer or distributors. We will use the following terminology to describe the distribution line: We will represent a distribution line a multipartite graph $G = (L, M_1, \ldots, M_q, R, E)$ with set of vertices $L \cup M_1 \cup \cdots \cup M_q \cup R$ and set of edges $E$. The set of vertices represents the actors (manufacturer, distributors, and retailers) of a supply chain. The set of nodes $L$ represents the manufacturers, $\{M_i\}$ represents the sets of distributors, and $L$ represents the retailers. The set of edges represents the flow of items and the flow of demand information. The edges $(v_i \rightarrow v_j) \in E$ represent flow of goods. Edges are such that,

(1) Vertices $L$ represent the manufacturing units and can only have edges with a set of vertices $M_1$.
(2) Vertices in the sets $M_1, \ldots, M_k$ are distribution centres. A vertex in $M_x$ can only have edges with vertices from the set $M_{x-1}$ with $x > 1$.

(3) Vertices $R$ represent the retail units and can only have edges with a set of vertices $M_q$.

We will use the discrete-time to represent the time it takes to move the products and the shelf life of the products. We will denote time as a sequence of uniformly increasing positive integer $t_1, t_2, t_3, \ldots$ with $t_i - t_{i-1} = dt$ for all $i$. We assume that distances between a pair of vertices can be traveled by time $d(e)$ where $e \in E^S$ is the edge between the vertices. Products are assumed to have a shelf life given by the equation: Decay level of a product $i$ at time $t$ is: $1/(1 + \frac{(.4*e)^t}{a})$, where $t$ is the time since the product has been produced and $a \in [0, A]$ chosen uniformly at random from $[0, A]$ where $A$ is a positive integer. Note that the maximum value of $Q_i^t$ is 1 and the minimum is zero. If $a$ is increased the products degrades slowly. We assume that $Q^* \geq .01$ is the threshold quality of the product that can be safely consumed. Hence any product $i$ with quality less than $Q^*$ is discarded and considered as food waste. We will use a just-in-time inventory management policy for all actors where an actor will only place an order for an item if an item is moved from its inventory or it has received an order for an item.

(1) If an actor $v_x$ receives an order for an item with shelf life $S$ then it will select an item with shelf life $S + \epsilon$ such that $\epsilon$ (positive number) is minimum. Thus an actor will move a product that meets the minimum shelf life requirement.
(2) If an actor $v_x$ receives an order for an item with shelf life $S$ and it does not have any item in its inventory to fulfil the order then it will unable to fulfil the order.
(3) If an actor has sent an item with shelf life $S + \epsilon$ from its inventory then it will order an item with a shelf life of at least $S + \epsilon$.
(4) If an actor has sent $n$ items then it will order $n$ items.

In this supply chain network, consumption of items, orders for items are created as follows:

(1) At a time $t_1$, the number of items to be sold at a retailer is a random number from a probability density function.
(2) At the time $t_1$, a retailer $v_x$ will evaluate the number of items it has sold, let it has sold $n$ items. The retailer will chose a distributor $v_y$ such that $(v_y \rightarrow v_x) \in E$, i.e., product can move from $v_y$ to $v_x$. It will place an order of $n$ items with a shelf life of more than a threshold (estimated by the maximum time it takes to move a product from the manufacturer to the retailer).
(3) At the time $t_1$, a distributor $v_x$ will evaluate the number of items it has sold and orders it has received at time $t_1$. If $v_x$ has moved an item with shelf life $S$ as it fulfils the order of either a distributor or a retailer then, $v_x$ will place an order for an item with shelf life $S + \epsilon$ ($\epsilon$ is a positive number) to any distributor or manufacturer $v_y$ such that $v_y \rightarrow v_x \in E$. If $v_x$ has received an order for an item with shelf life $S$ as it fails to immediately fulfils it then, $v_x$ will place an order to either a distributor or a retailer for an item with a shelf life of at least $S + \epsilon$.

(3) At the time $t_1$, if a manufacturer received an order for an item with shelf life $S$ then it will choose an item with shelf life at least $S$ from its inventory and move the product. It will immediately produce a new product and replenish its inventory. If it can not immediately fulfils the order then it will wait until a new product is created with a shelf life of $S$.

(4) The order of an item from an actor $v_x$ to another actor $v_y$ with shelf life $S$ is represented by the tuple $O_i =< v_x, v_y, S >$.

(5) The movement of an item with serial number $ID_z$ from an actor $v_x$ to another actor $v_y$ with shelf life $S$ is represented by the tuple $T_i =< v_x, v_y, ID_z >$.

The above process of order placement and product movement represents a static distribution line as an actor $v_x$ who has received an order for an item from the actor $v_y$ then only $v_x$ can fulfil the order. The problem of designing a dynamic distribution line is as follows: (1) If an actor $v_x$ who has received an order for an item from the actor $v_y$ then $v_x$ can delegate the order to another actor $v_z$ to fulfil the order. Product quality information will be available to decide on such order delegation. We will explain in the next section how to collect the product quality information in real time with IoTs. (2) Let at the $k$'th layer of the supply chain network actors $D_k$ supplies to the actors $D_{k+1}$. $O^k$ be the collection of orders from $D_{k+1}$ to $D_k$ and $I^k$ be the set of inventory items for the actors $D^k$. A bipartite graph $H = (D_k, D_{k+1}, E^H)$ will be constructed where $E^H \subset E$ (It means a product can only move along any permitted distribution path indicated by $E$. It also represents the preference of an actor to send/receive from other actors. Only one item can be transferred along one edge.) and for any $(v_x, v_y) \in E^H$ such that the shelf life of the item in the inventory of $v_x \in D_k$ is at least the lead time of the order placed by $v_y \in D_{k+1}$. The weight of an edge $(v_x, v_y) \in E^H$ is $1/1 + (Shelf\_life(a_x) - O_y^3)$ where $Shelf\_life(a_x)$ is the shelf life of $v_x$'s item and $O_y^3$ is the lead time of order placed by $v_y$. (3) A dynamic distribution line algorithm produces a maximum weighted matching $\mathcal{M}$ for the graph $H$. If the edge $(v_x, v_y) \in \mathcal{M}$ then the item from $v_x$ will move to $v_y$. In the next section, we will describe how to execute this matching algorithm in blockchain offline channels.

## 3.1   Problem of Designing Decentralised Distribution Lines

A generic model of decentralised distribution line may include an additional entity called 'mediators'. Mediators will buy and sell from all entities of the above-mentioned distribution line. In such a distribution line, the buyers and the seller will reveal quality information of products they want to buy or sell. Let $(q_B^1, \ldots, q_B^x)$ and $(q_S^1, \ldots, q_S^y)$ be the sets of quality of products to be bought or sold. Let $Match : (q_B^1, \ldots, q_B^x) \mapsto (q_S^1, \ldots, q_S^y)$ maps a product from a buyer to a seller. The optimal matching solution will minimize the following $\sum(q_B^i - Match(q_B^i))$, where $q_B^i - Match(q_B^i) \geq 0$. The problems of designing a decentralised distribution line that satisfies such optimisation condition are as follows: (a) Buyers would like to hide their quality information of products to be bought. This is because such information may reveal the inventory state of a buyer. (b) A mediator may not match the buyers and the sellers according to the objective shown in Eq. 1. A mediator may match a product with a long shelf life to a buyer who is demanding products with short shelf life. In this case, the buyer will benefit by procuring a product with long shelf life. (c) As the buyers want to hide information about the shelf life of ordered products, it is problematic for a buyer to verify if matching is performed with objective of Eq. 1.

## 4    Dynamic Distribution with Blockchains

We assume that all actors of the supply chain network are part of public proof of work-based blockchains. The proposed decentralised distribution line management solution may be executed in any blockchain network that can execute Hashed-Time Locked Contracts or similar scripts. In a supply chain network $G = (M, D_1, \ldots, D_q, R, E)$ there are $q + 1$ sets of mediator nodes $X_1, \ldots, X_{q+1}$ where each set $X_i$ contains a fixed number (positive integer) peers of the blockchain network. The supply chain network is a multi-partite graph with q+2 layers where $M$ is the first layer, $D_1$ is the second layer, and so on. Product move from layer $x$ to layer $y$ where $x < y$. Actors in the $i - 1$th layer and $i + 1$th layer will establish offline channels with mediators of the set $X_i$. Let $t_1, t_2, \ldots$ are discrete-time instances such that $t_i - t_{i-1} = dt$ where $dt$ is the number of new blocks in the blockchain. We assume that new blocks are added to the blockchain after an approximately equal time interval.

At the beginning of a time interval $t_x$, all actors in the $i + 1$th layer submit their order to any mediator node in the set $X_i$ with the lead time of the order. These actors will be referred to as the buyers. The buyers will be allowed a time $dt' < dt$ to send their orders. Order with lead time can be placed by creating a new offline channel or by updating an offline channel. After time $dt'$, the actors in the set $i'$th layer will send their intent to move inventory to the mediators $X_i$. These actors will be referred to as the sellers. Sellers will inform the mediators about their intent to move inventory by sending the self-life of their products to the mediators. They can do so by creating or updating offline channels with the mediators. The sellers can decide to move products from their inventory at any time in the time duration $t_i + dt'$ to $t_i + dt$ to sell their product before the time instance $t_{i+1}$. The mediators will execute an online maximum weighted bipartite graph matching algorithm to match the products from the sellers to the buyers. Next, we will explain how to open and update offline channels and execute the graph matching algorithm.

Blockchain offline channels [10] uses multi-signature addresses to open an offline channel among peers of the blockchain. This offline channel [10] is bidirectional and potentially infinite, i.e., it can execute the infinite number of transfers between two peers provided they do not close the channel and each of them has sufficient funds. We construct an offline channel for proof of work-based public blockchain with the following properties: (a) We construct a uni-directional channel between two peers, i.e., only one peer can send funds to another peer of this channel, (b) we construct a uni-directional channel that can be used for a finite number of transfers from a designated peer to another peer. The procedure for creating the uni-directional channel from $A$ to $B$ ($A$ transfers token to $B$)is as follows: Let $A$ and $B$ are two peers of the channel network $H$. $M_{A,B}$ is a multi-signature address between $A$ and $B$. This is a unidirectional channel from $A$ to $B$.

1. $A$ creates a set of $k$ ($k$ is a positive even integer) random strings $S_A^1, \ldots, S_A^k$. Using these random strings $A$ creates a set of Hashes $h_A^1 = Hash(S_A^1)$, $h_A^2 = Hash(S_A^2), \ldots, h_A^k = Hash(S_A^k)$ where $Hash$ is Hash function (using SHA256). $A$ creates a Merkle tree height $D = Log_2 k$ using these Hashes. In this tree there are $k$ leaf nodes and $k - 1$ non-leaf nodes of this Merkle tree. We denote the non-leaf nodes as $H_A^1, \ldots, H^{(k} - 1)_A$.
2. $B$ creates a set of $D - 1$ random strings and corresponding Hashes $H_B^1$, $\ldots, H_B^{D-1}$ such that there is a lexicographic order among these Hashes with $H_B^i \leq H_B^{i+1}$. We will call $H_B^x$ is ranked more than $H_B^y$ if the lexicographic order of $H_B^x$ is more than $H_B^y$.
3. $A$ will create a Hashed time-locked contract as follows: (a) Let $A$ wants to transfer $1 - d/D$ tokens to $B$ where $d \leq D$. (b) From the multi-signature address $M_{A,B}$ 1 token will be given to $A$ after time $T$(will be measured as the number of new blocks to be created from the current block in the blockchain) if $B$ does not claim these tokens by producing the key to any non-leaf node $H_A^x$ of the Merkle tree created by $A$, which is at a distance $d$ from the root of the Merkle tree. (c) If $B$ produces the key to such a Hash $H_A^x$ at depth $d$ then $(1 - d/D)$ tokens will be given to $B$ and remaining $d/D$ tokens will be given to $A$ if it can produce the key to a hash $H_B^x$ in the set $H_B^1, \ldots, H_B^{D-1}$ which is ranked more than $d$ hashes in this set.
4. $A$ will sign this HTLC and send it to $B$.
5. Now $A$ will send a transaction to $M_{A,B}$ of amount $1 + \epsilon$ token with the Merkle tree mentioned in the transaction. $B$ will send $\epsilon$ tokens to $M_{A,B}$ with $H_B^1, \ldots, H_B^D$ in the transaction data field where $\epsilon$ is the transaction processing fee. More than 1 token can be transferred by $A$. We are using 1 token as an example.
6. Before the next transfer from $A$ to $B$, $A$ will send the keys to the subset of Hashes $h_A^1, \ldots, h_A^k$ which can generate the Hash $H_A^x$. It will be possible for $A$ to transfer multiples of $1/D - 1$ tokens to $B$, in such a case multiple subsets of non-leaf Hashes of the Merkle tree will be revealed by $A$.

Note that, (a) $B$ signs and publishes the HTLC to claim the tokens. It gets the most number of tokens by using the last known keys of the Merkle tree leaf nodes. Thus $B$ will always use the last known keys of this Merkle tree leaf node. (b) The channel is secure as $A$ can not transfer to $B$ more than the current balance of the channel as $B$ will know the current balance by finding the non-leaf nodes from the keys supplied by $A$ and depth of such a non-leaf node. (c) All peers of the blockchain will know the existence of this channel and Hashes used in creating the channel as transactions to $M_{A,B}$ are visible to all peers.

### 4.1   Channel Between a Buyer and a Mediator

A buyer will pay a mediator if the following holds: (1) If the mediator can provide a product with shelf life more than the requested lead time of product ordered by the buyer to the mediator. (2) Additionally, the buyer should ensure that it is receiving a unique product. The protocol (shown in Fig. 1(b)) with buyer is $A$ and the mediator is $B$ is as follows: We modify the unidirectional channel as follows: (1) $A$ sends $B$ a sequence of Hashes of qualities (indicating the lead time of products) $\{Q_A^i\}$. (2) $A$ sends $B$ an HTLC similar to the HTLC in Fig. 1(a). The only changes are $B$ can claim tokens of value $d/D$ tokens if it can provide $d$ proofs of quality and $d$ proof of product uniqueness. (3) After sending this HTLC, $A$ sends the first quality requirement $q_A^1$ (whose Hash is $Q_A^1$) to $B$ and a set of keys in the Merkle tree created by $A$ to claim $1/D$ tokens from $M_{A,B}$. $B$ can claim these tokens by presenting proof of quality and proof of uniqueness. (4) Proof of quality will verify the quality of a product as claimed by a seller. We will assume that there is an IoT network monitoring quality information of the products and such quality information is periodically uploaded to an IPFS blockchain. A buyer or a seller can find proof of quality by checking the IPFS blockchain for quality information. (5) If $B$ can present such proof of quality and proof of uniqueness to $A$ then $A$ will reveal the next set of keys of the Merkle created by $A$ to send the next set of $1/D$ tokens. Otherwise, $A$ will reveal quality information $q_A^2$ (whose Hash is $Q_A^2$) to $B$ and ask it to find an appropriate product.

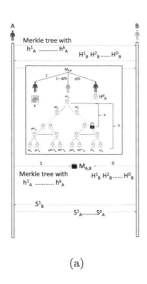

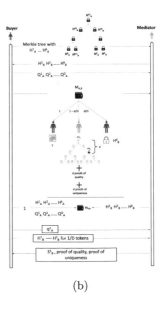

(a)                                                          (b)

**Fig. 1.** (a) Procedure of creating unidirectional offline channels. (b) Channel between a buyer and a mediator.

## 4.2    Channel Between a Seller and a Mediator

A channel between a seller and a mediator is similar to the uni-directional channel mentioned in the previous section with the following additional information:
(a) A seller will commit to a set of serial numbers to a mediator. It will store the
Hash of such serial numbers in the HTLC used to create the channel between
itself and a mediator. (b) Apart from revealing the quality of an item, the seller
must reveal a new serial number by producing the key to the Hash of a serial
number. Additionally, each seller may have a uni-directional channel (Fig. 1(a))
with at least one sink mediator with a Hash tree formed using the serial numbers
of products owned by the seller according to the channel construction protocol.
The sink mediators are mediators without channels from buyers.

## 4.3    Channel Among Mediators

The mediators are expected to build offline channels among themselves according
to the bi-directional offline channel protocol developed in [10].

## 4.4    Decentralised Product Matching

We will describe the protocol for decentralised distribution as a decentralised
matching algorithm that matches the order of a buyer to the inventory of a
supplier. First, we will explain how a mediator can match a product with the
order it has received from the buyers. Then we will explain how a seller can
sell a product via the mediators. The protocol of product sale from mediator to
buyers is as follows:

1. Assume that a mediator has received a product (of quality $Q^*$) from a seller
   or from another mediator and it wants to sell the product to the buyers who
   have established channels with it.
2. The buyers have established uni-directional channels with the mediator. Let
   there are 3 buyers $Buyer_A$, $Buyer_B$, and $Buyer_C$. They have established
   channels with quality information $(Q_A^1, Q_A^2, \ldots, Q_A^D)$, $(Q_B^1, A_B^2, \ldots, Q_B^D)$,
   and $(Q_C^1, A_C^2, \ldots, Q_C^D)$ respectively.
3. The mediator reveals $Q^*$ to $Buyer_A, Buyer_B$, and $Buyer_C$.
4. $Buyer_A, Buyer_B$, and $Buyer_C$ may reveal new reveal key to their respective
   quality Hashes (i.e., $Buyer_A$ may reveal the key to $Q_A^i$).
5. Mediator finds the buyer with the minimum difference between $Q^*$ and qualities revealed by the buyers. Say $Q^* - Q_B^i < Q^* - Q_A^i < Q^* - Q_C^i$. Mediator
   sells the product to $Buyer_B$ as it reveals the serial number to $Buyer_B$.
6. Next, the mediator will send the information $Q^* - Q_B^i < Q^* - Q_A^i < Q^* - Q_C^i$
   to all $Buyer_A, Buyer_B$, and $Buyer_C$.
7. Finally, the mediator will ask $Buyer_A, Buyer_B$, and $Buyer_C$ to reveal new
   quality information.

The protocol of product sale from the seller to mediator is as follows: (1) A seller $Seller_A$ will establish a unidirectional channel with a mediator $Mediator_1$ with serial number $Serial_1, \ldots, Serial_p$ according to the channel establishment protocol shown in Fig. 1(b). $Seller_A$ will establish a unidirectional channel with a sink mediator $Mediator_3$ with hash tree created by $Serial_1, \ldots, Serial_p$ according to the channel establishment protocol. (2) Now, $Seller_A$ will reveal key to $Serial_1$ to $Mediator_1$. (3) $Mediator_1$ will verify that there is a sink mediator $Mediator_3$ and there is a channel from $Seller_A$ to $Mediator_3$ with a hash tree that includes $Serial_1$. If $Mediator_1$ can verify this then it will pay $Seller_A$. (4) Now $Mediator_1$ will try to match the product with the serial number $Serial_1$ to the buyers who have a channel with it. If $Mediator_1$ can not do so (according to the previous protocol for product sale from a mediator to buyers) then it will try to resale this product to other mediators. (5) $Mediator_1$ can either sell it to another mediator or it can sell it to the sink mediator $Mediator_3$. (6) If it sells the product to sink mediator $Mediator_3$ by revealing the serial number then $Mediator_3$ can claim the payment from its with $Seller_A$. (7) If this sequence of product resale does not reach the sink mediator then it will indicate that the product is sold to a buyer otherwise it will indicate that the product can not be sold and it is returned to $Seller_A$.

## 5   Analysis

**Lemma 1.** *The decentralised product matching protocols are privacy-preserving.*

**Proof.** An adversary (who wants to know the quality information of products) will have access to the blockchain network and it has information on offline channels established among buyers, sellers, and mediators. However, it can not know the quality information because Quality information is kept in the channel from a buyer and a mediator or from a seller to a mediator. In both cases, Hashes of such information are included in the transaction used to establish the channels. Hence the adversary may not know the quality information of any buyer or seller if it can not control all mediators.

**Lemma 2.** *A mediator can not manipulate the matching protocol for product sale from mediator to buyers shown in Fig. 2(b).*

**Proof.** This holds because: (1) As shown in Fig. 2(b), $Buyer_A$ will verify if $Q^* - Q_A^1 > Q^* - Q_B^1$ as the mediator should produce the key to $Q^*$ and $Q_B^1$ to $Buyer_A$. (2) $Buyer_A$ has access to the transaction records and it can find the transactions from sellers to mediators and buyers to mediators that include $Q^*$ and $Q_B^1$. (3) $Buyer_A$ can verify that the above transactions are unspent and hence such quality information is not reused. (4) In case the mediator does not reveal the quality information, $Buyer_A$ may choose not to buy products via this mediator.

**Lemma 3.** *The mediators will not lose funds in the product sale protocol from a seller to a mediator shown in Fig. 2(c).*

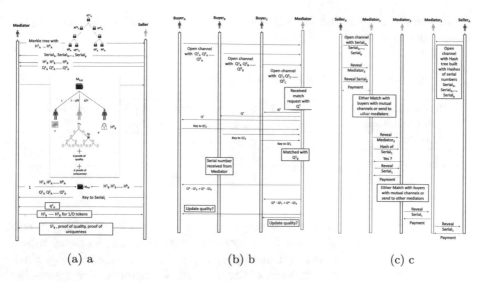

$$(a)\ a \qquad\qquad (b)\ b \qquad\qquad (c)\ c$$

**Fig. 2.** (a) Channel between a seller and a mediator, (b) protocol of product sale from a mediator to buyers, (c) protocol of product sale from a seller to mediators.

**Proof.** This holds because: (1) A mediator can verify the existence of a channel from the seller to a sink mediator by searching the transaction list of the blockchain. (2) Such a transaction denotes the existence of a channel from the seller to a sink mediator. (3) A mediator can verify that the above-mentioned channel is not closed as it can check the transaction list to ensure that the above transaction is unspent. (4) In such a live channel from the seller to the sink mediator, it can check the existence of the Hash of the serial number of product being sold by the seller. This will mean that the mediator can always resale the product to this sink mediator who can claim funds from the seller by producing the key to the Hash of the serial number.

## 6    Evaluation

We will evaluate the performance of the proposed with a simulation of supply chain and blockchains. We use proof of work-based simulation of blockchains used in [6]. We developed a supply chain simulator using agent-based modelling where all actors of a supply chain are modelled as autonomous agents. Figure 3(a) shows the workflow of a manufacturer as it executes two asynchronous processes. The first process simulates the product manufacturing rate as it manufacturer items (with a random decay) according to its demand. The second process simulates the order fulfilment procedure as it moves products from its inventory to the distributors. Figure 3(a) shows the workflow of a distributor, which is similar to the workflow of a manufacturer with an additional process that checks if an item in the transit have reached the distributor or if a product in transit should be discarded due to expiry of its shelf life. Figure 3(a) shows the workflow of a retailer who executes three asynchronous processes. We use the following supply chain network dataset (Table 1).

**Table 1.** Data for experiments

| $\|M\| = 20,\ \{\|D_1\| = 20\}_{i=1,2,3},\ \|R\| = 20$ | # of manufacturers, distributors, and retailers |
|---|---|
| $W(E) \in Random(1,6)$ | Distance among the actors |
| $\| \cup I_i \| = 10{,}000$ | Number of items |
| $a_j \in [5,15]$ | Decay rate |
| $\lambda_{demand},\lambda_{consume} = 6$ | PDF for retail demand and consumption |

We check the correctness of the supply chain simulation by executing it with increasing decay rate from 5 to 15. We observe (Fig. 3(a)) that as decay rate is increased waste is decreased and quality of items sold is increased (Fig. 4(a)). Next, we evaluate the performance of the dynamic distribution line generation with a static distribution line generation algorithm. As mentioned in Sect. 3, a static distribution line order can not be delegated and in a dynamic distribution, line order can be delegated using a graph matching algorithm executed via an offline channel network. We used an offline channel network with 5 matching nodes (nodes in the offline channels facilitating the matching process) between every pair of adjacent sets of nodes of the supply chain network (there are 5 layers in the supply chain network and we have 20 matching nodes). We execute the static distribution and dynamic distribution with the same sets of data by increasing the decay rate from 5 to 15. We observe that waste is significantly reduced using dynamic distribution lines (Fig. 4(b)) and the quality of items sold is increased by dynamic distribution lines (Fig. 4(c)).

## 7    Conclusion

In this paper, we proposed a decentralised distribution line protocol that dynamically routes perishable products to reduce food loss and waste. We used blockchain offline channels to implement such a distribution process for the high-scale execution of the distribution procedure. In the future, we will extend such distribution protocol for IPFS blockchains.

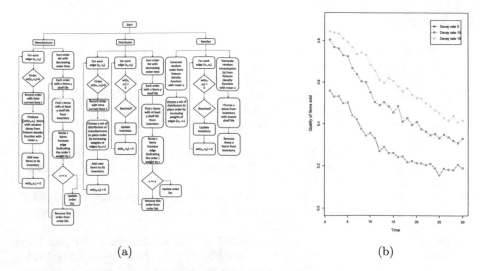

(a)                                              (b)

**Fig. 3.** (a) Workflow of the manufacturers, distributor, and retailers. (b) It shows the correctness of the simulator: as we increase the decay rate, waste is reduced and quality of products sold is increased.

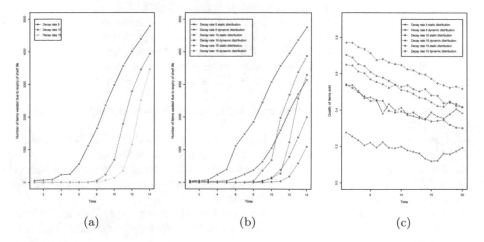

(a)                          (b)                          (c)

**Fig. 4.** (a) It shows the correctness of the simulator: as we increase the decay rate, waste is reduced and quality of products sold is increased, (b) the waste is significantly reduced using dynamic distribution lines, (c) quality of consumed product is significantly improved.

**Acknowledgement.** This publication has emanated from research conducted with the financial support of Science Foundation Ireland (SFI) under Grant Number SFI/12/RC/2289, co-funded by the European Regional Development Fund.

# References

1. Chen, W., Liu, H., Xu, D.: Dynamic pricing strategies for perishable product in a competitive multi-agent retailers market. J. Artif. Soc. Soc. Simul. **21**(2), 1–12 (2018). https://ideas.repec.org/a/jas/jasssj/2017-4-3.html
2. de Keizer, M., Akkerman, R., Grunow, M., Bloemhof, J.M., Haijema, R., van der Vorst, J.G.: Logistics network design for perishable products with heterogeneous quality decay. Eur. J. Oper. Res. **262**(2), 535–549 (2017). https://doi.org/10.1016/j.ejor.2017.03.049
3. Eom, K.H., Hyun, K.H., Lin, S., Kim, J.W.: The meat freshness monitoring system using the smart RFID tag. Int. J. Distrib. Sens. Netw. **10**(7), 591812 (2014). https://doi.org/10.1155/2014/591812
4. Eskandarpour, M., Dejax, P., Miemczyk, J., Péton, O.: Sustainable supply chain network design: an optimization-oriented review. Omega **54**, 11–32 (2015). https://doi.org/10.1016/j.omega.2015.01.006
5. Gaggero, M., Tonelli, F.: Optimal control of distribution chains for perishable goods. IFAC-PapersOnLine **48**(3), 1049–1054 (2015). https://doi.org/10.1016/j.ifacol.2015.06.222. 15th IFAC Symposium on Information Control Problems in Manufacturing
6. Hayes, B., Thakur, S., Breslin, J.: Co-simulation of electricity distribution networks and peer to peer energy trading platforms. Int. J. Electr. Power Energy Syst. **115**, 105419 (2020). https://doi.org/10.1016/j.ijepes.2019.105419
7. Reiner, J., Mike, N., Uysal, I., Lang, W.: Reducing food losses by intelligent food logistics. Philos. Trans. R. Soc. A (2014). https://doi.org/10.1098/rsta.2013.0302
8. La Scalia, G., Micale, R., Miglietta, P.P., Toma, P.: Reducing waste and ecological impacts through a sustainable and efficient management of perishable food based on the Monte Carlo simulation. Ecol. Indic. **97**, 363–371 (2019). https://doi.org/10.1016/j.ecolind.2018.10.041
9. Nakamoto, S.: Bitcoin: a peer-to-peer electronic cash system. https://www.bitcoin.org (2008)
10. Poon, J., Dryja, T.: The Bitcoin Lightning Network: Scalable Off-Chain Instant Payments. https://lightning.network/lightning-network-paper.pdf
11. Thakur, S., Breslin, J.G.: Scalable and secure product serialization for multi-party perishable good supply chains using blockchain. Internet Things **11**, 100253 (2020). https://doi.org/10.1016/j.iot.2020.100253
12. Tsao, Y.C.: Designing a fresh food supply chain network: an application of nonlinear programming. J. Appl. Math. **2013** (2013). https://doi.org/10.1155/2013/506531
13. Tykhonov, D., Jonker, C., Meijer, S., Verwaart, T.: Agent-based simulation of the trust and tracing game for supply chains and networks. J. Artif. Soc. Soc. Simul. **11**(3), 1 (2008). https://jasss.soc.surrey.ac.uk/11/3/1.html
14. Yan, B., Lee, D.: Application of RFID in cold chain temperature monitoring system. In: 2009 ISECS International Colloquium on Computing, Communication, Control, and Management, vol. 2, pp. 258–261 (2009). https://doi.org/10.1109/CCCM.2009.5270408

# ChronoEOS: Configuration Control System Based on EOSIO Blockchain for On-Running Forensic Analysis

Jose Alvaro Fernandez-Carrasco$^{(\boxtimes)}$, Telmo Egues-Arregui, Francesco Zola⬤, and Raul Orduna-Urrutia⬤

Vicomtech Foundation, Basque Research and Technology Alliance (BRTA), Paseo Mikeletegi, Donostia/San Sebastian 20009, Spain
{jafernandez,tegues,fzola,rorduna}@vicomtech.org

**Abstract.** In almost any industrial field, there are critical devices and elements that must be trusted. Therefore, it is meaningful to make a secure and reliable record of the events occurring in that environment, which may be used for a possible future forensic investigation in an industrial environment. Blockchain technology covers this need. This project presents ChronoEOS, an application that allows different versions of critical files to be stored securely and immutably, using EOSIO's blockchain technology. In this way, changes to the configuration files of a robotic arm are stored in the blockchain. In addition, a REST API has been deployed that allows the authorized user to make changes to the corresponding file by interacting with a Smart Contract deployed on the blockchain. Finally, there is a Lookout Agent pointing to the files that are being monitored. This code detects when the version stored in the blockchain is different from the original file, updating one or the other depending on where the last change has been applied.

**Keywords:** Forensic analysis · EOSIO · Smart industry · Control version

## 1 Introduction

In an increasingly technological world where the volume of devices, data and information has grown exponentially, being able to extract digital evidence out of the use of these elements has become very important for forensic investigations and trusted computing. This phenomenon is even more relevant in critical infrastructures such as energy plants or industrial environments. In fact, the management of sensitive and critical data arise different concerns about the security of information. In this scenario, a slight change in a configuration file of an industrial device can cause problems in the operation of the device, resulting in production fault, delay or even safety risks for plant operators.

For this reason, performing forensic analysis on industrial devices is becoming increasingly necessary to safeguard the legitimate interests of the owners of the

J. Prieto et al. (Eds.): BLOCKCHAIN 2022, LNNS 595, pp. 37–47, 2023.
https://doi.org/10.1007/978-3-031-21229-1_4

information and the systems that manage it. However, the current difficulties in extracting evidence make it very difficult to ensure that such information has not been modified. Furthermore, usually, these operations are performed offline, i.e., stopping the production plant, extracting the information and finally performing the analysis. For this reason, in this work, we propose a version control system based on EOSIO blockchain for on-running forensic analysis, called *ChronoEOS*.

Blockchain technology is one of the most innovative and disruptive concepts of the last few years [15]. This technology could be seen as a repository shared by the network participants, with the particularity that what is stored is immutable. Thus, when a transaction takes place on the blockchain, or any movement associated with it (creation of an account, call to a Smart Contract, etc.), it is stored in one of the blocks. When the block is full of information, it is attached to the blockchain, following the last existing block. Furthermore, blockchain information is copied among the network participants which are also involved in decision-making process, making the blockchain a distributed and decentralized system.

ChronoEOS represents a possible solution to monitor the changes that occur in sensitive repositories in the industrial environment to be used as a basis for the capture and analysis of digital evidence in a forensic analysis context. To this end, the proposed solution is based on blockchain technology to guarantee the integrity and immutability of the evidence collected for possible future investigations.

The rest of the paper is organized as follows. Section 2 introduces the used blockchain, forensic concepts and analyze related work. After that, Sect. 3 presents the problem description, whereas in Sect. 4 our solution and its structure are introduced. Section 5 describes the implementation and the obtained results. Finally, in Sect. 6, we draw conclusions and provide guidelines for future work.

## 2   Background

In this section, several concepts related to forensic analysis and blockchain are introduced. More specifically, in Sect. 2.1 an overview about forensic analysis is reported, whereas in Sect. 2.2, EOSIO blockchain is introduced, as well as information about forensic studies that directly use blockchain technology.

### 2.1   Forensic Analysis

The process involving the detection, investigation, and documentation of a digital device or system behind a security incident generated by misuse, an attack, or a crime can be defined as forensic analysis. This analysis is fundamental for Law Enforcement Officers (LEOs) to demonstrate the guilt or innocence of a user in court hearings. More specifically, forensic analysis is based on gathering digital information, which can be seen as everything that has to do with the evidence coming from computer systems such as PCs, mobile phones, or cameras [3]. However, the evidence cannot be handled under any conditions, and for it to be valid in a court, it should follow a precise procedure. That is why, chain of custody concept is relevant [11]. The chain of custody represents the trail or chronological steps that follow-up for the electronic evidence and are mandatory in order to preserve the integrity of this evidence. This process can be divided into five main parts as mentioned in [4]: *Collect evidence, Authenticate, Examine, Analyze* and *Report of the evidence*. The first part is related to the way in which the evidence is collected from the device; the second part involves techniques about how the evidence is authenticated and validated; the third phase includes examination of the evidence, followed by a deep analysis of the material. In the fifth part, a complete report of the whole process is written.

In this sense, forensic techniques cannot be applied only by LEOs for criminal hearings, but also by researchers [12] and dedicated practitioners to discover new information about techniques, failures and incidents in their infrastructure. Forensic analysis is even more relevant when incidents (fault or attack) affect critical infrastructures, such as supply chains, industrial systems, energy plants, and so on. In fact, in these cases, changes in the configurations, misuse or attacks, can have an impact not only on the economical aspect but also on human safety or national security.

## 2.2 Blockchain in Forensic Analysis

EOSIO is among the blockchains with the highest Potential [7], and it is specifically designed to facilitate the operation of decentralized applications, also called *dApps*. In terms of its architecture, EOSIO uses 3 key concepts common in computer science domain to explain how its blockchain works: *NET (Bandwidth)*, *CPU (Computation)* and *RAM (Memory or storage space)*. NET is related to the need to transmit information over the network. CPU contains information about processing power to deploy a dApp or to execute a Smart Contract function, whereas RAM gathers insight about the amount of blockchain information stored. CPU and NET are mostly used by ordinary users to interact with the blockchain's applications and contracts, while RAM is used almost exclusively by developers for implementing decentralized applications (dApps) [16]. Although, exists different consensus mechanisms [14], i.e., fault-tolerant policies used for coordinating the decision-making process, in EOSIO, Delegated Proof of Stake (DPoS) and Byzantine Fault Tolerance (BFT) algorithms are used [6]. These mechanisms are more sustainable and environmentally friendly, as they do not require an excessive consumption of computing resources [5]. More specifically, BFT and DPoS algorithms are employed at different EOSIO layers.[1] BFT is used in Layer 1, which aim is to monitor and confirm the execution of the blocks so that they are permanently recorded in the blockchain. On the other hand, DPoS is used in Layer 2, which is in charge of selecting the nodes to be authorized to sign blocks in the network [9]. In EOSIO, 21 producer nodes are elected through

---

[1] https://eos.io/news/dpos-bft-pipelined-byzantine-fault-tolerance/.

a voting system in which stakeholders can participate. These nodes share the blocks production, so that each producer signs 12 blocks, with a period of 0.5 s between blocks. This approach enables the possibility to work in environments with a very quick response, and, at the same time, keep control over the block production.

Having a record, such as a database or event logs, in a given environment is relatively straightforward, but ensuring that this evidence log has not been compromised in an unwanted and undetected manner is a complex task. Here is where blockchain technology can play a decisive role for the forensic community, as it allows for decentralized and immutable records of events and transactions occurring on the network. For example, in [10], authors propose the use the blockchain technology to ensure the integrity of a chain of custody process. In [2] the focus is on vehicle applications, wherein soon vehicles will collect a lot of information from their environment for different applications. Another solution for an IoT environment is presented in [1]. The authors propose Hyperledger Fabric, a permissioned blockchain, for storing metadata of important events related to Smart Homes applications. A work that ensure proof of existence and preservation for examination using encrypted IoT data is presented in [8]. Finally, in [13] is also applied blockchain technology to the IoT environment, to store device logs and ensure their immutability, preventing the existence of tampered logs.

In this work we propose a solution that combines blockchain technology (EOSIO) and forensic analysis to preserve traces of modifications in a critical industrial system. Moreover, once the modification is detected, our solutions allows authorized operators to revert such modifications, going back to a previous and safe system configuration. To the best of our knowledge, this is the first work toward the implementation of a more secure and reliable version control system that exploits EOSIO blockchain for on-running forensic analysis.

## 3   Use Case Description

"Traditional" techniques used for industrial forensic analysis, are based on analyzing and discovering the unauthorized modifications only after an incident or a failure as happened (ex post). This is due to the fact that, in these cases, forensic experts can exploits the incident for stopping the industrial system and gather all the available evidence, limiting their impact over the production. However, novel methods for performing the forensic analysis on-running and that at the same time can be use to detect configuration changes and so prevent incident and failures need to be investigated. For this reason, in this work, ChronoEOS, a configuration control system based on EOSIO blockchain for on-running forensic analysis is introduced. More specifically, we apply ChronoEOS in a concrete environment composed by a robotic arm for monitoring the modifications in its configuration files, as shown in Fig. 1.

**Robot Overview**. Universal Robots (UR) Ltd. is a company that builds robotic arms for industrial environments and the manufacturing world. UR developed

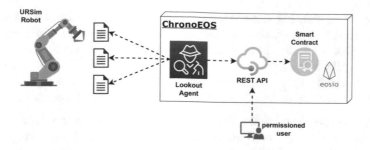

**Fig. 1.** ChronoEOS and robot application.

a simulator of these robots into software that can be installed on a Windows or Linux server. This program can simulate and visualize the movements performed by the robot, introduce personalized programs for the robots to execute, emulate the I/O analog and digital pins as well as monitor Modbus protocol interactions in the simulator. This environment enables the user to interact with many utilities and properties of a Universal Robots arm, even if not all of the functionalities of the real robots are available. In this work, **URSim Robot** is the chosen OT component containing sensitive files, i.e., files whose versions will be kept in the blockchain. The sensitive files are created in a directory where the programs run by the robots are stored.

## 4    ChronoEOS

In this paper, we implement ChronoEOS to keep track of changes in a configuration file (*sensitive or target information*) of an industrial robot (*target devices*). In this way, ChronoEOS gives the possibility to authorized users (operators) within the industrial environment to recover previous versions of the robot configuration or even to modify it according to specific needs.

ChronoEOS consists of the following three modules: Lookout Agent, Rest API and Blockchain.

**Lookout Agent**. This is deployed in the industrial robot and it is in charge of detecting changes in the configuration file as well as modifying the it directly when the API requires so. This code is executed in a scheduled process (**cron job**) so versions of the pointed target information are saved at fixed times. Since the calls are made to *upsert* function of the Smart Contract, the credentials of the authorized user (account name and private key) must be added to this code.

**Rest API**. With the idea of having an interface so that the operator (authorized user) can interact with both the Smart Contract and the target information, an API has been developed. The methods available in the API are as follows, which can be divided into two sub-blocks: *Blockchain EOS* and *Smart Contract*. Blockchain EOS contains the two necessary methods to start and stop the production of blocks in EOSIO, another for displaying the status of any of the monitored files, showing the different versions previously stored, and a method to obtain information on the transactions that have been recorded. Smart Contract contains methods to display the information for a given account, and a method to call the *upsert* function of the Smart Contract. This function receives as parameters the name of the account, its associated private key, the name of the configuration file to be modified and the new content (plain text).

**Blockchain**. The blockchain deployed in ChronoEOS is an EOSIO private blockchain, with three block-producing nodes. Moreover, the blockchain has an API node to listen to HTTP requests to the blockchain and a seed node that stores in memory the information of all the blocks, in the case at some point the blockchain crashes and you want to retrieve the information. The blockchain is the key element in this application to ensure the recording of evidence and to carry out forensic analysis with guarantees in this industrial environment.

Regarding the Smart Contract deployed, it uses multi-index tables for the registration of the target information. Multi-index tables[2] are a way of caching status or data in RAM for fast access. The blockchain records the transactions, but the multi-index tables allow the application data to be stored. In particular, in this Smart Contract, for each target information, the following information are stored:

1. **Version**: It starts with the initial version and increases as changes are made to the document.
2. **Timestamp**: Timestamp of the moment of the change.
3. **File**: Name of the document to be modified or observed.
4. **Content**: Content of the file in hexadecimal format.

The Smart Contract has an *upsert* function, which receives as inputs the values of the target information and *content* variables, which allows a new version of the data to be added to the table. The parameters version and timestamp are obtained directly in the function. An important particularity added to the contract is that the *upsert* function can only be called by the account that has deployed the contract. For this reason, an external account would not be able to modify the target information, which makes the proposed system more secure.

## 5    Experimental Results

### 5.1    Deployment and Evaluation

The steps to deploy ChronoEOS are the following:

1. To deploy EOSIO's private blockchain, with three producer nodes, one node to receive HTTP requests from the API, and one seed node to record the blocks information on the chain.
2. To create the account on the blockchain for the authorized user. Store the private key in a safe place, as it will be needed to make requests.

---

[2] https://developers.eos.io/manuals/eosio.cdt/v1.7/group_multiindex.

3. To deploy the Smart Contract for saving the multi-index tables for the created account.
4. To Deploy the REST API on some secure server, external to the robot environment, but accessible from it, in order to be able to make the API method calls.
5. On the device where the documents (target information) to be supervised are located, deploy the Lookout Agent, pointing to the objective files.
6. Execute the Lookout Agent in a scheduled process (for example, a cron).

Let $X_e$ the content of the robot configuration file taken at time $e$, and $X'_t$ the last version of this file taken at time $t$ and stored in the blockchain. The scheduled process (cron), in every execution, checks that $X_e = X'_t$. If the condition is verified, no actions are taken, otherwise ($X_e \neq X'_t$) the next step is to check in which environment the last change has occurred, using the timestamp. Specifically, if $e > t$, i.e., the timestamp of the file is greater than the timestamp of the last element stored in the blockchain, this means that $X_e$ has changed. Therefore, the Lookout Agent calls the *upsert* function of the Smart Contract, through the API, and indicates the new content to be saved in the blockchain as a new version of the file. On the other hand, if $e < t$, it means that $X'_t$ is changed. This can happen since the authorized user can directly use the API for this purpose, calling the *upsert* method. In this case, the Lookout Agent takes the *content* field and makes the change in the original file, updating the file with the content introduced through the API. Having all versions of the file on the blockchain as immutable data, this information can be used as evidence for future forensic investigations. In addition, the fact of saving it as Multi-Index Tables allows the authorized user to check past versions of the file, observing relevant data such as the content of the file and the timestamp. Thus, if this user requires to retrieve a past version of the file, he can do so without difficulty through the API and the Smart Contract.

Although the ChronoEOS application allows the requirements for forensic analysis to be met, it is equally important that the application is lightweight, with as little impact as possible in terms of resources consumed. For this reason, two measurements have been carried out:

- Measuring the impact of the Lookout Agent on the environment where it is deployed (the robot simulator), obtaining the computational resources it consumes when running (CPU, RAM and network packet transmission when calling the API).
- Measure the memory required to store the information generated in the blockchain.

The robot simulator specifications have been obtained from the official robot virtual machine image. The deployment has been made in a ESXi server with the mentioned parameters and the Lookout Agent is installed as a cron process. In order to measure the impact of the execution of this piece of code, *psutil* python library is used, capable of retrieving information on running processes and resource consumption.

During the test, changes have been made both in the original file and from the "upsert" function of the API, to analyse how many resources are required by the system to perform the corresponding actions. Thus, the calls to the Lookout Agent are in seconds (see Figs. 2, 3 and 4): 20, 80, 140, 200, 260. In seconds 5 and 120 the changes have been made from the API (function "upsert"), while in seconds 50, 160 and 220 the changes have been made directly in the original file. The graphs show how every minute (when the Lookout Agent process is executed) there are peaks in resources, both CPU, disk memory and packets received and sent. This is because when the Lookout Agent is called, it consults the blockchain to check if there has been any modification and applies the corresponding steps, depending on where the change has occurred. Another noteworthy aspect is that, when the changes are produced from the blockchain (before the first and third peak), the data sent is significantly less than when the change is produced in the original file, since in this case the new content must be sent to the blockchain, calling the upsert function, which requires a greater amount of information to be sent. On the other hand, it can be seen that RAM is hardly affected by this process.

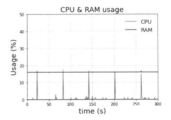

**Fig. 2.** CPU and RAM usage.

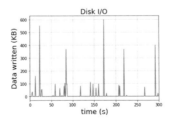

**Fig. 3.** Disk usage.

As for the memory required to store blockchain events, it is necessary to have some storage capacity, especially if you want to deploy the application in an environment where you do not want to stop block production at any time, as every half second a block is generated and added to the chain. So, for every

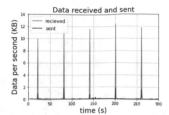

**Fig. 4.** Data I/O.

100,000 blocks, around 125 MB of storage is required, depending on the number of transactions that have been recorded. Below is a graph (Fig. 5) showing the disk space used by the blockchain, depending on the number of blocks it has. Note that the space is highly variable, as it depends on the number of transactions that take place (the more transactions, the more space the blocks need).

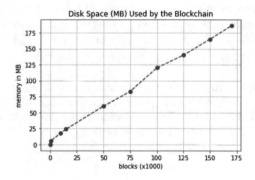

**Fig. 5.** Memory used by the blockchain.

## 5.2 Discussion

The fact of having a REST API from which a authorized user can manage and visualize the different critical documents is a relevant point in favor of the ChronoEOS approach, as all the tasks related to the traceability of these documents can be carried out from a single command post. On the other hand, although the Smart Contract is public and can be deployed by any account on the blockchain, it is designed so that only the account that has deployed the contract can call the *upsert* function. This ensures that critical files can only be modified from the REST API by authorized accounts. The most sensitive and vulnerable point of ChronoEOS is the Lookout Agent, as it must be guaranteed that the code is always running. If it is no longer running, a change in any file would not be detected, so the traceability of the code would be lost.

## 6   Conclusion

The main conclusion of this project is that ChronoEOS is an application that can help in Forensic Analysis with guarantees within a given industrial environment. In addition, the built-in Rest API provides flexibility and convenience to authorized users, as they can make controlled modifications to files, monitor changes, and start or stop the production of blocks on the blockchain and so on, from a user-friendly and easy-to-use command window. On the other hand, the EOSIO blockchain is well suited for this type of task, thanks to its consensus protocol (dPOS), which allows it to have a high rate of block production, and the fact that the programming language of the Smart Contracts is C++, which is widely used. However, logically, other blockchain technologies, such as Ethereum, can also be valid. As future lines for future work, work is expected to be done to improve the scalability of the application. Thus, the aim is that ChronoEOS can be applied to a more complex industrial environment, with a multitude of devices and files. This could mean having to adjust the Smart Contract and the Lookout Agent to adapt it to new needs. Another issue to be improved in future lines of work is the use of the cron process. The cron process can only be run at a minimum of every minute, resulting in malfunction issues, for example, two changes taking place within the same minute results in the first one not being registered in the blockchain, missing versions on the way.

**Acknowledgements.** This work has been partially supported by the Basque Country Government under the ELKARTEK program, project REMEDY (KK-2021/00091), and by the Spanish Centre for the Development of Industrial Technology (CDTI) under the project ÉGIDA (EXP 00122721/CER-20191012).

## References

1. Brotsis, S., Kolokotronis, N., Limniotis, K., Shiaeles, S., Kavallieros, D., Bellini, E., Pavué, C.: Blockchain solutions for forensic evidence preservation in IoT environments. In: 2019 IEEE Conference on Network Softwarization (NetSoft), pp. 110–114. IEEE (2019)
2. Cebe, M., Erdin, E., Akkaya, K., Aksu, H., Uluagac, S.: Block4forensic: an integrated lightweight blockchain framework for forensics applications of connected vehicles. IEEE Commun. Mag. **56**(10), 50–57 (2018)
3. Divith Devaiah, M., Metre, P.B.: Survey on current digital forensic practices (2017)
4. Giova, G., et al.: Improving chain of custody in forensic investigation of electronic digital systems. Int. J. Comput. Sci. Netw. Secur. **11**(1), 1–9 (2011)
5. He, N., Zhang, R., Wang, H., Wu, L., Luo, X., Guo, Y., Yu, T., Jiang, X.: {EOSAFE}: security analysis of {EOSIO} smart contracts. In: 30th USENIX Security Symposium (USENIX Security 21), pp. 1271–1288 (2021)
6. Huang, Y., Wang, H., Wu, L., Tyson, G., Luo, X., Zhang, R., Liu, X., Huang, G., Jiang, X.: Understanding (mis) behavior on the EOSIO blockchain. Proc. ACM Meas. Anal. Comput. Syst. **4**(2), 1–28 (2020)
7. Leiponen, A., Thomas, L.D., Wang, Q.: The dApp economy: a new platform for distributed innovation? Innovation 1–19 (2021)

8. Li, S., Qin, T., Min, G.: Blockchain-based digital forensics investigation framework in the internet of things and social systems. IEEE Trans. Comput. Soc. Syst. **6**(6), 1433–1441 (2019). https://doi.org/10.1109/TCSS.2019.2927431
9. Liu, J., Zheng, W., Lu, D., Wu, J., Zheng, Z.: Understanding the decentralization of DPoS: perspectives from data-driven analysis on EOSIO. arXiv preprint arXiv:2201.06187 (2022)
10. Lone, A.H., Mir, R.N.: Forensic-chain: Ethereum blockchain based digital forensics chain of custody. Sci. Pract. Cyber Secur. J. **1**, 21–27 (2018)
11. Prayudi, Y., Sn, A.: Digital chain of custody: state of the art. Int. J. Comput. Appl. **114**(5) (2015)
12. Rafique, M., Khan, M.: Exploring static and live digital forensics: methods, practices and tools. Int. J. Sci. Eng. Res. **4**(10), 1048–1056 (2013)
13. Ricardo, P., Jurcut, A.: BLOFF: a blockchain based forensic model in IoT (2021). https://doi.org/10.4018/978-1-7998-7589-5.ch003
14. Wang, Y., Cai, S., Lin, C., Chen, Z., Wang, T., Gao, Z., Zhou, C.: Study of blockchains's consensus mechanism based on credit. IEEE Access **7**, 10224–10231 (2019)
15. Xu, M., Chen, X., Kou, G.: A systematic review of blockchain. Financ. Innov. **5**(1), 1–14 (2019). https://doi.org/10.1186/s40854-019-0147-z
16. Zheng, W., Zheng, Z., Dai, H.N., Chen, X., Zheng, P.: Xblock-EOS: extracting and exploring blockchain data from EOSIO. Inf. Process. Manag. **58**(3), 102477 (2021). https://doi.org/10.1016/j.ipm.2020.102477

# Sharding-Based Proof-of-Stake Blockchain Protocol: Security Analysis

Abdelatif Hafid[1]([⊠])[ID], Abdelhakim Hafid[1][ID], and Adil Senhaji[2][ID]

[1] University of Montreal, Montréal, QC H3T 1N8, Canada
abdelatif.hafid@umontreal.ca, ahafid@iro.umontreal.ca
[2] Mizuho Securities, New York, NY 10022, USA
adil.senhaji@mizuhogroup.com

**Abstract.** Blockchain technology has been gaining great interest from a variety of sectors including healthcare, supply chain, and cryptocurrencies. However, Blockchain suffers from its limited ability to scale (i.e., low throughput and high latency). Several solutions have been proposed to tackle this issue. In particular, sharding proved that it is one of the most promising solutions to Blockchain scalability. Sharding can be divided into two major categories: (1) Sharding-based Proof-of-Work (PoW) Blockchain protocols, and (2) Sharding-based Proof-of-Stake (PoS) Blockchain protocols. The two categories achieve a good performance (i.e., good throughput with a reasonable latency), but raise security issues. This article focuses on the second category. In this paper, we provide a probabilistic model to analyze the security of these protocols. More specifically, we compute the probability of committing a faulty block and measure the security by computing the number of years to fail. Finally, we evaluate the effectiveness of the proposed model via a numerical analysis.

**Keywords:** Blockchain scalability · Sharding · Security analysis · Proof-of-Stake · Practical Byzantine fault tolerance

## 1 Introduction

With the rise of Bitcoin [8], Blockchain has attracted significant attention from both industry and academia. More specifically, it has been adopted in different industry segments including healthcare [7], finance [9], and public sector [1]. However, the capacity of Blockchain to scale is very limited [4]. For example, in the case of cryptocurrencies, Bitcoin [8] handles between 3 and 7 transactions per second (tx/s), which is very limited compared to traditional payment systems (e.g., PayPal [10]). Several solutions were proposed to scale Blockchain. In

Supported by CIRRELT – Interuniversity Research Centre on Enterprise Networks, Logistics and Transportation.

particular, sharding has emerged as a promising solution [4]. Sharding consists of partitioning the network into sub-networks, called shards; all shards work in parallel to enhance the performance of the network. More specifically, each shard processes a sub-set of transactions instead of the entire network processing all the transactions. While sharding considerably improves scalability, it decreases the level of Blockchain security. More specifically, in sharding-based Blockchains, it is easy for a malicious user (aka, malicious/Byzantine node) to conquer and attack a single shard compared to the whole network; this attack is well-known as a shard takeover attack (aka, 1% attack) [5].

Blockchain networks are susceptible to Sybil attacks by malicious nodes (called Sybil nodes). Several consensus mechanisms (e.g., **P**roof-of-**W**ork (PoW), **P**roof-of-**S**take (PoS), and **p**ractical **B**yzantine **F**ault **T**olerance (pBFT)) have been proposed to defend against these Sybil nodes. Sharding-based Blockchain protocols [4,12] can be classified into two classes: sharding-based PoW and sharding-based PoS Blockchain protocols.

Recently, Hafid et al. [2,3,5] proposed mathematical models to analyze the security of sharding-based PoW Blockchain protocols. In this paper, we focus on the sharding-based PoS Blockchain protocols. We propose a probabilistic model to analyze the security of these protocols by computing the probability of committing a faulty block. Based on these probabilities, we calculate the number of years to fail for the purpose of quantifying and measuring the security of the network.

The rest of the paper is organized as follows. Section 2 presents the proposed probabilistic model. Section 3 presents numerical results and evaluates the proposed model. Section 4 concludes the paper.

## 2   Probabilistic Model

In this section, we propose a probabilistic model to analyze the security of sharding-based PoS Blockchain protocols. Generally, a sharding-based PoS Blockchain protocol (e.g., Incognito [6]) consists of one specific chain, called the beacon chain and many shard chains. The beacon chain synchronizes all shard chains in the network.

First, we start by computing the probability of a shard to commit a faulty block. Second, we calculate the probability of the beacon chain to commit a faulty block. Third, we compute the probability of all shards committing a faulty block. Finally, based on all these probabilities, we compute the probability of committing/adding a faulty block to the blockchain.

### 2.1   Notations and Definitions

Table 1 shows the list of symbols and variables that are used to describe the proposed probabilistic model.

**Definition 1.** (*Faulty Block*) A faulty block is a block that contains fraudulent transactions.

**Table 1.** Notations and symbols.

| Notation | Description |
|---|---|
| $\mathcal{N}$ | Number of users |
| $n$ | Committee size of a shard |
| $n'$ | Committee size of the beacon chain |
| $H$ | Number of honest validators in a shard |
| $M$ | Number of malicious validators in a shard |
| $\mathcal{V}$ | Number of validators in a shard ($\mathcal{V} = H + M$) |
| $\zeta$ | Number of shards |
| $X$ | Random variable that computes the number of malicious nodes in the committee of a shard |
| $H'$ | Number of honest validators in the beacon chain |
| $M'$ | Number of malicious validators in the beacon chain |
| $\mathcal{V}'$ | Number of validators in the beacon chain ($\mathcal{V}' = H' + M'$) |
| $X'$ | Random variable that computes the number of malicious nodes in the committee of the beacon chain |
| $r$ | Resiliency of the shard committee |
| $r'$ | Resiliency of the beacon committee |
| $R$ | Percentage of malicious validators in a shard chain |
| $R'$ | Percentage of malicious validators in the beacon chain |
| $\mathcal{P}_f$ | Probability of conquering the protocol |
| $\mathcal{P}$ | Probability of a shard to commit a faulty block |
| $\mathcal{P}'$ | Probability of the beacon chain to commit a faulty block |
| $\mathcal{P}''$ | Probability of all shards committing a faulty block |
| $\mathcal{Y}_f$ | Number of years to fail |

**Definition 2.** (*Conquering the Protocol*) A protocol is said to be conquered if the malicious nodes success to add a faulty block to the blockchain.

**Definition 3.** (*Committee Resiliency of a Shard*) The maximum percentage of malicious nodes that the committee of the shard chain can support whereas still being secure.

**Definition 4.** (*Committee Resiliency of the Beacon Chain*) The maximum percentage of malicious nodes that the committee of the beacon chain can support whereas still being secure.

## 2.2   Architecture

In this section, we present a sample architecture of sharding-based PoS Blockchain protocols. This scheme is similar to that of Incognito [6].

Figure 1 shows a sample sharding-based PoS Blockchain protocol, which contains a single beacon chain and $\zeta$ shard chains. Shard chains produce blocks in parallel. All shard chains are synchronized by the beacon chain. More specifically, each shard has its own committee (i.e., a subset of the network nodes), which is randomly assigned by the beacon chain. Each shard chain processes a subset of the transactions submitted to the network. When a shard block is created, the beacon committee verifies the block; if it is valid, it adds the block header to the beacon chain. Otherwise, it drops it and sends the proof to other shards for a vote to slash the misbehaving shard committee. Furthermore, in each epoch, the beacon chain shuffles committees, of the shards, to increase the security of the blockchain. For Incognito [6], when a new random number is generated, the beacon chain shuffles the committees; one epoch, for Incognito, corresponds to generating a new random number. This number is generated periodically in a round-robin fashion [6,11].

## 2.3   Probability distributions

Generally, to add a faulty block to a sharding-based PoS Blockchain protocol (e.g., Incognito [6]), it must be confirmed by at least $\beta$ $(0 < \beta < 1; \beta = r)$ of the shard committee members, by at least $\beta$ of the beacon committee members $(\beta = r')$, and by at least $\beta$ of all shards' committees. For *Incognito* [6], $\beta = r = r' = \frac{2}{3}$.

**Lemma 1.** *The probability of a shard to commit a faulty block $(\mathcal{P})$ can be expressed as follows:*

$$P(X \geq \beta n) = \sum_{j=\beta n}^{n} \frac{\binom{M}{j}\binom{H}{n-j}}{\binom{V}{n}} \tag{1}$$

Proof of Lemma 1 results directly from the cumulative hypergeometric distribution [3,5].

**Lemma 2.** *The probability of at least $\beta$ of all shards committees committing a faulty block $(\mathcal{P}')$ can be computed as follows:*

$$\sum_{i=\frac{2\zeta}{3}}^{\zeta} \left( P(X \geq \beta n) \right)^i = \sum_{i=\beta\zeta}^{\zeta} \sum_{\alpha=\beta n}^{n} \left( \frac{\binom{M}{\alpha}\binom{H}{n-\alpha}}{\binom{V}{n}} \right)^i \tag{2}$$

***Proof.*** The minimum number of committees to commit a faulty block is $\beta\zeta$, where $\zeta$ is the number of shards. The probability of exactly $\beta\zeta$ committees confirm/agree to add a faulty block can be expressed as follows:

$$P_{\beta\zeta} = \left( P(X \geq \beta n) \right)^{\beta\zeta} \tag{3}$$

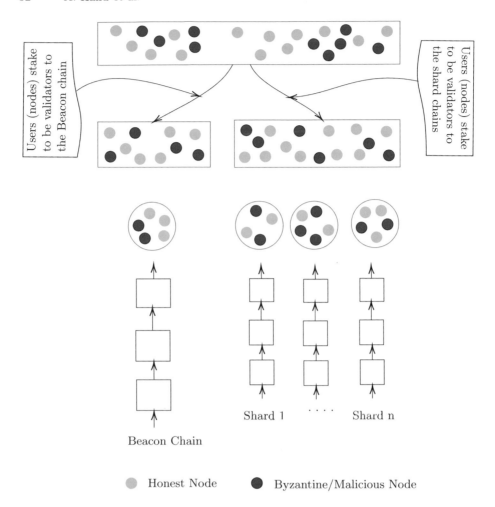

Beacon Chain

Shard 1 · · · · Shard n

Honest Node          Byzantine/Malicious Node

**Fig. 1.** A sharding-based PoS and pBFT blockchain protocol.

The probability to commit a faulty block by exactly $\beta\zeta + 1$ committees can be expressed as follows.

$$P_{\beta\zeta+1} = \left( P(X \geq \beta n) \right)^{\beta\zeta+1} \tag{4}$$

Similarly, the probability of exactly $\zeta$ committees (the entire number of shards in this case) agreeing to add a faulty block can be expressed as follows:

$$P_{\zeta} = \left( P(X \geq \beta n) \right)^{\zeta} \tag{5}$$

A faulty block can be committed if $\beta\zeta$ or $\beta\zeta + 1$ or $\beta\zeta + 2, \cdots,$ or $\zeta$ committees agree to add this block. This can be mathematically computed by the sum over all these probabilities and can be expressed as follows:

$$\mathcal{P}'' = P_{\beta\zeta} + P_{\beta\zeta+1} + \cdots + P_{\zeta} \tag{6}$$

**Lemma 3.** *The probability of the beacon's committee committing a faulty block ($\mathcal{P}'$) can be expressed as follows:*

$$P(X' \geq \beta n') = \sum_{j=\beta n'}^{n'} \frac{\binom{M'}{j}\binom{H'}{n'-j}}{\binom{\mathcal{V}'}{n'}} \tag{7}$$

Proof of Lemma 3 results directly from the cumulative hypergeometric distribution [3,5].

**Theorem 1.** (Committing a Faulty Block) *The probability of committing a faulty block ($\mathcal{P}_f$) by a given shard can be expressed as follows:*

$$\mathcal{P}_f = \sum_{k=\beta n}^{n} \sum_{i=\beta\zeta}^{\zeta} \sum_{\alpha=\beta n}^{n} \sum_{j=\beta n'}^{n'} \frac{\binom{M}{k}\binom{H}{n-k}\binom{M}{\alpha}^i\binom{H}{n-\alpha}^i\binom{M'}{j}\binom{H'}{n'-j}}{\binom{\mathcal{V}}{n}\binom{H}{n-\alpha}^i\binom{\mathcal{V}'}{n'}} \tag{8}$$

***Proof.*** To commit a faulty block, it must be confirmed/verified by at least $\beta$ of the shard committee members, by at least $\beta$ of the beacon committee members, and by at least $\beta$ of all shards' committees. This can be expressed by the product over the three probabilities (the calculated probabilities in Lemmas 1–3).

### 2.4   Years to Fail

To make the measurement of the security more readable, we propose to compute the number of years to fail ($\mathcal{Y}_f$) based on the calculated failure probability (i.e., the probability of conquering the protocol). This number can be expressed as follows:

$$\mathcal{Y}_f = 1/\mathcal{P}_f/\mathcal{N}_s \tag{9}$$

where $\mathcal{P}_f$ is the probability of committing (adding) a faulty block to the blockchain and $\mathcal{N}_s$ is the number of sharding rounds per year (aka, number of epochs per year).

## 3   Results and Evaluation

In this section, we evaluate the effectiveness of the proposed probabilistic model via numerical simulations.

### 3.1   Simulation Setup

In order to implement the proposed model, we make use of a built-in Python library called **SciPy**. Particularly, we import **hypergeom** from **scipy.stats** model.

## 3.2 Results and Analysis

In Fig. 2, we assume a network with $\mathcal{N} = 2000$ nodes, $\mathcal{V} = 200$, $\mathcal{V}' = 400$, $\zeta = 8$, $r = r' = 0.5$.

Figure 2a shows the probability of a shard to commit a faulty block versus the size of the committee. We observe that the probability $\mathcal{P}$ decreases when the size of the committee increases. More specifically, we observe that the probability corresponding to $\mathcal{R} = 0.2$ (i.e., 20% of malicious nodes in each shard) decreases rapidly compared to those of $\mathcal{R} = 0.25$ and $\mathcal{R} = 0.3$; this can be explained by the small percentage of malicious nodes. In other words, as the percentage of malicious nodes gets smaller the probability decreases and vice versa.

Figure 2b shows the probability of all shards committing a faulty block versus the size of the committee. We observe that the probability $\mathcal{P}'$ decreases when the size of the committee increases. Similarly, as the percentage of malicious nodes slightly increases in the shard, the probability of committing a faulty block increases.

Figure 2c shows the probability of the beacon chain to commit a faulty block $(\mathcal{P}'')$ versus the size committee of the beacon chain $(n')$. We also observe that

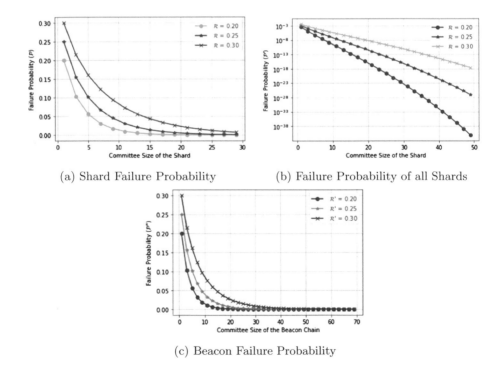

(a) Shard Failure Probability    (b) Failure Probability of all Shards

(c) Beacon Failure Probability

**Fig. 2.** (a) Probability of a shard to commit a faulty block $(\mathcal{P})$ versus the committee size of the shard $(n)$, (b) log-scale plot of the probability of all shards committing a faulty block $(\mathcal{P}')$ versus the size of the committee $(n)$, and (c) probability of the beacon chain to commit a faulty block $(\mathcal{P}'')$ versus the size committee of the beacon chain $(n')$.

the probability $\mathcal{P}''$ decreases when the size of the committee increases. More specifically, we observe that the probability corresponding to $\mathcal{R} = 0.2$ (i.e. 20% of malicious nodes in the beacon chain) decreases sharply compared to those of $\mathcal{R} = 0.25$ and $\mathcal{R} = 0.3$.

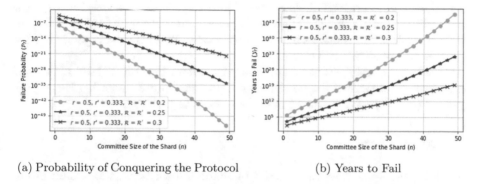

(a) Probability of Conquering the Protocol          (b) Years to Fail

**Fig. 3.** Log-scale plot: (a) probability of conquering the protocol ($\mathcal{P}_f$) versus the committee size of the shard ($n$), (b) number of years to fail ($\mathcal{Y}_f$) versus the committee size of the shard ($n$).

In Figure 3, we assume a network with 2000 nodes, $\mathcal{V} = 200$, $\mathcal{V}' = 400$, $\zeta = 8$, $n' = 100$. Figure 3a shows the probability of the conquering the protocol when varying the committee size of the shard. We observe that as the committee size of the shard increases the probability of conquering the protocol decreases. Figure 3b shows the number of years to fail ($\mathcal{Y}_f$) versus the committee size of the shard. We observe that when the committee size of the shard increases the number of years to fail increases.

**Table 2.** Probability of conquering the protocol.

| $\mathcal{R} = \mathcal{R}'$ | 10% | 15% | 20% | 30% |
|---|---|---|---|---|
| $\mathcal{P}_f^{\mathrm{a}}$ | 3.63E−66 | 2.10E−34 | 1.58E−18 | 1.70E−04 |
| $Y_f^{\mathrm{a}}$ | 7.56E+62 | 1.30E+31 | 1.74E+17 | 16.12 |
| $\mathcal{P}_f^{\mathrm{b}}$ | 0.0 | 5.14E−80 | 2.01E−41 | 5.30E−07 |
| $Y_f^{\mathrm{b}}$ | inf | 5.33E+76 | 1.36E+38 | 5171.32 |

[a] Scenario 1
[b] Scenario 2

In Table 2, we assume two scenarios to show the effectiveness and the feasibility of the proposed model: Scenario 1 proposes a network with $\mathcal{N} = 2000$, $\zeta = 8$, $\mathcal{V} = 200$, $\mathcal{V}' = 400$, and $r = r' = 0.333$ whereas Scenario 2 proposes a network with $\mathcal{N} = 4000$, $\zeta = 8$, $\mathcal{V} = 400$, $\mathcal{V}' = 800$, and $r = r' = 0.333$. It is noteworthy that the proposed model can be adopted to any scenario.

Table 2 shows the probability of conquering the chain (i.e., the probability of committing a faulty block; it is calculated based on Theorem 1) for different percentages of malicious nodes in the shards as well as in the beacon chain. Moreover, Table 2 shows the number of years to fail corresponding to these probabilities. We observe that as the percentage of malicious nodes increases the number of years to fail decreases. More specifically, we observe that the probability of conquering the chain is extremely low even with 20% of malicious nodes in each shard as well as in the beacon chain. This achieves a good security, which is about 1.74E+17 years to fail.

Finally, we conclude that by adjusting the committee size of the shard as well as the committee size of the beacon chain, we could protect sharded Blockchain systems (based on PoS) against malicious nodes (e.g., Sybil nodes).

## 4    Conclusion

In this paper, we address the security of sharding-based PoS Blockchain protocols. In particular, we provide a probabilistic model to compute the probability of committing a faulty block. Based on this probability, we compute the number of years to fail. Furthermore, this article depicts that we can control the number of years to fail by adjusting the committee size of the shard as well as the committee size of the beacon chain. Our future work includes the computation of the failure probability across-shard transaction.

## References

1. Abou Jaoude, J., George Saade, R.: Blockchain applications - usage in different domains. IEEE Access **7**, 45360–45381 (2019). https://doi.org/10.1109/ACCESS. 2019.2902501
2. Hafid, A., Hafid, A.S., Samih, M.: New mathematical model to analyze security of sharding-based blockchain protocols. IEEE Access **7**(19232899), 185447–185457 (2019). https://doi.org/10.1109/ACCESS.2019.2961065
3. Hafid, A., Hafid, A.S., Samih, M.: A novel methodology-based joint hypergeometric distribution to analyze the security of sharded blockchains. IEEE Access **8**(20000968), 179389–179399 (2020). https://doi.org/10.1109/ACCESS. 2020.3027952
4. Hafid, A., Hafid, A.S., Samih, M.: Scaling blockchains: a comprehensive survey. IEEE Access **8**(19800223), 125244–125262 (2020). https://doi.org/10.1109/ ACCESS.2020.3007251
5. Hafid, A., Hafid, A.S., Samih, M.: A tractable probabilistic approach to analyze Sybil attacks in sharding-based blockchain protocols. IEEE Trans. Emerg. Topics Comput. (2020). https://doi.org/10.1109/TETC.2022.3179638
6. Incognito. https://we.incognito.org/t/incognito-whitepaper-incognito-mode-for-cryptonetworks/168 (2021)
7. Kassab, M.H., DeFranco, J., Malas, T., Laplante, P., Destefanis, G., Graciano Neto, V.V.: Exploring research in blockchain for healthcare and a roadmap for the future. IEEE Trans. Emerg. Topics Comput. **9**, 1835–1852 (2019). https://doi.org/ 10.1109/TETC.2019.2936881

8. Nakamoto, S.: Bitcoin whitepaper. https://bitcoin.org/bitcoin.pdf (2008)
9. Nordgren, A., Weckström, E., Martikainen, M., Lehner, O.M.: Blockchain in the fields of finance and accounting: a disruptive technology or an overhyped phenomenon. ACRN Oxf. J. Finance Risk Perspect. **8**(1), 47–58 (2019)
10. PayPal. https://www.paypal.com/fr/webapps/mpp/home (2021)
11. Rasmussen, R.V., Trick, M.A.: Round robin scheduling—a survey. Eur. J. Oper. Res. **188**(3), 617–636 (2008)
12. Wang, G., Shi, Z.J., Nixon, M., Han, S.: Sok: Sharding on blockchain. In: Proceedings of the 1st ACM Conference on Advances in Financial Technologies, pp. 41–61 (2019)

# Cryptocurrencies, Survey on Legal Frameworks and Regulation Around the World

Yeray Mezquita[1(✉)], Dévika Pérez[2], Alfonso González-Briones[1],
and Javier Prieto[1]

[1] BISITE Research Group, University of Salamanca, Edificio Multiusos I+D+i, Calle
Espejo 2, 37007 Salamanca, Spain
[2] University of Cádiz, Cádiz, Spain
{yeraymm,alfonsogb,javierp}@usal.es
{devika.perezmedina}@gmail.com

**Abstract.** Cryptocurrencies have been in the spotlight for the past few
years. The centralization of the management of fiat currencies by banks
led to their emergence with Bitcoin in 2008 as a way to present truly
democratic money to the world. With the increase in the flow of capital to
this market, new challenges and problems have arisen that governments
must deal with. In this paper, a study of the legal frameworks that are
being implemented throughout the world with the intention of regulating
the cryptocurrency market is carried out. This study aims to understand
the different points of view and interests that each country has when
implementing each type of framework.

**Keywords:** Cryptocurrencies · Regulation · Review

## 1 Introduction

Cryptocurrencies have been in the spotlight for the past few years. The centralization of the management of fiat currencies by banks led to their emergence with
Bitcoin in 2008 as a way to present truly democratic money to the world [27].
After the inflationary crisis brought about by COVID, capitalists have begun to
look for alternatives to holding their wealth in cryptocurrencies.

With the increase in the flow of capital to this market, new challenges and
problems have arisen that governments must deal with [1, 29, 30]. For example,
cryptocurrencies have become a means of financing illegal activities, as well as
a means to evade taxes. On the other hand, investors feel unprotected against
the volatility of this market and manipulations by major capitals.

In this paper, a study of the legal frameworks that are being implemented
throughout the world with the intention of regulating the cryptocurrency market
is carried out. The study shows how the legal frameworks differ among the
world's economic powers. This study aims to understand the different points

of view and interests that each country has when implementing each type of framework.

Following this introduction, Sect. 2 shows the study of the legal frameworks proposed by each of the world's economic powers. The study has been divided into three major world economic areas: the European Union (Sect. 2.1), the United States (Sect. 2.2, and the BRICS economic-trade union (acronym of an economic-trade association that has been carried out by the states of Brazil, Russia, India, China, and South Africa since 2008) (Sect. 2.3). Finally, the work ends with a conclusion in Sect. 3.

## 2   Legal Frameworks

In this decade of cryptocurrencies' operation, the money invested until 2020 amounted to $221K million worldwide[1]. According to some studies, 44% of these transactions in 2018 were intended to facilitate illicit activities [8]. Some of the characteristics of cryptocurrencies, especially their anonymity, have made them an attractive element to be used in criminal conduct. In the literature studied, it's been pointed out that the main crimes that have been linked to this phenomenon are the financing of terrorism, money laundering, tax evasions, and the purchase of illicit material or services. Therefore, the criminal aspect should be a focus of concern for States or supranational organizations [7,11,12,15,17,28,33,35].

Thus, for authors such as Fernando Navarro, cryptocurrencies are considered an excellent way for money laundering. The author points out that according to some studies, the rise in the market in December 2017 was due to a speculative movement resulting from a large-scale money laundering [2]. Similarly, Patricia Saldaña points to Bitcoin as a protagonist in money laundering through the purchase of cryptocurrencies with money obtained from illicit gains [31].

On the other hand, the concern about tax evasion has been highlighted by another sector of the doctrine. Studies such as those carried out by Rain Xie, Mounteney, and García Sigman highlight that, in addition to money laundering, the use of these crypto-assets for transactions of illicit goods and services is very common. The most usual method is through the Dark Web, the Internet black market, for the purchase of narcotics, weapons, or consumption of child pornography [14,34]. García Sigman points out that cryptocurrencies are facilitating the purchase of narcotics on this dark internet wholesale [9].

As part of the study of the literature shows, one of the crimes that have generated greater concern to the institutions is tax evasions. Two characteristics facilitate the commission of the crime, the anonymity and the absence of financial intermediaries that control the movements as in the case of the [13] banks. Besides, terrorist groups have found cryptocurrencies a means of financing. This is pointed out by Dion-Schwarz in his study on the use of cryptocurrencies in terrorist activities [6]. The author highlights the use of this new technology for

---

[1] Data obtained from the total capitalization of the cryptomarket in real-time through www.tradingview.com. Date accessed June 9, 2020

the financing of terrorist groups and also highlights the alternative cryptocurrencies to Bitcoin that are most commonly used for these transactions such as Blackcoin or Monero [6].

Despite this high number of cases, most states do not have extensive regulation on the subject, not even on the nature of cryptocurrency [3,4,10,16,18–26]. This is not to say that cryptocurrencies have not currently attracted regulatory interest. The capital movements surrounding cryptocurrencies have led governments to become concerned about the fiscal and financial aspects of the phenomenon.

However, neither the lack of regulation nor the linking of the phenomenon with criminal acts has led to a decrease in the development of cryptocurrencies or the blockchain technology that hosts them. Therefore, finding the right balance between the regulation that the phenomenon needs to ensure its legal security and the flexibility linked to its constant development is a complex task. Even more so if we take into account that the approach to cryptocurrencies varies according to the state, the organizations, or the areas we are referring to [32].

A comparison between different economic spheres highlights the lack of common criteria mentioned above. To make this comparative approach, the situation of cryptocurrencies will be identified in three major world economic areas: the United States, the European Union, and the BRICS economic-trade union (acronym of an economic-trade association that has been carried out by the states of Brazil, Russia, India, China, and South Africa since 2008). As will be seen in this analysis, the positions range from the tolerance of cryptocurrency in the market to the prohibition of the operation of exchanges, showing a disparity between the different regulations or, directly, an absence of regulation.

## 2.1 European Union

The European Union presents one of the main examples of a lack of adequate or extensive regulation in this area. Its progress has been oriented to the presence of cryptocurrency as an instrument in the commission of crimes. In January 2019, Directive (EU)2018/843 of the European Parliament and of the Council of 30 May 2018 came into force. It was intended to amend Directive (EU) 2015/849 on the prevention of the use of the financial system for money laundering or terrorist financing.

The directive inclusion considers that the anonymity of virtual currencies allows their misuse for criminal purposes. To this end, the law established that it is needed control by the exchange service providers and the electronic wallet custodian service providers. The tools for this control must be provided by the National Financial Intelligence Units, which must obtain information that allows them to associate the addresses of virtual currencies with the identity of the owner of the same.

The European Union currently has a draft proposal for a European regulation on crypto-asset markets. This future regulation aims to regulate this type of market through a general framework for the entire European Union. This draft regulation is known as MiCA (Markets in Crypto-assets) [5]. The main objective pursued by the MiCA is the regulation of the market for crypto-assets which are currently not considered under EU financial legislation as financial assets or electronic money.

## 2.2   United States

The United States does not have developed regulations in criminal areas. On the other hand, it has been concerned to respond in the financial area. Considering its history with this phenomenon, the country has faced related problems in both sectors. Due to price manipulations in the cryptocurrency market, many US investors have lost millions of dollars. Faced with this situation, the New York Department of Financial Services proposed in 2014 to regulate the exchanges and being themselves the ones responsible to grant the licenses for the opening of new exchanges. In this way, investors can feel a state backing for their investments. Likewise, in tax matters, cryptocurrencies are included in the payment of taxes like the holding of any other property, as is the case with shares.

However, the same is not true for money laundering or financing of illicit activities. Currently, there is no regulation in this area despite having seen the relationship between cryptocurrencies and the criminal sphere that benefits from the anonymity they provide. Despite having so far this absence of regulations, in March 2020 Congressman Paul Gosar introduced the bill: "Cryptocurrency Act of 2020" to clarify and legitimize crypto-assets in the United States by differentiating digital assets and not granting a single definition and nature. In this proposal, there would be three categories: crypto-commodity, crypto-currency, and crypto-value, leaving the category of crypto-currency for those digital assets that are the representation of the currency of the United States to respect what is established in the Constitution. In this way, they try to classify cryptocurrencies to elucidate which regulation corresponds according to the function they have in the market.

The proposal of Paul Gosar is innovative and presents an interesting categorization of cryptocurrencies, but it still does not resolve the regulatory absence in the criminal field that has been highlighted on several occasions since the Silk Road investigation in the United States. It is not known whether this gap is due to a regulatory impossibility or a lack of initiative. What is certain is that the regulatory avenues that are being put forward advocate a tolerance through the control and regulation of crypto-assets and the blockchain system, not a ban on them.

## 2.3   BRICS

BRICS is the acronym of an economic-commercial association that has been carried out since 2008 by the states of Brazil, Russia, India, China, and South Africa (the latter since 2011). In the 2000s it was considered that these 5 countries could have an economic relevance in the future and it was proposed to create this union. The truth is that cryptocurrencies have a strong presence and influence in most of them. This fact was the reason why this union was chosen as the third pillar for our comparative study since it includes four countries whose form of regulation will be crucial for the development of cryptocurrencies. The states that are part of this union have had a complex development in the positions taken towards crypto-assets. Some of them even have gone through some moment of banning the sale and issuance of cryptocurrencies in their territories.

Brazil first associated the phenomenon with pyramid schemes and its possible prohibition was debated in Congress on several occasions. In the end, since 2015, a line focused on the regulation of these cryptoassets in terms of taxation was followed. Cryptocurrency is recognised as an economic good and a means of payment. In 2019, bill 2303/2015 was presented by federal deputy Aureo Ribeiro to try to regulate this phenomenon and its market more precisely. However, it was not until 29 September 2021 that the bill was approved, and now it only remains to be submitted to the Chamber of Deputies.

Russia and China share a very similar regulatory landscape. Both economic powers are major sources of cryptocurrency mining. However, given the energy expenditure involved along with the economic movement cryptocurrencies are causing, restrictive measures have been taken towards the activity.

In the case of Russia, cryptocurrency is considered a digital financial asset and not a currency. The Russian Central Bank has on several occasions lobbied the state powers against the acceptance of crypto-assets in the market as a possible risk to the economy. There is a draft law that envisages a very restrictive limitation of exchanges and the issuance and use of cryptocurrencies as stated by the Chairman of the State Duma Committee on Financial Markets, Anatoly Aksakov. This project will establish a definition of cryptocurrency and prohibit its use as a method of payment. However, this regulation does not envisage the same path for mining activity. According to Aksakov, the activity will be allowed as long as it is regulated and taxed for exercising it as it is considered a type of business that produces profit.

On the other hand, On the other hand, China was the birthplace of cryptocurrency mining and exchange platforms and it is an activity carried out by many users in the country. However, in September 2021 the Chinese government has proposed a ban on cryptocurrency-related activities. The reason that led the government to initiate these restrictions is not only for financial and economic reasons. Cryptocurrency mining farms are causing serious damage to the environment and electricity consumption. We will have to observe how this control of a transnational phenomenon evolves and whether they really succeed in banning the activity.

Finally, the case of India also reflects disparate positions on the phenomenon over the years. In April 2018, the Reserve Bank of India (RBI) banned the use of cryptocurrencies and any operations carried out with exchanges that handle these digital assets. Faced with this decision, the Supreme Court of India on 4 March 2020 struck down the measure adopted by the central bank, calling it disproportionate and unconstitutional. There are currently plans to regulate cryptocurrencies as commodities from February 2022.

## 2.4   Legal Issues Summary

The disparity of regulations shown in the analysis of the studied three territorial sectors highlights the difficulties faced by a transnational phenomenon such as cryptocurrencies. It is not only a concern for users but also for the states themselves in their intention to control the cryptocurrency activity. The analysis developed exposes that most of the states are favorable to the development of blockchain technology, which they do not perceive as a threat, but this is not the case with its best-known representation, the cryptocurrencies.

As assets that can be created voluntarily and are not subject to control to ensure financial security and consumer protection, cryptocurrencies cause rejection. However, some states have an intention to create their own state cryptocurrency. However, the proliferation of state cryptocurrencies generates uncertainty about the value they will have in other territories.

This observation is made from the perception of a wide range of conceptualization and denominations of digital assets that, as the study has shown, can be considered goods, digital securities, financial assets, or currencies. Therefore, the need for a global agreement on general standards for the Blockchain and cryptocurrencies should be considered, as is the case with other international regulations. The objective should be to increase confidence at financial and market levels and to address the criminal risks associated with cryptocurrencies.

At this point, it should be noted that crimes such as money laundering, terrorist financing, or drug trafficking are transnational criminal acts whose commission has benefited from the emergence of cryptocurrencies. In reality, this is not a novelty but a change in the means of commission as criminal groups take advantage of one of the main attractions of this phenomenon, their anonymity. Therefore, it is not so much a question of defining new crimes or modifying the definition of some of them, but rather of developing tools to minimize this characteristic. The international agreements required on this matter must therefore combine two elements: providing cryptocurrency with legal certainty and reducing its criminal potential.

In this digital era, regulation should not be oriented towards delegitimizing the innovation that cryptocurrencies and Blockchain technology represents. The path of international agreements on the matter must ensure a tandem between the financial and technological security of exchanges for the protection of consumers and the economic system and the development of this phenomenon.

## 3   Conclusion

In this paper it has been done a study in depth of the initiatives, regarding the cryptocurrencies ecosystem, of some of the most important states of the world. Although crimes such as money laundering, terrorist financing, or drug trafficking are transnational criminal acts whose commission has benefited from the emergence of cryptocurrencies. This is not a novelty but a change in the means of commission as criminal groups take advantage of one of the main attractions of this phenomenon, their anonymity. Therefore, it can be concluded that it is not so much a question of defining new crimes or modifying the definition of some of them, but rather of developing tools to minimize this characteristic. The international agreements required on this matter must therefore combine two elements: providing cryptocurrency with legal certainty and reducing its criminal potential.

**Acknowledgements.** The research of Yeray Mezquita is supported by the predoctoral fellowship from the University of Salamanca and co-funded by Banco Santander. This research was also partially supported by the project "Technological Consortium TO develop sustainability of underwater Cultural heritage (TECTONIC)", financed by the European Union (Horizon 2020 research and innovation programme under the Marie Skłodowska-Curie grant agreement No. 873132). Authors declare no conflicts of interest.

# References

1. Ahmad, P.: A review on blockchain's applications and implementations. ADCAIJ: Adv. Distrib. Comput. Artif. Intell. J. **10**(2) (2021)
2. Cardoso, F.N.: Criptomonedas (en especial, bitcóin) y blanqueo de dinero. Revista electrónica de ciencia penal y criminología **21**, 14 (2019)
3. Casado-Vara, R., Novais, P., Gil, A.B., Prieto, J., Corchado, J.M.: Distributed continuous-time fault estimation control for multiple devices in IoT networks. IEEE Access **7**, 11972–11984 (2019)
4. Castellanos-Garzón, J.A., Mezquita Martín, Y., Jaimes S, J.L., López G, S.M.: A data mining approach applied to wireless sensor networks in greenhouses. In: International Symposium on Distributed Computing and Artificial Intelligence, pp. 431–436. Springer (2018)
5. Commission, E.: Proposal for a regulation of the European parliament and of the council on markets in crypto-assets, and amending directive (EU) 2019/1937. https://eur-lex.europa.eu/legal-content/EN/TXT/?uri=CELEX:52020PC0593
6. Dion-Schwarz, C., Manheim, D., Johnston, P.B.: Terrorist Use of Cryptocurrencies: Technical and Organizational Barriers and Future Threats. Rand Corporation (2019)
7. Fatima, N.: Enhancing performance of a deep neural network by comparing optimizers experimentally. ADCAIJ: Adv. Distrib. Comput. Artif. Intell. J. (ISSN: 2255–2863) Salamanca **9**(2), 79–90 (2020)
8. Foley, S., Karlsen, J.R., Putniņš, T.J.: Sex, drugs, and bitcoin: How much illegal activity is financed through cryptocurrencies? Rev. Financ. Stud. **32**(5), 1798–1853 (2019)
9. García, L.I.: Narcotráfico en la darkweb: los criptomercados. URVIO Revista Latinoamericana de Estudios de Seguridad **21**, 191–206 (2017)
10. González-Briones, A., Castellanos-Garzón, J.A., Mezquita Martín, Y., Prieto, J., Corchado, J.M.: A framework for knowledge discovery from wireless sensor networks in rural environments: a crop irrigation systems case study. Wirel. Commun. Mob. Comput. **2018** (2018)
11. Gupta, S., Meena, J., Gupta, O., et al.: Neural network based epileptic EEG detection and classification (2020)
12. Hussain, A., Hussain, T., Ali, I., Khan, M.R., et al.: Impact of sparse and dense deployment of nodes under different propagation models in MANETs (2020)

13. Jafari, S., Vo-Huu, T., Jabiyev, B., Mera, Farkhani, M.A.: Cryptocurrency: A Challenge to Legal System (2018)
14. Lavorgna, A.: How the use of the internet is affecting drug trafficking practices (2016)
15. Li, T., Chen, H., Sun, S., Corchado, J.M.: Joint smoothing and tracking based on continuous-time target trajectory function fitting. IEEE Trans. Autom. Sci. Eng. **16**(3), 1476–1483 (2018)
16. Li, T., Corchado, J.M., Sun, S.: Partial consensus and conservative fusion of Gaussian mixtures for distributed PhD fusion. IEEE Trans. Aerosp. Electron. Syst. **55**(5), 2150–2163 (2018)
17. López, A.B.: Deep learning in biometrics: a survey. ADCAIJ: Adv. Distrib. Comput. Artif. Intell. J. **8**(4), 19–32 (2019)
18. Martín, Y.M., Parra, J., Pérez, E., Prieto, J., Corchado, J.M.: Blockchain-based systems in land registry, a survey of their use and economic implications. CISIS **2020**, 13–22 (2020)
19. Mezquita, Y., Alonso, R.S., Casado-Vara, R., Prieto, J., Corchado, J.M.: A review of k-NN algorithm based on classical and quantum machine learning. In: International Symposium on Distributed Computing and Artificial Intelligence, pp. 189–198. Springer (2020)
20. Mezquita, Y., Casado-Vara, R., González Briones, A., Prieto, J., Corchado, J.M.: Blockchain-based architecture for the control of logistics activities: pharmaceutical utilities case study. Logic J. IGPL **29**(6), 974–985 (2021)
21. Mezquita, Y., Gil-González, A.B., Prieto, J., Corchado, J.M.: Cryptocurrencies and price prediction: a survey. In: International Congress on Blockchain and Applications, pp. 339–346. Springer (2021)
22. Mezquita, Y., Gil-González, A.B., Martín del Rey, A., Prieto, J., Corchado, J.M.: Towards a blockchain-based peer-to-peer energy marketplace. Energies **15**(9), 3046 (2022)
23. Mezquita, Y., González-Briones, A., Casado-Vara, R., Chamoso, P., Prieto, J., Corchado, J.M.: Blockchain-based architecture: a mas proposal for efficient agri-food supply chains. In: International Symposium on Ambient Intelligence, pp. 89–96. Springer (2019)
24. Mezquita, Y., González-Briones, A., Casado-Vara, R., Wolf, P., Prieta, F.d.l., Gil-González, A.B.: Review of privacy preservation with blockchain technology in the context of smart cities. In: Sustainable Smart Cities and Territories International Conference, pp. 68–77. Springer (2021)
25. Mezquita, Y., Parra, J., Perez, E., Prieto, J., Corchado, J.M.: Blockchain-based systems in land registry, a survey of their use and economic implications. In: Computational Intelligence in Security for Information Systems Conference, pp. 13–22. Springer (2019)
26. Mezquita, Y., Valdeolmillos, D., González-Briones, A., Prieto, J., Corchado, J.M.: Legal aspects and emerging risks in the use of smart contracts based on blockchain. In: International Conference on Knowledge Management in Organizations, pp. 525–535. Springer (2019)
27. Nakamoto, S., et al.: Bitcoin: a peer-to-peer electronic cash system (2008)
28. Pimpalkar, A.P., Raj, R.J.R.: Influence of pre-processing strategies on the performance of ml classifiers exploiting TF-IDF and bow features. ADCAIJ: Adv. Distrib. Comput. Artif. Intell. J. **9**(2), 49 (2020)
29. Schmidt, W.C., González, A., et al.: Fintech and tokenization: a legislative study in Argentina and Spain about the application of blockchain in the field of properties (2020)

30. Srivastav, R.K., Agrawal, D., Shrivastava, A., et al.: A survey on vulnerabilities and performance evaluation criteria in blockchain technology (2020)
31. Taboada, P.S.: por qué las organizaciones criminales utilizan criptomonedas? los bitcoins en el crimen organizado. El Criminalista Digital. Papeles de Criminología **6**, 1–41 (2017)
32. Valdeolmillos, D., Mezquita, Y., González-Briones, A., Prieto, J., Corchado, J.M.: Blockchain technology: a review of the current challenges of cryptocurrency. In: International Congress on Blockchain and Applications, pp. 153–160. Springer (2019)
33. Vergara, D., Extremera, J., Rubio, M.P., Dávila, L.P.: The proliferation of virtual laboratories in educational fields. ADCAIJ: Adv. Distrib. Comput. Artif. Intell. J. **9**(1), 85 (2020)
34. Xie, R.: Why china had to ban cryptocurrency but the US did not: a comparative analysis of regulations on crypto-markets between the US and China. Wash. Univ. Global Stud. Law Rev. **18**, 457 (2019)
35. Zubair, S., Al Sabri, M.A.: Hybrid measurement of the similarity value based on a genetic algorithm to improve prediction in a collaborative filtering recommendation system. ADCAIJ: Adv. Distrib. Comput. Artif. Intell. J. **10**(2), 165–182 (2021)

# SmartTwin: A Blockchain-Based Software Framework for Digital Twins Using IoT

Miguel Pincheira$^{(\boxtimes)}$, Massimo Vecchio, and Fabio Antonelli

Fondazione Bruno Kessler, Trento, Italy
{mpincheiracaro,mvecchio,fantonelli}@fbk.eu

**Abstract.** A Digital Twin provides a virtual representation of a physical object, complementing theoretical models with real-world data provided by the Internet of Things. Currently, most cloud-based IoT platforms provide device virtualization with different levels of detail and complexity. Nonetheless, this approach has inherited the disadvantages and risks of cloud computing, such as centralizing and isolating the information. Thus, a blockchain-based Digital Twin could provide a unique combination of properties to address these challenges while ensuring availability, integrity, and confidentiality. Current research on blockchain-based Digital Twin has focused on the data-sharing functionalities. However, the role of the IoT sensors as secure, trustworthy data sources is still an issue that needs to be addressed in the Digital Twin domain. In this work, we propose SmartTwin as a software framework to develop blockchain-based Digital Twins. The architecture's key is considering IoT as direct actors on the blockchain system, taking advantage of the cryptographic capabilities of Blockchain to create a root-of-trust for the data feeding the Digital Twin. In addition, our proposal uses smart contracts as a software platform to define complex business logic. Finally, to illustrate the benefits of SmartTwin, we describe two use cases where the framework was used.

**Keywords:** Blockchain · Digital twin · Internet of things · Smart contracts

## 1 Introduction

The Digital Twin (DT) paradigm, a concept from the product lifecycle management, has migrated from engineering and manufacturing to several other domains. According to the authors of [1], a Digital Twin can be defined as a "virtual information construct" representing the actual or estimated state of a physical object. This virtual representation enabled a visible model to explain abnormal behaviors and even predict them, creating an aggregated value in the manufacturing chain. The engineering domain quickly embraced this capability

© The Author(s): under exclusive license to Springer Nature Switzerland AG 2023
J. Prieto et al. (Eds.): BLOCKCHAIN 2022, LNNS 595, pp. 67–77, 2023.
https://doi.org/10.1007/978-3-031-21229-1_7

of complementing theoretical models with real-world data to design, produce, and analyze complex systems, such as cars, aircraft, engines, and control units. Here, the Internet of Things (IoT) is one of the key enablers for developing DTs, as it provides the link between real devices and their virtual representation [2]. IoT sensors monitor several parameters of the physical object, acting as the data source for the virtualized information model. Furthermore, the application of IoT and DT in an industrial environment is gaining a fair amount of interest under the umbrella of the "Industry 4.0" concept [3] for monitoring production systems across the entire value chain. For this reason, most cloud-based IoT platforms provide device virtualization with different levels of detail and complexity, ranging from a simple "off-line model" based on the last know state to "predictive models" fueled by large amounts of data produced by IoT and powered by machine learning algorithms [4]. Nevertheless, this approach has inherited the disadvantages and risks of cloud computing, such as centralizing and isolating the information while creating additional barriers restricting interactions between interested parties. Moreover, these centralized cloud services act as a black box for IoT, where the stakeholders do not have control and total trust in how the data is managed and shared [5]. Therefore, blockchain-based applications provide a unique combination of properties to address the data-sharing challenges while ensuring availability, integrity, and confidentiality [3]. Further, the programming capabilities of Blockchain, such as smart contracts on Ethereum, can also provide a platform that relies on algorithms for friction-less automation and direct interaction of DTs and stakeholders [6].

However, existing research on blockchain-based DTs has focused on the data-sharing capabilities [3,7], or the use of Blockchain as an enabler for additive manufacturing [8]. Consequently, the role of the IoT sensors as secure, trustworthy data sources for the DTs is still an issue that needs to be addressed [9]. In this perspective, we propose SmartTwin as an architecture to enable the integration of IoT and blockchain technology to create blockchain-based DTs. The architecture's key is considering IoT as direct actors on the blockchain system, taking advantage of the cryptographic capabilities of Blockchain to create a root-of-trust for the data feeding the DTs. Using a layered architecture, we overcome several challenges that integrating IoT and Blockchain presents while providing different levels of abstractions. Furthermore, these different levels of abstraction could provide the integration of several additional functionalities, such as decentralized storage, privacy-protective mechanism, and secure off-chain computing. In addition, this architecture could enable new use cases where physical objects are not only represented in the virtual world but can also have the capabilities to interact autonomously in a decentralized, trusted way. To this end, our proposal uses smart contracts as a software platform to define complex business logic. The architecture is the product of a design science research approach (DSR) [10], validated in different use-cases, including Traceability applications [11], Water management systems [12], and Mining Inspections [13].

The rest of this paper is structured as follows: Sect. 2 briefly summarizes related works. Then, Sect. 3 describes in detail our proposed framework.

Section 4 describes the application of the framework on two use cases. Lastly, we draw our conclusions and plans for future works in Sect. 5.

## 2   Related Work

The concept of the Digital Twin is gaining much momentum, as shown by surveys such as [2]. Several researchers are proposing frameworks to build Digital Twins. For instance, the authors of [4] proposed a layered framework written in Java that aims to be application-independent. Similarly, the authors of [14] present a framework for data-driven digital twins, focusing on integrating ML approaches for extracting knowledge from the DTs data. Although these approaches do not consider Blockchain as a technological enabler, they highlight the importance of data collection and related issues. Moreover, they emphasize the need for multi-domain frameworks to implement the DTs.

Mandolla et al. presented one of the first works to address the combination of DTs and Blockchain [8]. Briefly, they study the metal additive manufacturing process for the aircraft industry using a DT approach with Blockchain. They focus on securing and organizing the data generated through the entire manufacturing process and emphasize how Blockchain could help build secure and connected manufacturing infrastructure. Similarly, the authors of [7] proposed a data management method for DT of products based on blockchain technology. They present the case study of a turbine and analyze their proposal to manage the data of the product's lifecycle, from the design to production and maintenance. Recently, the authors of [3] proposed a blockchain-based DTs focusing on data-sharing. They present and implement an owner-centric framework using DTs on the Ethereum blockchain. Their approach considers several functionalities and provides a separation between on-chain and off-chain elements. Furthermore, they use several software design patterns [15] for smart contracts, highlighting the need for structured software frameworks to develop DTs and decentralized applications using Blockchain technology.

Despite this interest, all previous work missed to address the integration of the IoT sensors in a blockchain-based DT software framework [16]. Furthermore, current works are inconsistent with the use of Smart Contracts as a foundation to develop blockchain-based DTs applications.

## 3   The SmartTwin Framework

We propose the SmartTwin architecture to guarantee a root-of-trust for the information provided by the IoT devices as the foundation for a blockchain-based Digital Twin. Furthermore, cryptography as a core component of the blockchain provides a secure base for the data. Finally, since the Digital Twin is no longer in a cloud server, it benefits from a peer-to-peer network with distributed replicated storage.

The architecture presents two novel characteristics: (i) it considers constrained sensing devices as direct actors on the blockchain system, and (ii) provides a basic layer of services to develop new types of blockchain-based Digital Twins. Foremost, the architecture assumes that each device securely manages its cryptographic keys. Although this is a strong assumption, it is aligned with the state-of-the-art [16]. Furthermore, blockchain holds the potential for more complex identity schemes as key-building blocks for realizing completely decentralized public key infrastructures [16].

In the proposed architecture, sensing devices are direct actors on the blockchain system, have a unique blockchain identity, and generate digitally signed transactions. The use of cryptography at the core of the architecture aligns with the current security challenges in IoT and guarantees a root of trust for the sensed data, bringing the IoT devices a step closer to trustworthy blockchain oracles [9].

Furthermore, the proposed architecture uses smart contracts to provide the basic services for decentralized applications. Smart contracts provide a platform to define complex business logic, autonomously enforcing agreements between untrusted actors based on trusted values coming from ubiquitous IoT devices. This approach allows true IoT interoperability among multi-vendor devices and applications, with increased security offered by the underlying blockchain network.

We conceptualized the proposed architecture as the layered framework shown in Fig. 1. The framework comprises three main modules, i.e., Blockchain, Gateway, and Device. These modules resemble the typical 3-layer architecture of a modern IoT system composed of Cloud, Edge, and Device [17]. In this way, the framework can be easily mapped to existing IoT systems.

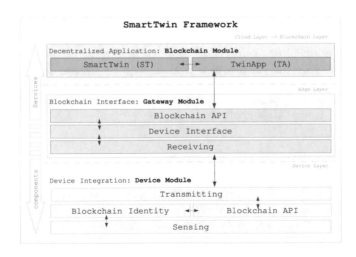

**Fig. 1.** The SmartTwin framework

The **Device module** is located at the Device layer and provides the necessary tools to transform a constrained sensing device into a direct actor on a blockchain system. The **Gateway module** is located at the Edge layer and is a simple relay component between the lower and upper modules. The **Blockchain module** replaces the centralized Cloud Layer with a decentralized Blockchain Layer, providing the interface for developing new types of decentralized IoT applications following the Digital Twin paradigm. From a service-oriented perspective, the lower modules address the low-level requirements of constrained sensing devices. Conversely, upper modules provide high-level services to develop new types of decentralized IoT applications. The following paragraphs, describes the modules in detail.

## 3.1   M1: Device Module

The device module enables the integration of constrained sensing devices as direct actors on a blockchain-based system. This module provides the standard functionalities of an IoT device (i.e., sensing and transmitting) and the cryptographic functionalities to interact with the blockchain system (i.e., identity and APIs). The following sub-modules describe the four functionalities of the device module.

- **M1.1) Sensing**: Controls the sensor that transforms a reading of the physical world into a sensed data value $v$. For better compatibility with existing IoT systems, this sub-module works as traditional sensing operations in IoT devices.
- **M1.2) Blockchain API'**: Provides a limited set of methods to interact with high-level functionalities of the blockchain network, i.e., access and retrieve information and interact with the smart contracts. In particular, this module specifies the transaction format $Tx$ to match the corresponding $ST$. Additionally, the module also provides the operations to transform $Tx$ into a network-compatible version $Tx'$ (e.g., compressing/decompressing, encoding/decoding) according to the communications capabilities of the device (M1.4).
- **M1.3) Blockchain Identity**: Provides all the cryptographic functions required to create a direct interaction with the blockchain (i.e., a valid transaction $Tx$). The functions include Elliptic Curve Digital Signature Algorithm (ECDSA) and hashing algorithms optimized for constrained devices. The key pair $K$ is managed in this sub-module and provides a unique identity $dI$ to the device $d$.
- **M1.4) Transmitting**: Provides the methods to transmit $Tx$ (or $Tx'$) to the gateway, using the communication interface available in the sensor. For better compatibility with existing IoT systems, this sub-module works as traditional transmitting operations in IoT devices.

## 3.2   M2: Gateway Module

The gateway module is a transparent gateway connecting two (or more) communication interfaces as traditional IoT architectures at the edge layer. The gateway module can not modify the transactions $Tx$ as is signed by the sensing devices, guaranteeing data integrity through this layer. The two sub-modules of this module work as follows:

- **M2.1) Receiving**: Receives $Tx$ (or $Tx'$) using the communication interface that connects with the sensing device. For better compatibility with existing IoT systems, this sub-module works as usual receiving operations in edge devices.
- **M2.2) Device Interface**: It provides the functionalities to adapt the information from the device to a format compatible with the blockchain API, according to the communications interfaces between the gateway and the device (e.g., compression, encoding).
- **M2.3) Blockchain API**: Provides all the methods to interact with high-level functionalities of the blockchain system, i.e., access and retrieve information and interact with the smart contracts. The module also provides the operations to recover $Tx$ from $Tx'$ (e.g., compressing/decompressing, encoding/decoding). The sub-module M1.2 (Blockchain API') is a reduced version of this component.

## 3.3   M3: Blockchain Module

The blockchain module enables the creation of the blockchain-based Digital Twin. It relies on the scripting capabilities of the blockchain platform (i.e., smart contracts). The framework defines two models as high-level smart contract abstractions that specify a minimum set of primitives. Applications can extend these models to fill the requirement of each particular use case. The base models are SmartTwin and TwinApp, as described below.

- **M3.1) Smart Twin**: Defines a base smart contract to create a SmartTwin (ST) representing an IoT device on the blockchain. The SmartTwin is composed of properties, which are handled by two methods: *setValue()* to register a value (or property) from the IoT device and *getValue()* to return the registered value. The main objective of these properties is to describe the characteristics of interest of the object represented by the Digital Twin (e.g., physical measures). However, these properties can also represent advanced properties using other technological components. For instance, one property represents private information stored off-chain. In this case, the property's value would be the hash of an encrypted file stored on a distributed file system (DFS), such as IPFS, or Swarm. Similarly, the property could be a link to secure computations or zero-knowledge proof.
  The contract also stores an internal variable called device, representing the identity $dI$ of the device $d$ that can change the internal state of the Smart-Twin. A call to *setValue()* is only accepted if it comes from the device $d$.

Finally, the SmartTwin implements an abstract method called *callApp()* used
to interact with the TwinApp apps.

- **M3.2) Twin App**: Defines a base smart contract to create a TwinApp
  (TA) that interacts with a SmartTwin contract. The TwinApp implements
  two methods, namely *registerTwin()* and *queryTwin()* and stores an internal
  variable called *owner*, representing the user *uinU* that created the TwinApp.
  The method *queryTwin()* obtains the latest value stored on a SmartTwin.
  This method is only valid if SmartTwin has previously registered with the
  TwinApp using the *registerTwin()* method. Registering the Twin does not
  aim to impose restrictions, rather it aims to initialize the system.

The blockchain module aims to provide high-level components for developing
applications and thus, UML diagrams can provide a better description. Figure 2
shows a class diagram representing the SmartTwin and the TwinApp. Figure 3
shows a sequence diagram of SmartTwin and a TwinApp interacting in the
context of a sample application.

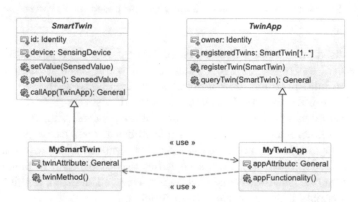

**Fig. 2.** Class diagram of a sample application using Smart Twins, and Twin Apps.

## 4   Use Cases

The architecture is the outcome of a design research approach, validated in
different use-cases, including Traceability applications [11], Water management
systems [12], and Mining Inspections [13]. For space limitation, here we describe
only two applications, highlighting the benefits that the SmartTwin framework
provided to develop the use case.

### 4.1   Agri-food Traceability

In [11], we used our framework for a fully decentralized, blockchain-based traceability IoT application for Agri-Food supply chain management. As shown in

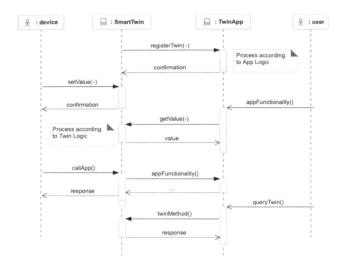

**Fig. 3.** Example interactions over the proposed architecture.

Fig. 4, we used our framework to create Digital Twins representing several items across the entire process (e.g., plants at the farm, cases during transportation, and the final product). Furthermore, our architecture allowed to develop decentralized applications that directly interacted with the SmartTwin. These applications implemented steps in the traceability process (e.g., transportation, deliver) and implemented additional functionalities such as organic certifications to further increase the business value of the system. The proposed solution was developed, deployed, and evaluated using two blockchain implementations (i.e., Ethereum and Sawtooth) and several low-cost IoT sensors, highlighting the framework's capabilities to be blockchain independent.

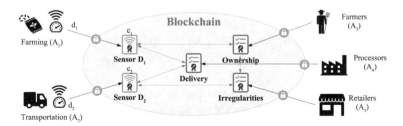

**Fig. 4.** Architecture using the SmartTwin framework for an agri-food traceability application (extracted from [11])

## 4.2    Water Management

In [12] we explored how the energy-efficient IoT devices and blockchain could be used to incentivize virtuous behaviors in agricultural practices. As shown in Fig. 5, we used our framework to create Digital Twins of water valves actuated by low-power IoT devices. Moreover, our architecture allowed to implement an application to interact with the SmartTwin for several functionalities, such as billing, certifications, and rewards. By implementing a complete proof-of-concept in the Ethereum network, we evaluated the impact of the architecture in terms of memory, program size, communications, and power consumption on six different IoT sensors, highlighting the framework's small footprint for the IoT sensors.

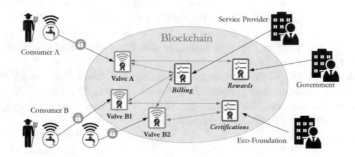

**Fig. 5.** Architecture using the SmartTwin framework for an water management system (extracted from [12]).

# 5    Conclusions and Future Work

In this work, we presented the first version of SmartTwin, a software framework to develop blockchain-based Digital Twins. The framework's key is considering IoT as direct actors on the blockchain, i.e., the devices interact with the blockchain without any additional system component. Furthermore, using a layered architecture, we overcome several challenges that integrating IoT and Blockchain presents while providing different levels of abstractions. The framework is the outcome of an iterative DSR approach, designing, building, and evaluating blockchain-based applications in different case studies. Compared to the current research on blockchain-based DT, which focuses on data-sharing, our framework focuses on the IoT devices as data sources, making them direct actors on the blockchain network. This approach guarantees a root of trust for the sensed data, which is the base for the Digital Twin. Furthermore, in the proposed framework, smart contracts provide a platform to define complex business logic, enabling the development of new types of blockchain-based Digital Twin applications.

Future works for the proposed framework include evaluating metrics related to smart contracts and formal security analysis. Exploring the use of layer-2 architectures on public blockchains (e.g., lightning network, state channels, plasma network) and hybrid approaches is another intriguing research path. Finally, evaluating methods for integrating secure off-chain computing into the framework is ongoing work.

# References

1. Grieves, M., Vickers, J.: Digital twin: mitigating unpredictable, undesirable emergent behavior in complex systems. In: Kahlen, F.-J., Flumerfelt, S., Alves, A. (eds.) Transdisciplinary Perspectives on Complex Systems: New Findings and Approaches, pp. 85–113. Springer International Publishing, Cham (2017)
2. Tao, F., Qi, Q., Wang, L., Nee, A.Y.C.: Digital twins and cyber–physical systems toward smart manufacturing and industry 4.0: correlation and comparison. Engineering **5**(4), 653–661 (2019)
3. Putz, B., Dietz, M., Empl, P., Pernul, G.: Ethertwin: blockchain-based secure digital twin information management. Inf. Process. Manag. **58**(1), 102425 (2021)
4. Picone, M., Mamei, M., Zambonelli, F.: WLDT: a general purpose library to build IoT digital twins. SoftwareX **13**, 100661 (2021)
5. Reyna, A., Martín, C., Chen, J., Soler, E., Díaz, M.: On blockchain and its integration with IoT. Challenges and opportunities. Future Gener. Comput. Syst. **88**, 173–190 (2018)
6. Seebacher, S., Schüritz, R.: Blockchain technology as an enabler of service systems: a structured literature review. In: Proceedings of the International Conference on Exploring Services Science, pp. 12–23. Springer, Berlin (2017)
7. Huang, S., Wang, G., Yan, Y., Fang, X.: Blockchain-based data management for digital twin of product. J. Manuf. Syst. **54**, 361–371 (2020)
8. Mandolla, C., Petruzzelli, A.M., Percoco, G., Urbinati, A.: Building a digital twin for additive manufacturing through the exploitation of blockchain: a case analysis of the aircraft industry. Comput. Ind. **109**, 134–152 (2019)
9. Heiss, J., Eberhardt, J., Tai, S.: From oracles to trustworthy data on-chaining systems. In: Proceedings of the IEEE International Conference on Blockchain (2019)
10. Hevner, A., Chatterjee, S.: Design science research in information systems. In: Design Research in Information Systems, pp. 9–22. Springer, Berlin (2010)
11. Pincheira, M., Vecchio, M., Giaffreda, R.: Benchmarking constrained IoT devices in blockchain-based agri-food traceability applications. In: Proceedings of the International Congress on Blockchain and Applications, pp. 212–221. Springer, Berlin (2021)
12. Pincheira, M., Vecchio, M., Giaffreda, R., Kanhere, S.S.: Exploiting constrained IoT devices in a trustless blockchain-based water management system. In: Proceedings of the IEEE International Conference on Blockchain and Cryptocurrency (ICBC), pp. 1–7 (2020)
13. Pincheira, M., Antonini, M., Vecchio, M.: Integrating the IoT and blockchain technology for the next generation of mining inspection systems. Sensors **22**(3) (2022)
14. Friederich, J., Francis, D.P., Lazarova-Molnar, S., Mohamed, N.: A framework for data-driven digital twins for smart manufacturing. Comput. Ind. **136**, 103586 (2022)

15. Gamma, E., Helm, R., Johnson, R., Vlissides, J.M.: Design Patterns: Elements of Reusable Object-Oriented Software. Addison-Wesley Professional (1994)
16. Ali, M.S., Vecchio, M., Pincheira, M., Dolui, K., Antonelli, F., Rehmani, M.H.: Applications of blockchains in the internet of things: a comprehensive survey. IEEE Commun. Surv. Tutorials **21**(2), 1676–1717 (2018)
17. Viriyasitavat, W., Xu, L.D., Bi, Z., Hoonsopon, D.: Blockchain technology for applications in internet of things-mapping from system design perspective. IEEE Internet Things J. **6**(5), 8155–8168 (2019)

# Proofs and Limitations of the Pathway Protocol

Marc Jansen[1(✉)], Ilya Sapranidi[2], and Aleksei Pupyshev[2]

[1] University of Applied Sciences Ruhr West, Mülheim, Germany
marc.jansen@hs-ruhrwest.de
[2] GC DAO, Dubai, UAE
{sapranidi,alexp}@gton.capital

**Abstract.** Blockchain technology offers efficient solutions for a multitude of use-cases, but one of the most successful areas in which this technology has been applied is the field of decentralized finance (DeFi). Decentralized exchanges allow a large number of users to exchange one cryptocurrency for another, without the need for any central authority with any degree of control over the process. Although many different models of decentralized exchanges have been proposed, the Automated Market Maker model currently seems to be the most widely applicable and, therefore, successful. As in the classical financial world, companies and projects may seek to solicit or provide market making services for various trading pairs on exchanges. While for orderbook-based exchanges, the execution of such services is studied deeply and rather straightforward, for AMMs the process becomes more complicated. To solve this, different approaches to market making are offered by the Pathway protocol, focused on two major goals: first, to bring the token price for a particular trading pair to a certain target price, and second, to keep the liquidity of that trading pair intact. The original publication on the Pathway protocol describes its design and goals in detail, but lacks mathematical evidence that the protocol is formally capable of achieving these goals. Therefore, this paper delves a little deeper into the mathematical background of the protocol and formally proves its most important characteristics claimed by the original paper. Last but not least, this paper provides a more in depth discussion of the benefits and limitations of the Pathway protocol, based on the presented mathematical foundations.

**Keywords:** Blockchain · Automated market makers · Cryptocurrencies · Market making

## 1 Introduction

Blockchain [1] has become a key decentralization technology, and decentralized finance (DeFi) [2] is one of the main drivers of its success. In recent years, numerous protocols have emerged that allow users to control their finances independently of large banks or other intermediaries. Various use-cases of traditional

J. Prieto et al. (Eds.): BLOCKCHAIN 2022, LNNS 595, pp. 78–87, 2023.
https://doi.org/10.1007/978-3-031-21229-1_8

finance, such as borrowing and lending funds, exchanging different cryptocurrencies for each other, derivatives, and even insurance, to name a few, have been implemented on decentralized infrastructure. With Ethereum and other EVM based systems leading the space, other platforms, with slightly less powerful languages for the implementation of Smart Contracts [3], provide a currently still growing infrastructure for the development of different financial protocols.

One of the largest fields of interest has been decentralized exchanges (DEX) [8], which provide a decentralized way to exchanging different blockchain assets for each other. While exchanges in traditional finance largely rely on centralized orderbooks to create markets and to price various assets, new pricing models have been developed for the decentralized equivalents of centralized exchanges in DeFi. Here, the most prominent model has been the Automated Market Maker (AMM) [7]. Yet, the name is a bit misleading: the AMM model itself does not provide market maker functionality in the traditional financial sense, but rather allows one to price a given asset relative to another asset. AMMs rely on pre-defined mathematical relations that can vary with each protocol, to determine asset pricing. A classic approach was pioneered by Uniswap, which defines the price relation between tokens based on their amounts in the pool.

While traditional market making is fairly straightforward on orderbook-based exchanges, where you as the market maker can simply add new orders to provide liquidity to the market, the model is quite complex for AMM-based exchanges. This is where the Pathway protocol comes into play [6]. The protocol offers a two-step process performed on AMM-based pools that allows for a new price to be determined for an asset traded on a DEX, while at the same time ensuring that certain conditions on the exchanges are met, such as, for instance, liquidity not being affected by Pathway protocol operations. Although the original description of the protocol clearly demonstrates the necessary steps that need to be taken by a market maker on an AMM DEX, mathematical proofs of the model are missing from the original publication. Therefore, this paper deals exactly with just that missing evidence, to demonstrate that the Pathway protocol can be trusted and used reliably.

The remainder of the paper is organized as follows: after the introduction, an overview of the state of the art is presented. This is followed by a brief description of the basic operations of Pathway protocol, including their mathematical representations used throughout the paper. Immediately following this, the impact of Pathway protocol operations on the exchange (and its corresponding pools) is revisited in light of the mathematical representation used in this paper, also raising the question of what the most important characteristics of the Pathway protocol are that need to be proven. The next chapter provides evidence for three of those most fundamental characteristics, essentially showing that the protocol can be trusted. After the fundamental trustworthiness is established, a discussion of the protocol is presented, together with some of its limitations. Last, an outlook and a brief description of potential topics for future works conclude the paper.

## 2   State of the Art

AMMs are usually organized into so-called pools. For the sake of simplicity, the rest of the paper is based on the assumption that each pool consists of two assets that can be exchanged for each other via pool operations. In order to allow tokens to be exchanged for each other, pools must have a certain number of tokens. This is where liquidity providers come into play: users can provide liquidity (in the form of tokens) to a pool, allowing other users to exchange the tokens of the token pair for each other. In order to incentivize liquidity provision, a fee resulting from the exchange of tokens is distributed among liquidity providers in proportion to the amount of liquidity provided.

There are various AMM models on the market that users, both liquidity providers and traders, can choose from. These models mainly differ in the methodology used to determine the price of tokens in a pool, often referred to as the invariant of a given pool. While some pools link the price of tokens only to the ratio of tokens in the pool, others may also consider external parameters, such as prices on other exchanges, to determine the price. Nevertheless, the basic idea is that whenever the price of a token in the pool differs from the market price, it constitutes an arbitrage opportunity, which arbitrageurs may use to make a certain risk-free profit and thus return the price in the pool to the market price, essentially relying on the no-arbitrage theory [4].

Despite the different implementations of AMMs, two general rules apply to all AMMs considered in this paper [8]:

1. Operations that work with the pool's liquidity, either providing liquidity or removing liquidity, do not affect the price of the assets in the pool.
2. The invariant of the pool remains constant for exchange activities between the tokens of the given pool.

These two properties are integral for the Pathway protocol to work properly, mainly because the operations proposed by the protocol include removing a certain amount of liquidity from the pool and exchanging tokens in the pool.

The second important idea on which the Pathway protocol is based is the idea of market making. In traditional finance, this term usually refers to the party that provides liquidity in a given market [5]. In our case, the market would be established by an AMM pool. While market making in traditional finance can make use of different strategies, such as placing orders to a traditional orderbook, following a certain strategy, these operations are not possible in AMM-based pools, simply because there is no equivalent of an order book. Therefore, new strategies need to be developed based on the aforementioned AMM functionality, i.e. liquidity provision/removal and token exchange. Here, the Pathway protocol provides a solution by defining two simple steps that allow to manage the price of tokens in a given pool.

Market making is deemed necessary for DeFi projects or DAOs with governance tokens to allow token holders to enter and exit positions and increase their exposure to the market. Passive market making strategies usually employed by DAOs are not robust enough and therefore sensitive to market fluctuations and external factors. On the other hand, active approaches to market making create a separate class of algorithmic governance tokens, with price stability achieved by linking it to a predictable target value. As stated by the authors of Pathway, the implementation of such protocols can highly increase the market efficiency of DAO systems.

## 3   Pathway Operations

The Pathway protocol describes an approach to active market making on a decentralized exchange (DEX), implemented as an automated market maker (AMM). While this may seem strange, it makes sense if we think about what AMM actually does. Contrary to its name, its main purpose is not to create a market in the traditional sense of the term, but to provide a deterministic way of identifying prices of the assets traded on it. This is where the Pathway protocol comes in, by providing a set of operations that can be used to influence token pricing.

Classical AMMs like Uniswap [9], Curve [10], DoDo [11] and the like, are usually organized in so called pools. Here, for the sake of simplicity, and since this corresponds to the original publication of the Pathway protocol, we assume that each pool consists exactly of two tokens $T_1$ and $T_2$ which are traded against each other.

The Pathway protocol describes two different steps that, when executed one after the other, allow to control the price of the two tokens traded in the pool. In the first step, liquidity is removed from the pool. In the second step, liquidity is exchanged from one token to another through the pool. Essentially, the purpose of performing these two steps is to change the price p1 of token $T_2$ calculated in token $T_1$ to the new price $p_2$.

Assuming $t_1$ is the amount of token $T_1$ in the pool and $t_2$ the amount of token $T_2$, the price $p_1$ for token $T_2$ expressed in $T_1$ would then be:

$$p_1 = \frac{t_1}{t_2}$$

Under the assumption that the liquidity $l$ in the pool needs to be kept constant, as well as the number of tokens $T_1$, we can calculate the amount of liquidity, expressed in $T_2$, that needs to be removed from the pool in order to achieve the price of $p_2$ as following:

$$l = l_r * p_2 + t_1$$
$$\Leftrightarrow l_r = \frac{l - t_1}{p_2}$$

## 3.1   Step 1: Removing Liquidity from the Pool

Since liquidity should be removed in equal shares for each token $T_1$ and $T_2$, the amount of tokens that need to remain in the pool could be calculated as:

$$\frac{l_r}{2} * \frac{t_2}{t_1} \text{ of token } T_1, \text{ and}$$

$$\frac{l_r}{2} \text{ of token } T_2$$

## 3.2   Step 2: Swapping $T_1$ Tokens to $T_2$

In the second step the amount of:

$$\frac{l_r}{2} * \frac{t_2}{t_1}$$

$T_1$ tokens resulting from the removal of the liquidity will now be swapped to $T_2$ tokens via the pool.

# 4   Impact of Pathway Operations on the Pool

In order to better understand the evolution of the pool state, in terms of the amount of tokens $t_1$ and $t_2$ in it, we can use the following calculations. The number $t_2'$ of tokens $T_2$ after step 1 and step 2 in the pool can be calculated as:

$$t_2' = t_2 - \frac{l_r}{2}$$

Similarly, the number $t_1'$ of tokens $T_1$ after step 1 and step 2 in the pool can be calculated as:

$$t_1' = t_1 - \frac{l_r}{2} + \frac{l_r}{2} = t_1$$

Therefore, the amount of tokens $t_1$ stays intact after the operations. What remains is the question of how the Pathway operations described above actually influence the price of the tokens in the pool. Here, it is important to consider three different questions for the proof:

1. What is the impact of step one on the price of token $T_2$ calculated in $T_1$?
2. Is the price of the token $T_2$ calculated in $T_1$ really $p_2$?
3. Does a round of Pathway operations retain the original liquidity of the pool?

# 5   Proofs

As described above, three different points need to be proven in order to demonstrate that the Pathway protocol is working as intended. We start by proving that the first step does not impact the price: this is important in case where the different steps proposed by the protocol cannot be performed atomically. Therefore, here is the first theorem:

**Theorem 1.** *The price of the token $T_2$ calculated in $T_1$ after step 1 is still $p_1$.*

**Proof.** The amount of tokens $t_2$ that needs to removed from the pool can be calculated as:

$$t_2 - \frac{l - t_1}{p_2}$$

And since half of this liquidity needs to be removed in $T_2$ tokens, the amount of tokens $T_2$ removed from the pool could be calculated as:

$$\frac{1}{2} * (t_2 - \frac{l - t_1}{p_2})$$

The amount of tokens $t_1$ that will be removed from the pool is then calculated as:

$$\frac{1}{2} * \frac{t_1}{t_2} * (t_2 - \frac{l - t_1}{p_2})$$

We can then calculate the price after step 1 as:

$$\frac{t_1 - \frac{1}{2} * (\frac{t_1}{t_2} * (t_2 * \frac{l-t_1}{p_2}))}{t_2 - \frac{1}{2} * (t_2 - \frac{l-t_1}{p_2})} =$$

$$= \frac{t_1 - (\frac{1}{2} * \frac{t_1}{t_2} * t_2 - \frac{1}{2} * \frac{t_1}{t_2} * \frac{l-t_1}{p_2})}{t_2 - (\frac{1}{2} * t_2 - \frac{l-t_1}{2*p_2})} =$$

$$= \frac{t_1 - (\frac{t_1*t_2}{2*t_2} - \frac{t_1*l-t_1^2}{2*t_2*p_2})}{t_2 - \frac{t_2}{2} + \frac{l-t_1}{2*p_2}} =$$

$$= \frac{t_1 - \frac{t_1}{2} + \frac{t_1*(l-t_1)}{2*t_2*p_2}}{t_2 - \frac{t_2}{2} + \frac{l-t_1}{2*p_2}} =$$

$$= \frac{\frac{t_1}{2} + \frac{t_1*(l-t_1)}{2*t_2*p_2}}{\frac{t_2}{2} + \frac{l-t_1}{2*p_2}} =$$

$$= \frac{\frac{1}{2} * (t_1 + \frac{t_1*(l-t_1)}{t_2*p_2})}{\frac{1}{2} * (t_2 + \frac{l-t_1}{p_2})} =$$

$$= \frac{t_1 + \frac{t_1*(l-t_1)}{t_2*p_2}}{t_2 + \frac{l-t_1}{p_2}} =$$

$$= \frac{t_1 * (1 + \frac{l-t_1}{t_2*p_2})}{t_2 * (1 + \frac{l-t_1}{t_2*p_2})} =$$

$$= \frac{t_1}{t_2} =$$

$$= p_1$$

In the second theorem, we show that after the second step the price is exactly equal to $p_2$, or the target price, to demonstrate that the protocol can achieve its primary goal:

**Theorem 2.** *The price of the token $T_2$ calculated in $T_1$ after step 2 is $p_2$.*

**Proof.** The amount of tokens $T_1$ in the pool after step 2 are still $t_1$ (see above), while the amount $t_2'$ of tokens $T_2$ in the pool can be calculated as:

$$t_2 - (t_2 - \frac{l - t_1}{p_2})$$

Therefore, the price of the token $T_2$ expressed in $T_1$ is calculated as follows:

$$\frac{t_1}{t_2-(t_2-\frac{l-t_1}{p_2})} =$$
$$= \frac{t_1}{t_2-t_2+\frac{l-t_1}{p_2}} =$$
$$= \frac{t_1}{\frac{l-t_1}{p_2}} =$$
$$= \frac{p_2*t_1}{l-t_1} =$$
$$= \frac{p_2*t_1}{2*t_1-t_1} =$$
$$= \frac{p_2*t_1}{t_1} =$$
$$= p_2$$

since $l$ calculates to:

$$l = t_1 + t_2 * p_1 = t_1 + t_2 * \frac{t_1}{t_2} = 2 * t_1$$

Finally, we need to prove that the operations performed by the Pathway protocol do not change the liquidity of the pool. In cases of single-sided pools like on Dodo PMM DEX, token pricing can be regulated by providing additional liquidity to or removing it from the pool, e.g. by liquidity providers. However, the Pathway protocol provides a generic solution, which can be deployed with different AMM setups, not only single sided pools, while still allowing to execute effective market making strategies. Therefore, we now need to prove that the liquidity of the pool remains intact after performing both steps of the Pathway protocol:

**Theorem 3.** *The liquidity $l'$ of the pool after step 2 is still $l$.*

**Proof.** The liquidity $l'$ of the pool after step 2 can be calculated as:

$$l' = t_1 + p_2 * t_2'$$

since, as already described in the proof of Lemma 2, $t_2'$ can be calculated as:

$$t_2' = t_2 - (t_2 - \frac{l-t_1}{p_2})$$

this leads to:

$$l' = t_1 + p_2 * t_2' =$$
$$= t_1 + p_2 * (t_2 - (t_2 - \frac{l-t_1}{p_2})) =$$
$$= t_1 + p_2 * (t_2 - t_2 + \frac{l-t_1}{p_2}) =$$
$$= t_1 + p_2 * \frac{l-t_1}{p_2} =$$
$$= t_1 + l - t_1 =$$
$$= l$$

## 6    Discussion and Limitations

While the Pathway protocol seems to be efficient in its goal to control the price of a given asset in a pool, it also suffers from several potential shortcomings and intricacies that a market maker should have in mind while deploying the protocol into production.

## 6.1    A Potential Attack Vector: A Sandwich Attack

One potential attack vector is a so-called sandwich attack [12] which can arise when Pathway is used to maintain a stable target price. As described in the original paper [6], malicious actors can take advantage of token selling that can sporadically happen in the pool. To do so, they can buy the tokens back for a lower price, expecting a set of Pathway liquidity interventions to occur. This allows those actors to sell their tokens for a definite profit. This attack is possible because with a general implementation of Pathway, the price should deterministically return to a stable target at some point, which means that using Pathway interventions for maintaining a stable price "peg" on AMMs has a fundamental weakness. As proposed in the original Pathway article, a so-called Proactive Market Maker (PMM) DEX can be used if maintaining a stable peg is necessary.

Proactive Market Makers (PMM) make use of variable parameters such as asset ratio and curve slope, which allows to flexibly set the mathematical relation between assets. On PMM DEXes, such as DODO, market price discovery is usually guided by systems of oracles that can provide price, peg or any other data feeds into smart contracts.

## 6.2    Pathway Protocol Might Lead to a Decreased Influence on the Pool

Let us assume, for the sake of simplicity, that liquidity providers Alice and Bob each provide 50% of liquidity in the pool, resulting in 50 Liquidity tokens owned by each of them. Let us further assume that the total liquidity of the pool amounts to 100$. Hence, Alice's and Bob's liquidity tokens each have a total value of $50. Now, Bob runs the Pathway protocol steps in order to change the price to a certain target level. Let's assume Bob needs 10% of all liquidity tokens in order to achieve the target price. The liquidity tokens used during the Pathway protocol steps are burned by the pool. Therefore, after the execution of the Pathway steps, the total number of liquidity tokens becomes 90, since 10% from 100 have been burned by the AMM, and hence the distribution between Alice and Bob calculates to:

Bob: $40/90 = 44.44\%$
Alice: $50/90 = 55.56\%$.

Since the liquidity in the pool is invariant towards the Pathway protocol steps (see Theorem 3), the pool's liquidity is still $100. Hence, Alice's share of the liquidity in the pool increased by 5.56%, while Bob's liquidity declined by 5.56%, resulting in an actual loss of exposure for Bob as liquidity provider.

On the other hand, Bob received tokens from the swap in step 2, which make up for his loss on the financial level. Yet, Bob sacrificed some influence on the pool liquidity, which in turn limits his ability to conduct larger liquidity interventions in the future.

This is essentially the reason why the authors of [6] state that: "In the perfect scenario, all assets contained in the AMM pool are owned by the DAO." Such requirements are easily achievable by protocols that own their complete liquidity on DEXes (POL) [13]. In a POL scenario, the decline in the amount of liquidity tokens does not have a negative impact because there is just one holder of those tokens anyway and their liquidity provider tokens always represent the complete liquidity of the pool.

## 7    Outlook and Future Work

The main innovation of the Pathway protocol is that it provides a completely defined algorithmic Market Making approach for AMM based Decentralized Exchanges. This paper has shown that the liquidity intervention steps proposed in the original paper are mathematically sound and achieving the corresponding goals.

Pathway provides various opportunities for market makers to implement different token pricing strategies in AMM based pools. Therefore, the protocol itself is defined independent of why the target price in question should be achieved in the pool. We therefore think that different market making goals and outcomes might be interesting subjects for further research.

For instance, the target price that should be achieved by the Pathway protocol could be defined by an oracle, mirroring a price of an external exchange. Another possibility for the target price to be defined is by holding a fixed peg, e.g. $1, to implement a dollar-based stablecoin. Here, the combination of Pathway protocol interventions and the concept of Protocol Owned Liquidity for the creation of an algorithmic stablecoin seems to be promising.

Additionally, further research is necessary with respect to how the protocol steps should be executed in practice. While the initial definition implies an externally-controlled execution of both steps, it is also possible to execute those steps directly by creating dedicated smart contracts, which results in an atomic implementation of the Pathway protocol.

## References

1. Nakamoto, S.: Bitcoin: a peer-to-peer electronic cash system. Decentralized Business Review, 21260 (2008)
2. Werner, S.M., Perez, D., Gudgeon, L., Klages-Mundt, A., Harz, D., Knottenbelt, W.J.: SoK: Decentralized Finance (DEFI) (2021) [online]. Available at: https://arxiv.org/abs/2101.08778
3. Jansen, M., Hdhili, F., Gouiaa, R., Qasem, Z.: Do smart contract languages need to be Turing complete? In: International Congress on Blockchain and Applications, pp. 19–26. Springer, Cham (2019)
4. Ross, S.A.: The arbitrage theory of capital asset pricing. J. Econ. Theory **13**, 341–360 (1976)
5. Dolgopolov, S.: Insider trading: informed trading, and market making: liquidity of securities markets in the zero-sum game. Wm. Mary Bus. Law Rev. **3**, 1 (2012)

6. Pupyshev, A., Sapranidi, I., Khalilov, S.: Pathway: a protocol for algorithmic pricing of a DAO governance token (2022) [online]. Available at: https://arxiv.org/abs/2202.06541
7. Buterin, V.: On Path Independence (2017) [online]. Available at: https://vitalik.ca/general/2017/06/22/marketmakers.html
8. Xu, J., Vavryk, N., Paruch, K., Cousaert, S.: SoK: Decentralized Exchanges (DEX) with Automated Market Maker (AMM) Protocols (2021) [online]. Available at: https://arxiv.org/abs/2103.12732
9. Adams, H.: Uniswap Whitepaper (v1) (2018) [online]. Available at: https://hackmd.io/C-DvwDSfSxuh-Gd4WKE_ig
10. Egorov, M.: StableSwap - efficient mechanism for Stablecoin liquidity (2019) [online]. Available at: https://curve.fi/files/stableswap-paper.pdf
11. DODO Team: DODO - A Next-Generation On-Chain Liquidity Provider Powered by Pro-active Market Maker Algorithm (2020) [online]. Available at: https://dodoex.github.io/docs/docs/whitepaper/
12. Zhou, L., Qin, K., Torres, C.F., Le, D.V., Gervais, A.: High-frequency trading on decentralized on-chain exchanges. In: 2021 IEEE Symposium on Security and Privacy (SP), pp. 428–445. IEEE (2021)
13. Chitra, T., Kulkarni, K., Angeris, G., Evans, A., Xu, V.: DeFi Liquidity Management via Optimal Control: Ohm as a Case Study (2022)

# Profitable Fee Controller for Payment Channel Networks

Anupa De Silva$^{(\boxtimes)}$ (iD), Subhasis Thakur(iD), and John Breslin(iD)

Data Science Institute, National University of Ireland Galway, Galway, Ireland
{anupa.shyamlal,subhasis.thakur,john.breslin}@insight-centre.org

**Abstract.** Payment channel networks (PCN) are one of the promising second layer solutions for the scalability issue of cryptocurrencies. They offer high throughput and low-cost transactions for the users. However, practical and design issues of PCN hinder users from embracing the technology. This paper focuses on how to increase user adoption by boosting revenue as transaction mediators. Our approach is twofold; reduce the cost of refilling channels due to channel exhaustion and collectively acquire a maximum gain from all channels belonging to the respective nodes. For that, we devise a novel fee controller based on Model Predictive Control (MPC), where we formulate the problem as an Optimal Control Problem (OCP). It is optimized to obtain maximum profit while proactively alarming the payers about the exhaustion through a dynamic fee mechanism. We evaluate the proposed fee controller's effectiveness compared with the conventional static fee controller and OptimizedFee controller. Our experimental simulations demonstrate that the proposed fee controller successfully improves the revenue and controls channel exhaustion simultaneously.

**Keywords:** Cryptocurrency · Payment channel networks · Layer-two · Model predictive control · Fee controller

## 1 Introduction

Industries are rapidly embracing blockchain technology regardless of what domain they are in, tech or non-tech. Primarily, they benefit from its traceable, decentralised, and immutable infrastructure. However, the Blockchain Trilemma points out that prioritising blockchain's security and decentralised nature restricts blockchain's scalability. Among many proposals [3] to control the scalability issue, layer-two protocols are prominent as an orthogonal scaling solution that can absorb the robustness of the underlying blockchain layer (i.e., layer-one) and operate independently. PCN emerged as a promising layer-two solution introduced to minimise blockchain latency and cost issues and increase the throughput. PCN facilitate unlimited miniature transfers with instant settlements, making it ideal for micropayments. Micropayments are not feasible on decentralised cryptocurrencies (e.g. Bitcoin) due to their slow transaction

J. Prieto et al. (Eds.): BLOCKCHAIN 2022, LNNS 595, pp. 88–99, 2023.
https://doi.org/10.1007/978-3-031-21229-1_9

confirmation and expensive fees incurred by time-consuming and high-cost computations to build the consensus. Theoretically, PCN are only bounded by the communication overhead between participants. Hence, it can operate fast and economically with high throughput yet provides the same security as if the transfer happened on the chain.

The cryptocurrency community is increasingly embracing PCN due to the cryptocurrency price surge. However, the general comment on PCN in academia is that PCN have built-in design issues [7]. In this paper, we mainly focus on channel exhaustion (see Sect. 2.2) in PCN caused by the imbalanced flow of payments. PCN channel has a fixed capacity, and when the balance drops, it becomes limited to cater to new transactions. Hence, the channel should be restored by refilling. It demands an on-chain transaction causing channel downtime and an additional cost to the channel owners. Such channel exhaustions primarily affect the intermediaries (see Sect. 2.1) who joined the network to make a profit through transaction mediation. Controlling this issue enables intermediaries to sustain the channel for an extended period and maximise revenue.

This paper aims to strengthen the PCN user base by empowering intermediary nodes in the network. We devise an automated fee controller to manage the inbound and outbound cash flow, which avoids or instantly recovers from channel exhaustion. It is mainly beneficial for intermediary nodes to survive in the network for an extended period and reduce on-chain fees incurred by refilling. At the same time, it is designed to achieve the fundamental objective of an intermediary, gaining maximum utility of the channels, thus, yielding the maximum possible profit. The rest of the paper is organised as follows. First, we present the operations of PCN in general and relevant issues, followed by existing research related to fee policies and channel exhaustion. Then, we formulate the optimisation problem for Model Predictive Control and present the design of the proposed controller. Using Sala, we perform experiments on the controller's effectiveness and evaluate the proposed controller comparing it with two other controllers.

## 2   Background

There are several implementations of PCN for different platforms (e.g. Lightning Network for Bitcoin and Raiden for Ethereum). In general, PCN [7] operate by creating channels between peer-to-peer couples. Each couple collectively stores a locked deposit on the blockchain using a Multisig Smart Contract. The smart contract is programmed in such a way that this deposit can be released either with the consent of both participants (both should digitally sign) or after the expiration of the lock. The speciality is that the participants can keep updating their respective funds adhering to the same agreement but refrain from recording it on the chain. Every update is publishable; hence, at any point in time, both can publish the latest state on-chain individually (i.e., after a dispute) or collaboratively and close the channel. Hence, the offline consensus is built through the threat to publish. It is, in a way, double spending where the latest one gets rejected if one participant pushes an old state to the chain. Rationally, it is a state which is more beneficial for the publisher. To avoid that, we can include a time-lock that makes each state update valid after a specific time; hence the victim gets a time window to push the latest on the chain.

Creating a channel for every pair is also a costly operation on the chain. Therefore, when a couple has not established a direct payment channel, we can route the payment through a set of intermediaries who have already opened channels. This synchronisation of channels can further reduce the creation of new on-chain transactions. For that, PCN extend the same smart contract using an additional lock using the hashed value of a secret. The process starts with the payment receiver generating a secret and sending its hash value to the sender. The sender finds a path to the receiver through connected channels. Each participant, including the original sender, must compensate the peer in return for the secret of the hash value before a set expiry time. They should follow the path to the original receiver to obtain the secret from the original receiver, who is compensated by the actual amount at the end of the path. Intermediaries return to the original sender with the preimage, where they can unlock a fee for the brokerage in addition to the actual amount. In general, there is a base fee plus a fee proportional to the transferred amount. Once the sender receives the secret, they can verify and prove the transaction completion.

## 2.1 Participants of PCN

We identify three types of entities in PCN with different intentions. Current literature has given less or no attention to recognising their perspectives distinctively. They are the payment participants, defined as payers, payees and intermediaries, who can also overlap. Payers and payees actuate the coin liquidation within the network. The liveness of the network depends on their existence. Payers initiate transactions through the nodes and expect security, reliability, and privacy from PCN. A low transaction fee is also an essential requirement. Payees are primarily merchants who expect the payments through the network and intend to engage with a larger audience. Merchants can also cause an imbalance in the network and channel exhaustion in neighbouring nodes because of continuous payment in-flows. Intermediaries intend to earn by providing brokerage services to the network. They are profitable based on transaction fees and the position of the network.

## 2.2 Issues in PCN

Specific PCN design considerations cause inevitable reliability and availability issues for the users. PCN nodes are not revealing their actual balances in real-time but the initial capacities (on-chain deposit from both participants). However, it is a requirement for all payment participants to hold sufficient funds to forward a transaction. Payers have to guess from the initial capacities when the actual balances are concealed or learn it from trial and error before a successful transaction. For example, the channel selection criteria can be set so that the transaction amount must be less than the channel capacity. Although it is necessary to be true, it is not sufficient to qualify a channel because of the disparity of balances. If failed, the payer has to choose another path, probably with higher fees. The empirical study on a real-world PCN in [10] reveals that the transactions fail mainly due to insufficient balance. The conventional view on this issue is that disclosing the balances is a threat to the privacy of the nodes. However, there are practical attacks [4,5] to expose the channel balances by manipulating the free rides offered by PCN. It is possible to hop from payer to payee at no cost and gain usable insights about the nodes and transactions. For example, the attacker can initiate a payment with an invalid hash lock for a known or unknown recipient. This bogus payment inevitably fails, and refunds (i.e., free attack); still comes back with manipulatable information.

Apart from the design issues, there are practical complications in PCN. Imbalance transaction flow of the network is one of them. When the payment flow gets biased in one direction, a disparity is created between the inbound and outbound of the node, and the outbound flow eventually becomes unusable. Since they are on two separate transactions on the chain, they cannot be adjusted offline, although both belong to a single person. It takes at least one on-chain transaction unless the channels are created within a channel factory [1]. An unbalanced network is the root cause of most PCN complications. It intensifies transaction failures as a consequence of the balance concealment. Nodes in an unbalanced transaction flow have to face frequent re-fillings or closures of channels and bear the respective on-chain fees. It deteriorates the revenue of the nodes, particularly the intermediary nodes. Further, it affects the availability of the nodes considering the overall network. Another inter-related issue is the traffic congestion in channels. Congestion occurs when there are unsettled transactions piled up, and as a result, the channel cannot accommodate other transactions. In the payment execution, participants must wait for the secret to arrive. Till then, the locked-up funds are also not usable. On-hold time is a crucial factor as it is a missed opportunity to engage in a different payment execution which causes a loss of revenue. An attacker can also leverage this along with the free rides mentioned above and deploy an attack to handicap a channel.

## 3   Related Works

Current literature consists of several techniques to handle channel exhaustion issues in PCN, including refilling and building channel factories. When the channel is depleting or depleted, the channel participants have to record the updated amounts on the chain and close the channel. If they are willing to restart, two on-chain operations will require closing and re-opening the channel. However, the splicing technique can limit them to one operation, yet it still costs one on-chain transaction fee. Instead, they can adjust their balances by making circular payments to themselves. Pickhardt et al. [8] device an optimisation algorithm to maintain a balanced network by making circular payments. It still costs the intermediary fees. REVIVE by Khalil et al. [6] introduces a free of charge(subject to the responsiveness of the nodes) re-balancing mechanism without closing the channels. Here, PCN nodes collectively shift their funds in a circular path and refill their balances. However, this method introduces a centralised party to the network.

Creating channel factories [1] is a safety measure to handle the depleted channels in the channel creation stage. In channel factories, more than two parties collaborate in the initial funding and later perform bilateral transactions reducing the space taken on-chain by offline adjustments to the funds. However, when the number of involved parties increases, the number of disputes surges and the channel factory begins to operate as a set of normal bilateral channels.

The skewness of transaction flow is the primary cause of channel exhaustion. Therefore we can control the channel exhaustion by preventing skewness. The current research works realise this by introducing different routing protocols and fee policies. Thakur et al. propose a flocking based routing protocol for a balanced network and empirically analyse its effectiveness in [9]. Stasi et al. device a fee policy for a sustainable and balanced network focusing on multi-part payments [2]. They measure the imbalance of a channel by the difference in channel balances and devise a fee calculation algorithm, OptimizedFee. Channel imbalance decides which proportion of payment amount to be charged by the payer. They intend to maintain equal balances in both peers, making payers effortlessly determine if a node holds sufficient funds. As a result, it increases the chances of payment execution being successful. They observe a significant reduction in network imbalance when using the proposed policy in a simulated network.

In summary, we can find preventive (i.e. Channel Factories), proactive(i.e. balanced routing), and reactive (i.e. re-balancing) solutions to handle exhausted channels. In the bidirectional payment networks, asymmetric flow is a recurrent event. Therefore, preventive or reactive methods are not always sufficient as a long term answer. In contrast, proactive strategies can demonstrate promising results for an extended period. On the other hand, the existing works to improve the reliability of PCN are based on minimising the balance difference (i.e. a central balance) to maintain a balanced network. However, we argue that a central balance also implies limited channel liquidity, which could affect the income of the intermediary nodes.

## 4    MPC Based Fee Controller (MPC-FC)

Intermediary nodes stake their coins to join the network to gain financial earnings by facilitating PCN payment execution. In designing the fee controller, we prioritise maximising their profit while controlling the channel exhaustion.

The process of controlling the channel exhaustion is intuitive, such that we manage the outbound coin flow using the fee rate adjustment. When the outbound balance depletes, we raise the fee to discourage users from choosing the channel in their path selection until inbound traffic reinforces and restores the depleted node. In reality, it is more complex as a node can hold several channels. For instance, recovery from depletion expects to satisfy both outbound flow restrictions and attract inbound. The node cannot do the latter as it has no control over the inbound flow as it does in the outbound flow (its peer controls the fee). However, the flow should exit from one of the channels owned by the same node, which means lowering the fees of the rest of the channels can incentivise inbound transactions. Still, it is a trade-off as reducing fees also affects the total income of the node. We design the proposed fee controller, MPC-FC, as an optimisation problem where channels collaborate to maximise the revenue while controlling the channel exhaustion.

## 4.1   Model

Assume a node engages with $n$ other nodes, creating $n$ number of channels with capacities denoted by the vector $C$. At time step $k$, we define the following states of the node.

**Definition 1.** Let $M_k$ be an $n \times n$ matrix denoting monetary values of incoming and outgoing transactions to and from each peer through the node at time step $k$.

For example, $m_{i,j} \in M$ holds the total monetary value of the transactions which comes from $j$th peer and exits to $i$th peer through the node.

**Definition 2.** Let $n \times 1$ matrix, $F_k$ denotes the fees charged by the node for each channel as a portion of the transaction amount being mediated and $\forall f \in F_k$, $0 \leq f_{min} \leq f \leq f_{max} \leq 1$.

If we denote the movable balances with each peer by the matrix $B_k$ at time step $k$,

$$B_{k+1} = B_k + \Delta B \tag{1}$$

where $\Delta B = sumv(M_k)^T + sumh(M \times (F_{k-1} - J_{n,1}))$ which includes incoming payments and fees and exclude outgoing payments. $sumv$, $sumh$ and $sum$ are the functions to calculate vertical, horizontal and total sums of a matrix respectively. $J$ indicates a matrix of ones.

Let the profit of the node denoted by $p_k$ at time step $k$.

$$p_{k+1} = p_k + sum(F_{k-1} \times M) \tag{2}$$

In this paper, we define $M_{k+1}$ as follows where, $norm(F)$ is the normalized fee rates and $O_{max}$ is the maximum outbound values $(J_{n,n}B)$. $w_m$ is the weight we set depending on the traffic. In simulations, we have set the $w_m = s/n$ where $s$ denotes the frequency of fee controller is run.

$$M_{k+1} = w_m \times O_{max} \times norm(F) \tag{3}$$

Here, we manipulate $M_{k+1}$ such that the controller has to raise the fee when the outbound balance is low and vice versa to maximize the profit.

## 4.2   Optimization Problem

We design the following optimal control problem based on the above.

$$min \sum_{k=0}^{H} -p_k + w_f \times \Delta F \qquad (4)$$

such that: Eqs. 1–3, and

$$J_{n,1} \times f_{min} \leq F \leq J_{n,1} \times f_{max} \qquad (5)$$

$$O_{n,n} \leq M \leq J_{n,1} \times C \qquad (6)$$

$$O_{n,1} + a_{min} \leq B \leq C - a_{min} \qquad (7)$$

$$B_0 = J_{n,1} \times C, p_0 = 0, M_0 = O_{n,n} \qquad (8)$$

Here we maximize the profit which is subjected to satisfy the inequalities in 5–7. $w_f$ is the weight that can prioritize the fee fluctuations. $O$ denotes the matrix of zeros, and we use $a_{min}$ to set lower and upper margins for the balances to avoid exhaustion.

## 4.3   Model Predictive Control

The fee controller uses the MPC technique to repeatedly solve the above OCP and obtain the optimal fee rates for the channels of the selected node. It starts with initial states as mentioned in Eq. 8 where initial balances are the individual channel capacity, profit, and transaction amounts are all zero. Then, for a given frequency and states, MPC computes the control input, $F$, by solving the OCP for the predicted $H$ number of future steps. Then it applies the first fee vector of the computed control sequence and observes the states ($M$, $B$, and $r$), which are fed back to the system to recompute the OCP. This continues in every $s$ step of the simulator.

## 5   Implementation

We occupy an agent based PCN simulator, Sala.[1] It can simulate the generic functionality of PCN, including establishing and maintaining payment channels, initiating, receiving, and mediating the transactions. It also allows configuring nodes with different fee strategies. We implemented MPC-FC using the do-mpc library,[2] which is based on CasAdi[3] and IPot Solver.[4] We integrated the proposed controller, MPC-FC into the Sala simulator along with other existing controllers; Static Fee Controller (S-FC) which is the general fee strategy in PCN implementations and OptimizedFee Controller (OF-FC) proposed in [2].

---

[1] www.github.com/anupasm/SALA.

[2] www.do-mpc.com.

[3] web.casadi.org.

[4] www.coin-or.github.io/Ipopt.

# 6    Experiments and Results

Our objective is to investigate how efficiently the proposed fee controller reduces channel exhaustion and calculate the cost incurred as a consequence. We empirically examine the proposed fee controller using the Sala simulator. We do not use real-world PCN transaction data (which is not available as well) for the experiments; instead, we use extensive simulations to simulate PCN behaviour by randomising transaction parameters (i.e. payer, payee, and value). Those simulations were carried out at a fixed number of nodes, capacities, limits of transaction values and fee percentages. However, our result evaluation is independent of such parameters as we evaluate fee controllers proportionally.

## 6.1    Setup

In the overall setup, we randomly generated networks of 50 nodes connected with six other peers on average. Each channel capacity is identical, and both peers contribute equally ($c = 100$). Also, we assign the three identified roles, merchant, intermediary and client, to each node to the proportion 1:2:2. For a realistic simulation, we assume that initiating and receiving transactions depends on the node's role. For example, we set a high probability for merchants to receive more incoming payments. And those payments are randomly generated in between $m_{min} = 4$ and $m_{max} = 10$. For simplicity, we assume the base fee is negligible and common for all nodes. Then, we set the static fee rate to 10% of the transaction value whereas MPC-FC and OF-FC charges $f_{min} = 5\%$ to $f_{max} = 15\%$. We also assume that payers always choose the cheapest and shortest path, which is intuitive.

In each simulation step, we monitor the collected fees, balances held by each node from the completed transactions, and the values of the incoming and outgoing transactions that pass through the node. We derive three variables from those data to analyse channel exhaustion. The first one is the total exhausted steps in the form of the step counts when the balances of related channels drop below $m_{min}$. Setting the lower limit to zero is impractical as the balance does not drop below that. We define average recovery time, extending the total exhausted steps where we calculate the average of exhausted segments to gain an insight into the recovery time. The last variable is the dropped fees from the transactions that could not pass because of insufficient balance. All values were extracted after $s = 500$ number of steps.

## 6.2    Exhaustion and Revenue

We compare how the three controllers manage exhaustion and revenue generation by repeating the same simulation three times with one of the three controllers each. As described in the overall setup, the first run generates the network and transactions and lets all nodes charge a static percentage of transacted value from all transactions. In the second and third runs, we embed the MPC-FC and OF-FC controllers inside the exhausted node, respectively and simulated the same network and the same set of transactions, excluding the transaction paths. We let payers determine the transaction paths independently in real-time to pick the cheapest and avoid expensive paths.

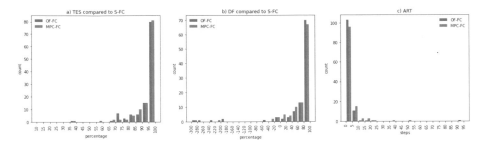

**Fig. 1.** Histograms of three parameters, indicating different aspects of exhaustion of 119 simulations. (a) Percentage distribution of total exhausted steps reduction compared to the respective S-FC based simulation. (b) Percentage distribution of dropped fees compared to the respective S-FC based simulation. (c) Distribution of average recovery time in steps.

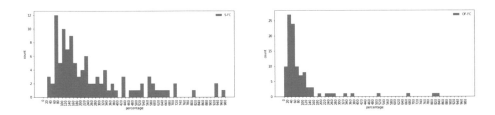

**Fig. 2.** MPC-FC collected fees compared to S-FC and OF-FC in 119 simulations.

Figure 1 presents the distribution of percentages of total exhausted steps reduction, dropped fees reduction, and average recovery time figures derived from 119 number simulations. Total exhausted steps and dropped fees values of OF-FC and MPC-FC simulations are given in proportion to S-FC based runs. OF-FC and MPC-FC demonstrate nearly similar behaviour in all three sections. Both limit the exhaustion to around 5% compared to S-FC, and most importantly, average recovery time values indicate that most nodes could recover from exhaustion after around five steps. The same applies to dropped fees figures, where both controllers reduce the dropped transactions amid exhaustion by approximately 80%. Figure 2 indicates the fees collected by the proposed MPC-FC compared to the S-FC and OF-FC, respectively. MPC-FC achieves higher collected fees in all 119 simulations, and on average, it is a 231.32% and 56.34% raise in profit compared to S-FC and OF-FC, respectively.

Overall, MPC-FC surpasses the other two controllers in terms of collecting fees. Both OF-FC and MPC-FC perform significantly better than the S-FC in managing exhaustion. We observe a similar behaviour between OF-FC and MPC-FC in all aspects considered.

## 7  Evaluation and Discussion

This section evaluates the proposed controller from different aspects by benchmarking with the other two; S-FC, the most common fee mechanism and the OF-FC, which promotes central balance to keep the network balanced.

The intermediaries in PCN expect to make the most of its limited investment. Here, we proposed an approach similar to OF-FC to adjust fees to control the balance. However, we refrain from the idea of keeping the balance central and considering channels in isolation. All the channels belonging to the relevant node collaboratively adjust the fees to maximise the profit. Therefore, according to the test run results of collected fees, the proposed MPC-FC surpasses the OF-FC by more than 50% margin, and it is more than two times what the static controller collected. The reason for the significantly low figures given by S-FC is that its channels remained exhausted for an extended period. However, we did not refill the channel upon exhaustion for simplicity; otherwise, S-FC based node should bear an additional cost (i.e. on-chain fee) which should be deducted from the profit.

The next and the main objective of this research is to minimise channel exhaustion. For that, total exhausted steps indicate how well the channel avoids exhaustion. In solving the OCP of MPC-FC, we have set constraints to maintain a non zero balance. According to the outcome of the simulations, OF-FC and MPC-FC equally perform well in dropping the total exhausted steps figures, even though there is a subtle difference. There are occasions where it is inevitable to avoid balance hitting bottom. However, it is not an issue as long as it can recover. Average recovery time indicates how fast a channel can recover when it gets exhausted. Fees are raised in OF-FC and MPC-FC when the channel balance deteriorates. In the meantime, the input stream flows in the reverse direction and raises funds. The average recovery time results in Fig. 1 indicates that both can recover from exhaustion within around five steps. OF-FC and MPC-FC demonstrate nearly similar performance. It does not occur with a fixed fee as in S-FC, and as a result, the channel rarely recovers from the exhausted state. Putting both outcomes together, we can derive that both OF-FC and MPC-FC can lessen the exhaustion and instantly recover from exhaustion.

Even though a channel is not exhausted, it can remain at a low balance and drop the incoming transactions. Therefore, dropped fees represent the lost income because of insufficient balance from a node's perspective. Dropped fees also imply the Transaction Failure Rate (TFR) in the view of the network reliability. The literature review found research works promoting PCN's balancedness to reduce the TFR, including the OF-FC. OF-FC tries to keep the balance central by adjusting the fees. In contrast, the proposed MPC-FC tries to stir the channel balances to secure maximum profit. Therefore, it is intuitive to expect higher TFR (in this case, dropped fees) in MPC-FC. However, dropped fees results in Fig. 1 suggest otherwise where we cannot see a significant difference between OF-FC and MPC-FC. The primary reason is that MPC-FC signals PCN senders using the channel by increasing the fees and vice versa. It is similar to revealing balances, but, in this case, higher fees actively discourage the senders. Hence, MPC-FC managed to control dropped fees to a reasonable extent. To summarise, MPC-FC has simultaneously achieved the main objectives of an intermediary, higher collected fees and lower total exhausted steps, average recovery time, and dropped fees compared to S-FC and OF-FC.

Fee policy communication is crucial with MPC-FC and OF-FC because of the decentralised nature of PCN. Both MPC-FC and OF-FC are dynamic controllers that alter the fees according to the dynamic states of the system. Therefore, the frequency of communicating the fee rates should be designed to avoid flooding the system. Moreover, nodes with MPC-FC and S-FC do not necessitate exposing their balances to the network, whereas OF-FC implicitly expects private balances to be public. Hence, MPC-FC can reduce transaction failures, protecting the privacy of the PCN nodes simultaneously.

The future direction includes inspecting the network-wide effect of using the proposed controller. Here we have explored individual usage of the fee controller. However, the same concept is applicable in a distributed way to achieve network-wide goals. It is important to examine how it will affect the transaction reliability and payers' cost in PCN, in general.

## 8   Conclusion

Payment Channel Networks have the potential to become the solution for micro-payments, while mainstream cryptocurrencies have become almost infeasible to perform day-to-day transactions without compromising security. However, there are few design and practical issues in PCN that hinder user adoption and consequently limit PCN's true potential. This paper attempts to mitigate one of such barriers: we empower users by reducing their costs caused by exhaustion and improving the gain by operating an intermediary node. We propose a novel technique to dynamically adjust the fee rates in PCN using Model Predictive Control. Our simulation results suggest that the proposed controller elevates user revenue compared to existing fee controllers and considerably decreases channel exhaustion and transaction failures. This result empirically attests that central balance reduces the node revenue. On the other hand, central balance is not the only solution to reduce transaction failures caused by insufficient balance. Future work intends to improve the fee controller to operate collaboratively with other nodes in a distributed manner, achieving user and network-wide objectives.

**Acknowledgements.** This publication has emanated from research supported in part by a grant from Science Foundation Ireland and the Department of Agriculture, Food and the Marine on behalf of the Government of Ireland under Grant Number

SFI/16/RC/3835 (VistaMilk), and also by a grant from SFI under Grant Number SFI/12/RC/2289_P2 (Insight). For the purpose of Open Access, the author has applied a CC BY public copyright licence to any Author Accepted Manuscript version arising from this submission.

# References

1. Burchert, C., Decker, C., Wattenhofer, R.: Scalable funding of bitcoin micropayment channel networks. Roy. Soc. Open Sci. **5**(8), 180089 (2018)
2. Di Stasi, G., Avallone, S., Canonico, R., Ventre, G.: Routing payments on the lightning network. In: 2018 IEEE International Conference on Internet of Things (iThings) and IEEE Green Computing and Communications (GreenCom) and IEEE Cyber, Physical and Social Computing (CPSCom) and IEEE Smart Data (SmartData), pp. 1161–1170. IEEE (2018)
3. Hafid, A., Hafid, A.S., Samih, M.: Scaling blockchains: a comprehensive survey. IEEE Access **8**, 125244–125262 (2020)
4. Herrera-Joancomartí, J., Navarro-Arribas, G., Ranchal-Pedrosa, A., Pérez-Solà, C., Garcia-Alfaro, J.: On the difficulty of hiding the balance of lightning network channels. In: Proceedings of the 2019 ACM Asia Conference on Computer and Communications Security, pp. 602–612 (2019)
5. Kappos, G., Yousaf, H., Piotrowska, A., Kanjalkar, S., Delgado-Segura, S., Miller, A., Meiklejohn, S.: An empirical analysis of privacy in the lightning network. In: International Conference on Financial Cryptography and Data Security, pp. 167–186. Springer (2021)
6. Khalil, R., Gervais, A.: Revive: Rebalancing off-blockchain payment networks. In: Proceedings of the 2017 ACM SIGSAC Conference on Computer and Communications Security, pp. 439–453 (2017)
7. Papadis, N., Tassiulas, L.: Blockchain-based payment channel networks: challenges and recent advances. IEEE Access **8**, 227596–227609 (2020)
8. Pickhardt, R., Nowostawski, M.: Imbalance measure and proactive channel rebalancing algorithm for the lightning network. In: 2020 IEEE International Conference on Blockchain and Cryptocurrency (ICBC), pp. 1–5. IEEE (2020)
9. Thakur, S., Breslin, J.G.: A balanced routing algorithm for blockchain offline channels using flocking. In: International Congress on Blockchain and Applications, pp. 79–86. Springer (2019)
10. Waugh, F., Holz, R.: An empirical study of availability and reliability properties of the bitcoin lightning network. arXiv preprint arXiv:2006.14358 (2020)

# R?ddle: A Fully Decentralized Mobile Game for Fun and Profit

Athanasia Maria Papathanasiou[✉], Chalima Dimitra Nassar Kyriakidou,
Iakovos Pittaras, and George C. Polyzos

Mobile Multimedia Laboratory, Department of Informatics, School of Information
Sciences and Technology, Athens University of Economics and Business, Athens,
Greece
papathanasiou@aueb.gr, nassarkyriakidou@aueb.gr, pittaras@aueb.gr,
polyzos@aueb.gr

**Abstract.** We explore the adoption of Web 3.0 technologies in gaming,
focusing on creating new business models for novel decentralized games
that attract the attention of all actors involved: players, designers, et al.
In particular, we leverage advances in Distributed Ledger Technologies
(DLTs) and the InterPlanetary File System (IPFS) to build a flexible,
transparent, and fully decentralized mobile game, called R?ddle, in which
players, potentially around the world, compete in posing and solving
mathematical riddles, designed by a gaming company or anyone, in order
to get rewarded accordingly. Furthermore, our presented game supports
a fully decentralized version of the Play-to-Earn (P2E) gaming model
that allows players to earn money, while they are playing and enjoying
the game. Finally, with our proposed architecture, we manage to achieve
gas and cost minimization on Ethereum, by using the IPFS effectively.

**Keywords:** InterPlanetary File System (IPFS) · Distributed Ledger
Technologies (DLTs) · Blockchains · Smart contracts · Ethereum ·
Mobile gaming · Crypto-games · Play-to-Earn (P2E) · Web 3.0

## 1 Introduction

The growing popularity of Distributed Ledger Technologies (DLTs) has made
them one of the most used technologies for several applications in any domain,
such as in energy sector, Internet of Things (IoT), supply chains, etc., as they
come with many benefits and intriguing properties. One of these domains that
can be significantly benefited is the mobile gaming industry. Indeed, many gaming companies have already utilized the blockchain technology within gaming
and they have developed and published a variety of games [6]. In addition to
these games, the use of blockchains within gaming has brought new business
opportunities in the gaming industry. In particular, a new gaming model has

J. Prieto et al. (Eds.): BLOCKCHAIN 2022, LNNS 595, pp. 100–109, 2023.
https://doi.org/10.1007/978-3-031-21229-1_10

been enabled, called Play-to-Earn (P2E), which allows players to generate revenues by just playing games. With this new model, blockchain-based games have attracted a lot of attention (more than 800,000 unique users) [2].

In a previous work by some of the authors [7], it was studied how blockchains can be utilized in mobile gaming ecosystems, and the benefits they provide were demonstrated, i.e., *decentralization, immutability, availability, transparency, and auditability*, among others. In [3], a fully decentralized system was envisioned that uses the InterPlanetary File System (IPFS) [1]. The presented system enables games having evolvable trading assets, represented as Non-Fungible Tokens (NFTs), completely owned by the users.

In this paper, we use the IPFS and the Ethereum blockchain [10] in a different way. In particular, we present *R?ddle*, a novel, decentralized mobile game. In this game, players solve mathematical riddles at three levels of difficulty; easy, intermediate, and advanced. The riddles are set by the gaming company. However, players can add their own riddles too. Furthermore, players can use hints to solve the riddles, by paying the defined amount of money, in cryptocurrency. A ranking system is used to keep track of the top players, based on their in-game level, in order to get rewarded. Players level up by solving riddles. Finally, players can also get rewards by adding their own riddles in the game.

To develop a proof of concept implementation of the aforementioned game, we leverage the Ethereum blockchain and its support for smart contracts to build a fair, trusted baseline system for the game. Furthermore, we use the IPFS to reliably host, in a decentralized manner, the riddles and the corresponding hints of the game. In this paper, we make the following contributions:

- We design and implement a fully decentralized and transparent baseline system for mobile games. We store the riddles and the hints of the game on IPFS, which provides tamper-proof, decentralized storage, to minimize cost, respecting decentralization.
- We guarantee that the rules of the game (e.g., the ranking system, etc.) will be respected, by using the Ethereum blockchain and smart contracts.
- We enable the P2E gaming model, by allowing users to get rewarded either by playing the game, or by adding their own riddles, and all this in a decentralized manner.
- We achieve gas minimization by using effectively the IPFS.

The remainder of this paper is organized as follows: In Sect. 2, we present the related work in the area. In Sect. 3, we present our proposed system design, as well as, its implementation. In Sect. 4, we quantitatively evaluate our system, while in Sect. 5, we present and discuss its advantages. Finally, in Sect. 6, we conclude the paper.

## 2   Related Work

We first introduce some typical examples of popular blockchain-based games. A simple one is BetDice, which runs on the EOS platform. In this game, each player selects a number between 1 and 100 and then rolls a dice. In case the target number is less than the number of the dice, the player wins. Due to the fact that it runs on the EOS chain, it does not introduce any gas fee or transaction delay and is therefore preferred from casinos on the Ethereum platform. Another game, which introduced the topic of asset ownership is CryptoKitties [8]. This game uses ERC-721 tokens [9] (kitties), which can be collected and traded among players. It uses Ethereum smart contracts and has a web application interface, so that players can interact easily with Ethereum. Another notable reference is Gods Unchained, a card game, where players purchase cards provided by the game operators, with the use of smart contracts, in limited quantity.

In addition, it is worth mentioning a few of the most important blockchain-based games that encorporate the P2E gaming model, as introduced in [4]. These games take place in Metaverses, which are virtual environments, where users create avatars to duplicate their physical-world experiences on virtual platforms. They also use NFTs, which represent in-game property. One of these games is Decentraland (MANA), which is a virtual game platform that runs on the Ethereum blockchain. Through its features, it allows users to create and monetize their content and property in the game. Another interesting reference is Axie Infinity (AXS) [5]. This game has gained general recognition in the industry. This is due to its seamless and dynamic P2E features. The goal of many players is to earn money, while they are having an entirely new level of enjoyment. Last but not least, one of the most played decentralized P2E games is the SandBox (SAND). This game allows users to create their own in-game avatars, as well as trade and sell content and property on NFT LAND. Like the aforementioned games, SandBox allows anybody to create, share, and monetize their in-game assets.

Thus, we observe from the numerous use cases and applications that blockchain technology can strengthen the gaming industry offering a variety of new interesting properties and features, such as transparency, immutability, availability, etc. Furthermore, using IPFS, we can create fully decentralized systems and games that do not have any trust relationships with centralized third-party servers.

## 3   System Overview and Implementation

In this section, we present an overview of our solution and its architecture, which is illustrated in Fig. 1. In our system, there is a *player* owning a mobile device, which contains our (Android) application and an Ethereum wallet (a public/private key pair) used for signing transactions sent to the blockchain, a *game administrator* having an Ethereum wallet, who creates the smart contracts, deploys them on the blockchain, and uploads the riddles on IPFS. Finally, there is the Ethereum infrastructure and the corresponding smart contracts, and the IPFS, where the riddles and the corresponding hints are stored.

In our system, there are two smart contracts, the `PlayerContract` and the `RiddleContract`. The PlayerContract implements all the actions for creating

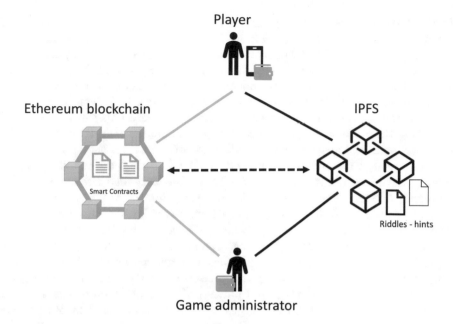

**Fig. 1.** An overview of the system's architecture.

and managing the players. Namely, the creation of an avatar, the retrieval of its attributes, and the modifier that correlates the avatar's ID with the user's account. Furthermore, it contains functions that check whether a user is eligible to level up, as well as, if she can use a hint. We should note here that it is essential to assure that some actions regarding a player, are only performed by the owner of the avatar. Due to blockchain's transparency every player in the game can view the functions inside the contract. Therefore, we do not want another player, except the owner of the avatar, to send transactions in order to level up, solve a riddle, or get a hint. Due to that fact, a modifier is used to ensure that only the owner of an avatar can call functions regarding their avatar. PlayerContract also includes functions such as play, solve, and reward, which are invoked, when a player tries to solve a riddle. Play function is called when a user chooses a riddle to solve. This function asks the player to pay a defined amount of money, which is transferred on the smart contract's address. Solve function checks whether the player should level up, while reward function shares to players the amount of ethers stored in the smart contract's address.

The latter contract is responsible for managing the riddles. It has one function, which is used for the creation of a riddle and is responsible for assigning each riddle to a specific level of difficulty; easy, medium, and hard. Both of the smart contracts implement appropriate checks to allow only the game administrator to perform administrative tasks.

The game administrator is responsible for adding and "pinning" the riddles of the game and the corresponding hints on IPFS. It is crucial to pin the content after adding it on IPFS, in order to guarantee that it will be live for long term and that it will not be garbage collected. When she uploads a riddle on the

IPFS, a CID is produced (in the form of the hash SHA-256), which is then used for fetching the uploaded riddle. These CIDs are stored on the blockchain, and in particular on the RiddleContract. The CIDs are used in the application, in order to extract the riddles from IPFS and have them appear on the user's mobile devices. The CIDs of the hints are not published anywhere, so the users cannot find them easily on IPFS. However, a malicious user can find them by observing what the node of the gaming company on IPFS uploads. This can be addressed by making the node of the gaming company changing its peer ID regularly. On the other hand, the answers of the riddles are stored within the application, so as to ensure that nobody will find them on IPFS. But, if the mobile phone is rooted, then a malicious user will be able to find and read the answers from the mobile phone. We can address this, by encrypting the answers. To store the keys, we can use the Android Keystore system. Nevertheless, the solution of this issue is out of the scope of our paper, and we have not implemented it.

From a high-level perspective, the system works as follows. Initially, the player launches the application. Then, she can register to create a new account or log in to an existing one. During this phase, the player is required to provide her Secret Recovery Phrase (or seed phrase) of her blockchain wallet, in order for her keys to be extracted from it (the seed phrase is secured by the gaming company and it is not stored on IPFS nor the blockchain). This is to find the user's account credentials. If she does not own an Ethereum wallet, then she has to create one, in order to be able to perform transactions within the application. After the registration phase, the player's avatar is created and stored in the PlayerContract. If the player has already created an account, she logs in using her user name and her seed phrase. Then, her avatar is retrieved from the blockchain. After that phase, the player can browse the application and choose one of the following options; play, settings, or ranking. By selecting "play", the difficulty level menu appears. *R?ddle* is inclusive of players of any mathematical background and provides three levels of increasing difficulty; easy, intermediate, and advanced.

To solve a riddle or request a hint, the player has to pay a defined amount of money in ethers. When the player chooses to solve a riddle, the unique URL that correlates to it, it is extracted from IPFS and the riddle is fetched on the application. Similarly, the same process applies for the hints. In addition, players can track their own, as well as, all the other players' progress in the game by selecting the ranking option from the menu. The ranking system keeps track of the three top players in the game and it is based on the players level.

As previously mentioned, we enable the P2E gaming model. Thus, the amount of money collected from the players, who try to solve a riddle or request a hint for that riddle, is stored on the contract's balance. Then, it is transferred to the first player's wallet in the ranking system. This is done in the case that the riddle is placed by the game administrator. Finally, players can contribute to the game by creating their own riddles, uploading them on IPFS, and setting the corresponding fees. By doing so, they can get rewarded with half of the amount of money that are stored in the contract's balance for that specific riddle, while the rest of it is transferred to the first player of the ranking.

The smart contracts of the system are implemented using Solidity programming language, and they are deployed on the Ropsten Ethereum test network.[1] The mobile application is developed using the Android studio and the Java programming language. To interact with the Ethereum blockchain, we used the Web3j java and android library,[2] while for the interaction with the IPFS, we have used the io.ipfs HTTP API.

## 4    Evaluation

To evaluate our proof of concept implementation of the *R?ddle* game, we measured the gas used for the key functions, as well as, we calculated the time needed to perform the most important game actions, i.e., when a user solves a riddle. The transaction cost (computational and storage overhead), which in Ethereum is expressed as the cost of gas for executing transactions is calculated every time a function is called. Gas is denoted to gwei.[3] Regarding the execution time, we are considering the time needed to send the transactions to the blockchain by calculating the execution time, which is required for all the functions that take part in the scenario of solving a riddle.

In Table 1, measurements of the basic functions are shown. The UploadIPFS row refers to the execution time needed for a riddle to be added and pinned on IPFS, while the ExtractIPFS row refers to the time needed for a riddle to be extracted from IPFS and appear on the application. LoadWallet row refers to the execution time, which is required to load the Ethereum wallet credentials for a player. Moreover, execution time is calculated for the basic activities; register, ranking, login, and solving a riddle, which are all shown by the corresponding rows of Table 1. The average execution time of 20 executions, for each action, is shown in the second column of Table 1. The deviation in the measurements can be attributed to the load and the traffic of the network and is not affected by the system's implementation.

**Table 1.** Average execution time of key functions measured in seconds.

| Functions | Execution time |
| --- | --- |
| UploadIPFS | 2.05 |
| LoadWallet | 1.25 |
| Register | 90.27 |
| Ranking | 5.04 |
| Login | 1.75 |
| Solve | 1.71 |
| ExtractIPFS | 1.22 |

---

[1] https://ropsten.etherscan.io/.
[2] https://docs.web3j.io/4.8.7/.
[3] A gwei is equal to $0.000000001\,\text{ETH} = 10^{-9}\,\text{ETH}$

As Table 1 indicates, most actions require just a few seconds, except from the register action. Considering Table 2, which measures gas price, the function createNewAvatar, which is called in register activity, has one of the most higher prices, and the highest among non-payable methods. Note that gas is calculated based on the complexity of the function in the smart contract, and createNewAvatar is one of the most complex functions. Due to that fact, it is reasonable to say that the execution time, which is required for register activity will be higher than the rest of the functions, which are less complex. However, the aforementioned issue does not affect the player's gaming experience, as createNewAvatar is only called once, when a player registers for the first time in the application.

Regarding the cost of the game, the gas needed for the key functions in the smart contracts is calculated. The contract functions giveHint, play, and reward are payable and therefore the corresponding gas price on Table 2 takes into account both the gas needed to execute the functions and the amount of ether required to be sent to the contract, as these functions are payable methods. Function createNewAvatar is only called in register activity, while the rest of the functions are called in solve activity, meaning that they are executed every time a player tries to solve a riddle. The rest of the activities from Table 1, which does not appear on Table 2, use only view functions, which are read-only and thus, they do not introduce any cost.

Therefore, as calculated from Table 2, in order for a user to solve a riddle, which is the main aim of the game, an amount of gas equal to 0.00207247 ether is required in case the player does not request a hint, whereas in the other case, an amount of gas equal to 0.00299355 is needed.

Table 2. Transaction gas fee of key actions, measured in gwei.

| Functions | Cost (gwei) |
| --- | --- |
| CreateNewAvatar | 854,930 |
| GiveHint | 921,080 |
| Play | 720,570 |
| Reward | 681,160 |
| Solve | 670,740 |

However, in our P2E gaming model, players can create riddles on their own or try to get a higher place in the ranking and hence receive rewards. In the first case, they receive half of the money for each riddle a player solves, while in the second one, they may even receive the whole amount of money for a riddle. In either cases, players can receive by far more money than the amount of ether they used to play the game. As illustrated in Fig. 2, if six players try to solve a riddle placed by a specific user, the user receives 0.00216171 ether, which is more money than the amount of ether he used to play the game. Consequently, for a number of players greater than six, the user, who created the riddle can make a profit.

In addition, a player can earn more money by getting the first place in the ranking system. In this case, the user will receive half of the amount of money

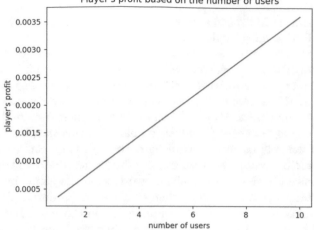

**Fig. 2.** Player's profit based on the number of users.

for riddles created by other players, in which scenario, as we discussed earlier, for a number of players greater or equal to six, he receives more money than the amount he used to play the game. On the other hand, if the riddle is placed by the game administrator, the player who ranked first receives the whole amount of money that each user paid to solve the riddle. Again, the profit increases linearly with the number of players. In this case, the top player can make a profit even if only three players try to solve a riddle created by the game administrator. If the aforementioned cases are combined, the player, who ranked first can earn a significant amount of money, which gives a motivation to players to try solving riddles and continue playing the game.

## 5  Discussion

In this paper, we presented a system that benefits from the blockchain technology, as well as enables the P2E gaming model, in which players can make a profit by creating their own riddles or solving riddles and ranking first in the game. Moreover, we used the Ethereum's capability for smart contracts in order to build a totally decentralized mobile game, open to all players, who just own an Ethereum wallet. We also took advantage of IPFS, as a decentralized storage unit, in which riddles and their corresponding hints are uploaded.

Our game has several advantages compared to legacy (blockchain-based) games. First of all, smart contract execution is deterministic and a smart contract once deployed, it cannot be modified. In this way, it is guaranteed that the rules of the game will not change with the time, since transactions do not require a central authority and their records are immutable and distributed across the network, meaning that they are visible by all participants at all times. Thus, our system is transparent, in the sense that players can be sure that the rules of the game will always be respected and no one, not even the gaming company, can modify them. Moreover, the use of smart contracts allows us to easily implement automated decentralized payments without the need of trusted intermediaries, enabling the P2E gaming model.

Furthermore, our system is fully decentralized as it is based on the Ethereum blockchain and IPFS. The riddles and their corresponding hints are stored on IPFS and not on centralized trusted servers owned by the gaming company, as it happens in traditional games that use relational databases owned and managed by the gaming companies. Players can also upload their own riddles on IPFS, without the interception of any centralized party. In this way, trust is promoted as players do not rely on any centralized authority. They can try to solve any riddle they want or create their own as the rules in the smart contracts indicate. Players also have the chance to receive rewards from the game, by contributing to it. Contributions entail uploading their own riddles. However, players can also earn an amount of money in ether by ranking first in the game. In this way, they have the opportunity to solve riddles based on their preferences on riddles, which can increase their probability to receive rewards.

Finally, by using IPFS, we reduce the chances of hardware overload for the gaming company by decreasing the storage requirements of our game. Moreover, with the use of IPFS as a storage unit (and pinning services), we ensure that the riddles and the corresponding hints will be available for long term. Even if the gaming company stops supporting the game, or if the company suffers from network outages, the game will continue to be available and playable.

Despite the numerous (security) advantages in the (mobile) gaming industry that blockchain technology can offer, as we have already mentioned above, it introduces a cost, which is based on the transactions that occur inside the blockchain. Due to the fact that gas is calculated based on the computational cost of a function, complex functions consume more gas. In our system, the function that is responsible for the creation of a player's avatar requires a significant amount of gas. However, it does not affect the player's gaming experience, as the method is only called once, when the player registers in the application.

In order to reduce the amount of gas that is consumed when a function is called, functions should be less complex, which means that smart contracts should consist of only the attributes that are important for the players or the game administrator in order to interact with the blockchain. More specifically, in our design, smart contracts contain only necessary information (e.g., player's account, riddle's ID, etc.) resulting in the minimization of the gas used for all actions. IPFS also plays a significant role in gas minimization, as it stores both riddles and their hints, while only the hashes of the riddles are stored on the blockchain.

## 6   Conclusions

In this paper, we presented *R?ddle*, a game relying on the Ethereum blockchain and the IPFS for efficiency, enabling a fully decentralized ecosystem for posing and solving (typically mathematical) puzzles. Anyone with access to the blockchain can pose riddles (and set economic parameters) and anyone can choose the riddles to play and solve, being rewarded accordingly. Introducing puzzles, one can be rewarded but also, more importantly, she can decide on their own the game parameters, such as cost to play. On the other hand, players can decide to play games they just like, or games based on their expectation to benefit from. Our design offers transparency, high availability, full decentralization, and enables the P2E gaming model.

**Acknowledgments.** The work reported in this paper has been funded in part by the Research Center of the Athens University of Economics and Business.

# References

1. Benet, J.: Ipfs—content addressed, versioned, p2p file system (2014)
2. Herrera, P.: State of the blockchain game sector (2021). https://dappradar.com/blog/bga-blockchain-game-report-july-2021
3. Karapapas, C., Pittaras, I., Polyzos, G.C.: Fully decentralized trading games with evolvable characters using NFTs and IPFS. In: Workshop on Decentralizing the Internet with IPFS and Filecoin (DI2F), in Conjuction with IFIP Networking Conference 2021, pp. 1–2 (2021)
4. Kaur, M., Gupta, B.: Metaverse technology and the current market (2021)
5. Mavis, S.: Official axie infinity whitepaper. In: Axie Infinity (2021). https://whitepaper.axieinfinity.com/
6. Min, T., Wang, H., Guo, Y., Cai, W.: Blockchain games: a survey. In: 2019 IEEE Conference on Games (CoG), pp. 1–8. IEEE (2019)
7. Pittaras, I., Fotiou, N., Siris, V.A., Polyzos, G.C.: Beacons and blockchains in the mobile gaming ecosystem: a feasibility analysis. Sensors **21**(3) (2021)
8. Sako, K., Matsuo, S., Meier, S.: Fairness in ERC token markets: a case study of cryptokittics. In: International Conference on Financial Cryptography and Data Security, pp. 595–610. Springer, Berlin (2021)
9. Entriken, W., Shirley, D., Evans, J., Sachs, N.: EIP-721: ERC-721 Non-fungible token standard (2018). https://eips.ethereum.org/EIPS/eip-721
10. Wood, G.: Ethereum: a secure decentralised generalised transaction ledger. Ethereum Project Yellow Paper 151 (2014)

# Mitigation of Scaling Effects in NTF-Based Ticketing Systems

Dan Heilmann[✉], Daniel Muschiol, Lars Karbach, Moritz Korte, Nils Orbat, Vincenzo Schulte am Hülse, and Marc Jansen

University of Applied Sciences Ruhr West, Mülheim, Germany
dan.heilmann@stud.hs-ruhrwest.de

**Abstract.** While cryptocurrencies have seen an enormous upswing in recent years, other technical advantages of the blockchain are also being used more frequently. Particularly in focus is the "Non-Fungible Token" (NFT), which stands out from conventional fungible goods and currencies because of its blockchain-wide uniqueness. Due to this characteristic, it is possible to represent personalized as well as limited assets, such as lots or tickets, by means of NFTs. This paper explores different technical approaches to implementing a ticketing system based on a blockchain, with a specific focus on NFTs. Based on the Waves Blockchain, three different possible solutions for the case study "Ticketing" were implemented using decentralized applications (smart contracts) and a JavaScript-based testing tool. Subsequently, various scenarios were designed, which simulated different event sizes and thus visitor numbers. The resulting data was examined with respect to possible scaling effects and general feasibility. The final evaluation of this work will answer the question whether NFTs can be used as a basis for a ticketing system and addresses any scaling related issues that arose during testing.

**Keywords:** Blockchain ticketing system · Ticketing · Non-fungible token · Scaling · Waves · Ticketing

## 1 Introduction

Cryptocurrencies and blockchain technologies have been established worldwide and are gaining more and more attention. In recent years, hundreds of companies and startups have been founded that work with blockchain technologies [8]. The trading volume of NFTs reached 34,530,649.86 USD in May 2021 [10]. There are hundreds of coins and tokens that can be used as digital currencies.

The previous form of event tickets relied on physical tickets. Because they usually use a system backed by traditional databases, the ticketing system can be scaled almost indefinitely. However, these tickets can be damaged or lost, as they usually have to be carried along as a printed ticket [4]. There are various problems associated with traditional ticketing systems. For example, large quantities of tickets can be purchased by a single person, thus creating a secondary market, as it is no longer possible to trace the transfer of the tickets. This problem also applies to visitors who should not have access to the event (stadium bans, etc.) [9]. Printed tickets are not particularly secure against forgery due to the fact that they are paper-based. This means that copies of the original ticket can be used at the entrance, provided it has not been cancelled beforehand, without the rightful ticket holder noticing. Organizers are currently trying to circumvent such forms of fraud by handing out personalized tickets [9]. In addition to longer admission controls, however, personalized tickets also prevent resale on a reasonable level [2].

The current design of event tickets relies on digital tickets. For this purpose, different algorithms need to be implemented, which at the same time also need to cover fraud, security and portability. The problem this work seeks to solve, becomes apparent when considering the maximum possible transactions per second on a blockchain. According to official data, the prime leader *Bitcoin* only performs seven transactions per second [5]. This number would not be sufficient to fill a high-capacity stadium within a few hours if admission control had to communicate with the blockchain for each visitor. This paper covers a research to find different algorithms, including a reference implementation and testing. The results will be discussed within Sect. 5.3.

This leads to the following question, which will be worked out within this paper:

*RQ: Can possible scaling effects in NFT-based ticketing systems, be mitigated by efficient algorithms?*

This work is split into four chapters. In addition to related work, architecture and implementation, the evaluation presents the lessons learned.

Especially in the evaluation chapter, all found algorithms are tested for resilience and reliability as well as evaluated for possible use cases. A conclusion and an outlook will close this work.

## 2   Related Work

Due to the continuous success of bitcoin as a cryptocurrency to this day, various other research areas of blockchain technology have been explored [6]. Although NFTs could also be used as stadium tickets, they are still mostly used as trading cards or digital artwork [10]. For example, in their paper, Ferdinand Regner et al. examined the usefulness of NFTs in relation to event tickets, as well as challenges and implementations. One of the biggest obstacles to implementation is scaling, an issue that will be explored in the context of this work [7]. Currently, the service of personalized NFT tickets and their validation is provided by *bam Ticketing*.[1] However, the company does not provide any information on any performance or scaling problems. It also remains unclear which blockchain is used by *bam Ticketing*.

---

[1] https://www.bam.fan/ (14.02.2022).

In their paper, Corsi et al. describe mechanics that can be used to solve the previously mentioned authorization problems [3]. However, this work does not take a position on general scaling problems.

To solve scaling problems, Adam Back et al. refer to creating a sidechain of the blockchain. This approach allows the mentioned problems to be distributed across several blockchains of the same type [1]. Zhou et al. mention an increased block count to solve scaling problems, but this doing so also increases block runtime [11].

# 3   Architecture

In the following chapter various algorithms will be used to analyze a blockchain for economies of scale. For the implementation, the blockchain *waves*[2] is used. It offers a high transaction speed and low costs compared to Bitcoin or Ethereum.[3] In this context, a system for two algorithms consisting of a test client and a decentralized application that resides on the blockchain was used. In case of the third algorithm, the decentralized application is substituted as a database cache. Since the focus of this work is to study economies of scale, it is assumed that the buyer already has a NFT ticket that gives him access to the desired event. The exact structure of an NFT is described in Sect. 4.1.

## 3.1   Algorithm—w/Return

In order to gain access to an event, the NFT must be sent back to a dApp of the vendor before the event begins. When the NFT is returned, a public function of the smart contract associated to the dApp is invoked. Within the function, the data that is stored in the data storage is fetched by using the unique ID of the received NFT. Furthermore, the status of the NFT stored at purchase and the check-in time are checked. Only in case the status of the NFT contains the value inactive and the time is within the period specified by the vendor, a successful execution of the transaction is granted. In case of success, the status checked-in and the transaction ID are then stored in the data storage, which then allows the buyer to generate a QR code, which is required access the event.

## 3.2   Algorithm—wo/Return

The second algorithm is constructed almost equivalent to the first algorithm, but works under the premise, that the visitor does not have to return the NFT to the organizer in advance. This substitution allows the visitor to decide much more dynamically when he wants to enter the event, and the strain on the blockchain is relieved, especially for large events, because every interaction with the blockchain takes up time.

---

[2] https://waves.tech/ (21.02.2022).
[3] https://docs.waves.tech/en/blockchain/waves-protocol/waves-ng-protocol (08.01.2022).

### 3.3   Algorithm—Cache

The third algorithm consists of a hybrid strategy. The blockchain continues to be the "single source of truth" and each visitor receives a dynamically generated NFT which they can us for authentication. While the first two algorithms require one or two transactions to be made on the blockchain for each authentication, this algorithm loads the entire data store into a local cache as soon as the purchase time window is closed. This local cache is then used to authenticate visitors at the point of entry. Since the visitor no longer needs to communicate directly with the blockchain at this point, much better scalability for large events can be expected here. The cache is not intended to be a replacement for the blockchain but is intended to assist with authentication prior to communicating back to the blockchain. Therefore the blockchain remains as the only single source of truth.

### 3.4   Test Scenarios

In order to predict the feasibility and scalability of the algorithms, each algorithm was simulated 10 times for each number of tickets specified in Table 1. To do this, the required number of NFT tickets were minted using a dApp and then sent to the test user wallets prior to the execution of each run. Subsequently, an execution of the algorithms presented in the previous parts took place using the test client and dApp or cache. The results were further analyzed in Sect. 5.2.

**Table 1.** Testcases.

| Algorithm | Number of tests | Number of ticket per testcase | | | | |
|---|---|---|---|---|---|---|
| 1. w/Return | 10 | 200 | 400 | 800 | 1600 | 3200 |
| 2. wo/Return | 10 | 200 | 400 | 800 | 1600 | 3200 |
| 3. Cache | 10 | 200 | 400 | 800 | 1600 | 3200 |

## 4   Implementation

During the implementation cycle, various problems had to be solved, which will now be discussed in the following.

### 4.1   NFT

For each NFT issued, the following entries are first written to the organizer's data store:

- Public key of the buyer
- Event ID
- Status of the NFT (*checked-in/used/unused*).

   This information is used to authenticate the visitor, as described in Sect. 4.3.

## 4.2 QR Code

Every visitor can generate a QR code based on his NFT. Since the focus of this work is not on the creation of an application that actually generates the QR code, a QR code is only simulated here with the help of the library *ts-lib-crypto* (Fig. 1).

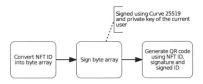

**Fig. 1.** QR code generation.

## 4.3 Authentication

In the first algorithm, the visitor's NFT is passed to the check-in function. The dApp then checks whether the attached NFT is valid and not checked-in. If this is the case, the status is changed to *checked-in* and the public address of the visitor in combination with the address of the NFT is written to the data memory of the dApp.

This process ensures that an NFT can only be assigned to exactly one visitor at the time of check-in. Afterwards, the visitor can authenticate himself with his QR code as described in Sect. 4.2. With the help of the public key, which is stored in the dApp data store, it can be checked if the scanned ticket contains a valid signature, that matches the signature within the QR code, which was previously created with the visitor's private key. If the signature is valid and the NFT referenced in the QR code matches the previously created link from Sect. 4.1, the visitor is allowed to attend the event. The other two algorithms work according to a similar principle, except that no check-in is performed there and a different data store is taken for processing. Figure 2 is intended to provide a holistic overview of the authentication flow.

## 4.4 Cache

For the simulation of a cache, a PostgreSQL database was used. This database is filled with the data store of the dApp before the event starts, and subsequently serves as the basis for verifying the visitor's QR code. Unlike the other scenarios, no interaction between the user and the blockchain is needed after the ticket is purchased.

When the visitor enters or leaves the event, the status of the NFT within the cache is set to "used/unused". With this technique, it is still possible to check whether a user is allowed to re-enter the event despite the lack of communication with the blockchain.

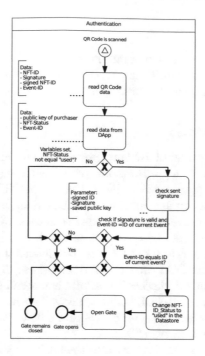

**Fig. 2.** Authentication procedure.

## 4.5  Test Scenario

In order to test the algorithms described above, all processes are mapped using JavaScript (Node.js). Since JavaScript is an asynchronous programming language, it is ideally suited for a scenario, that needs to cover extreme load testing.

For each algorithm and each scenario mentioned in Sect. 3.4, a separate test was written, which creates the required NFTs and QR codes. Then, a simulated user is authenticated to the organizer. In the first two algorithms, this is done by communicating directly with the blockchain; in the last algorithm, the cache is first prepared by pulling all relevant data from the blockchain, and then authentication to the cache takes place.

Since there are no other requirements for the testing framework, except for the basic programming language Javascript, the common unit testing framework $Jest^4$ was used here.

All tests were performed with the same test server (16 GB RAM, 8 CPU cores) under constant conditions within the Waves Testnet. The internet connection of the test server was a constant 253 Mbit/s, without any other disturbing factors in the internal network.

---

[4] https://jestjs.io/ (21.02.2022).

## 5 Evaluation

The purpose of this section is to examine the scalability and feasibility of the dApp and the associated algorithms. To this end, the evaluation strategy is first explained in more detail and the results are then discussed.

### 5.1 Evaluation Strategy

The algorithms are evaluated by implementing the three previously defined approaches. The different algorithms are tested with regard to their resilience and reliability by having to process different quantities. The test results for each algorithm will show an average runtime and the calculated standard deviations will show whether the previously established scenarios are feasible. Further scenarios will then be simulated by means of extrapolation.

There are two main criteria, that will be taken into consideration, when evaluating the algorithms. On the one hand, the speed of the admission time of the individual algorithms is checked for a different number of visitors. Secondly, the transaction density is checked, i.e. how many visitors can be admitted within a short period of time and how many visitors can be admitted over a longer period of time, such as at a trade fair.

Research on various scenarios serves as the basis for this classification (Table 2).

**Table 2.** Possible application scenarios.

| Event | Visitors | Distribution |
|---|---|---|
| Cinema | 3000 | Concentrated entry |
| Concert | 20,000 | Concentrated entry |
| Footballgame | 80,000 | Concentrated entry |
| American footballgame | 91,000 | Concentrated entry |
| Festival | 400,000 | Concentrated entry |
| Simultaneous footballgame | 500,000 | Concentrated entry |
| Industrial fair | 5,000,000 | Constant (per week) |
| Expo (Peking) | 730,000 | Constant (per week) |
| Traintickets (Long-distance) | 151,000,000 | Constant (per half year) |

### 5.2 Results

The analysis was performed as described in Sect. 5.1. As described in Sect. 3.4, the tests were simulated during the period from 01/27/2022 to 01/29/2022. The resulting times of the three algorithms are shown in Fig. 3, where each datapoint represents a duration. It can be seen that the duration of the algorithms in the largest test case with 3200 tickets is within an interval of 10–45 min. Here, the algorithm with cache shows the best value to be at 10 min, whereas the algorithm with early return takes up to 45 min to finish. Based on these results, it can be said that the duration is linear to the number of tickets. Therefore, extrapolation is also possible without further problems. Almost all runs could be completed within an average duration ± twice the standard deviation, with the exception of the cache algorithm with 3200 tickets.

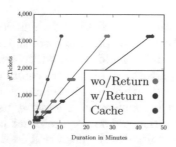

**Fig. 3.** Algorithm duration.

90% of the runs could be performed within this time. In order to be able to compare the algorithms with regard to the reliability and the speed, the mean standard deviation was calculated in percent (see Table 3). It is noticeable that the algorithm with the return of the ticket has the lowest average spread of the standard deviation. However, this algorithm also has the longest processing time and thus the highest standard deviation. As seen in Table 3, the percentage standard deviation reduces for all algorithms as the number of tickets increases, so it can be assumed that these values would increase linearly as the number of tickets increases.

**Table 3.** Standard deviation in %.

| #Ticket | Cache (%) | w/Return (%) | wo/Return (%) |
|---------|-----------|--------------|---------------|
| 100     | 18.12     | 8.82         | 25.03         |
| 200     | 10.81     | 5.27         | 16.43         |
| 400     | 5.64      | 3.70         | 10.90         |
| 800     | 2.49      | 1.75         | 5.27          |
| 1600    | 1.60      | 1.97         | 4.10          |
| 3200    | 1.32      | 1.58         | 1.51          |

For all scenarios that were implemented, the algorithms would probably be best suited for small events, with highly concentrated entry, or with equally distributed entry times like that of a trade show. For example, the fastest algorithm (Sect. 3.3) would require an admission time of 4 h and about 20 min for a stadium with 80,000 visitors. This admission time is clearly too high for an event where the audience should be admitted quickly. For example, the admission time at many stadiums ranges from $2.5^5$ to $3\,h.^6$ The same would apply to long-distance train tickets. To validate all tickets within a six-month period would require about 8278 h, whereas a full year has only 8760 h.

## 5.3    Discussion

This part of the paper will discussed the previous results. It was shown that even large volumes of tickets can be processed via the blockchain. Not every scenario could be implemented. It should be noted that positive scaling effects can only occur with the cache algorithm, since here communication takes place with the database and not directly with the blockchain, where the number of transactions per second cannot be influenced. Therefore, using the cache algorithm, an increase in ticket validation points for the event could also simultaneously increase the performance/throughput speed and lead to a reduced admission time. For example, although tickets with this algorithm are only valid at certain gates/entrances, the total number of admissions per database would again decrease and increase the overall performance of the system.

Overall, it can be concluded that all algorithms work, but are only suitable for different use cases. For events with high, simultaneous admission demand, the cache algorithm is best suited. For a uniformly distributed or small group of visitors, the purely blockchain-based algorithms are most suitable.

## 6    Conclusion and Outlook

Based on the previously described chapters, a conclusion as well as an outlook is now given. It is described how the application could be further developed and also which difficulties would have to be expected.

### 6.1    Conclusion

In this paper, three algorithms were presented that should cover NFT ticketing and at the same time be both very performant and as secure as possible. This work was able to show that even though blockchains in general do not offer particularly high transaction speeds, they can still be made usable for high scaling. This circumstance could be demonstrated primarily by means of a hybrid approach between a cache and the link to the blockchain (see Sect. 3.3). At the same time, not every algorithm is suited for every scenario, since both a particularly

---

[5] https://schalke04.de/veltins-arena/veltins-arena-hinweise-zu-einlasskontrollen/ (14.02.22).

[6] https://www.bvb.de/Tickets/Infos-AGBs/Stadionordnung (14.02.22).

high volume of visitors and a high concentration at the entrance lead to waiting times and thus to problems with acceptance. It can therefore be assumed that the described algorithms can only be used for small and medium-sized events with a high visitor density or for large events with a distributed number of visitors over time.

## 6.2   Outlook

As already described in Sect. 3, the Waves Blockchain has a high transaction speed, is at the same time inexpensive and delivers a good overall performance. If the performance of the algorithms needs to be increased further, this could most easily be done with a faster blockchain, which is currently the bottleneck of the algorithms. Thus, the development of a leaner, faster blockchain could be considered, creating potential for further research in this area. So it remains to be seen whether NFT tickets can be a potential alternative for large-scale events with high attendance and admission densities, while also solving problems such as ticket counterfeiting.

# References

1. Adam Back, M.C.: Enabling blockchain innovations with pegged (2014)
2. Bamakan, S.M.H., Nezhadsistani, N., Bodaghi, O., Qu, Q.: A Decentralized Framework for Patents and Intellectual Property as NFT in Blockchain Networks. Preprint, In Review (Oct 2021). https://doi.org/10.21203/rs.3.rs-951089/v1, https://www.researchsquare.com/article/rs-951089/v1
3. Corsi, P., Lagorio, G., Ribaudo, M.: TickEth, a ticketing system built on ethereum. In: Proceedings of the 34th ACM/SIGAPP Symposium on Applied Computing. ACM (Apr 2019). https://doi.org/10.1145/3297280.3297323
4. Glaap, R., Heilgenberg, M.C.: Digitales ticketing. In: Der digitale Kulturbetrieb, pp. 127–159. Springer Fachmedien Wiesbaden, Germany (2019)
5. Li, C., Li, P., Xu, W., Long, F., Yao, A.C.C.: Scaling Nakamoto consensus to thousands of transactions per second. arXiv arXiv:1805.03870 (2018)
6. Nakamoto, S.: Bitcoin: a peer-to-peer electronic cash system (2009). http://www.bitcoin.org/bitcoin.pdf
7. Regner, F., Schweizer, A., Urbach, N.: NFTs in Practice—Non-fungible Tokens as Core Component of a Blockchain-Based Event Ticketing Application. Technical report, München (2019). https://www.fim-rc.de/Paperbibliothek/Veroeffentlicht/1045/wi-1045.pdf
8. Statista: Anzahl der pro Jahr gegründeten Blockchain-Unternehmen und Start-ups weltweit von 2008 bis 2018 (2021). https://de.statista.com/statistik/daten/studie/1062771/umfrage/anzahl-der-pro-jahr-gegruendeten-blockchain-unternehmen-und-start-ups-weltweit/
9. Tackmann, B.: Secure event tickets on a blockchain. In: Data Privacy Management, Cryptocurrencies and Blockchain Technology, pp. 437–444. Springer, Berlin (2017)
10. Wang, Q., Li, R., Wang, Q., Chen, S.: Non-fungible token (NFT): overview, evaluation, opportunities and challenges. arXiv preprint arXiv:2105.07447 (2021)
11. Zhou, Q., Huang, H., Zheng, Z., Bian, J.: Solutions to scalability of blockchain: a survey. IEEE Access **8**, 16440–16455 (2020). https://doi.org/10.1109/access.2020.2967218

# Digital Content Verification Using Hyperledger BESU

Carmen María Alba⬨, Francisco Luis Benítez-Martínez$^{(\boxtimes)}$ ⬨,
Manuel Ventura-Duque⬨, and Rafael Muñoz-Román⬨

FIDESOL (Fundación I+D del Software Libre), Granada, Spain
{cmjimenez,flbenitez,maventura,rroman}@fidesol.org

**Abstract.** One of the greatest current social challenges is the certification of digital content to guarantee its veracity and authenticity, and the possibilities offered by distributed ledger technologies thanks to the properties of the blockchain in terms of security, immutability, traceability and transparency provide a specific framework for this. This work presents the design of a prototype model within the Ethereum BESU network that can be scalable and interoperable with other DLTs. We have, therefore, studied both the regulatory and technological framework to design a tool that is capable of certifying digital content. In this work, we will only focus on the technological aspects according to the main certification modalities that exist today, and the main applications of blockchain technology for the management of certificates. We will conclude by explaining how the described prototype works as one of the central elements for the deployment of a platform specially designed in blockchain to detect fake news and fight against disinformation.

**Keywords:** Blockchain · Digital content · Certification · Disinformation

## 1 Introduction

There can be no doubt that the amount of information currently available on Internet is enormous and that every day more processes are managed digitally. Such processes generate a vast number of documents that need to be certified in order to support their veracity and legality. In addition to verifying this documentation, it is also necessary to ensure the communication channels and verify that they are being used in the safest possible way. One of the main problems of the digital age is the veracity of the content that is posted on social media and it has been estimated that the year 2022 will mark the point when more false information will be posted than true. We are now facing a new pandemic: the pandemic of misinformation as "infodemic" [1] or "information pollution" as described by the World Economic Forum [2].

Given the varied nature of electronically stored information, it is necessary to establish a tool which is capable of unifying different digital format types and conducting a single type of certification. This was a particular concern in the pre-Internet era and was highlighted by Russell [3]. Many institutions are aware of the need to certify, encrypt and sign digital content. In our modern lives, it is becoming increasingly common to require a certified email or a sealed and legalized document or even to generate a document that certifies that documentation has been delivered to a user, all online.

One constant problem has always been to certify digital content using traditional mechanisms before distributed registration technologies. Although the most commonly used methods for certifying any type of online content by signing documents is to use a personal digital certificate or some type of QR code, these are not entirely secure or immutable. Third parties have also provided alternative certification mechanisms but their certification mechanisms are not always completely transparent. For this reason, the best option for the development of a tool which is able to certify different formats in a single certification may well be the use of blockchain technology for digital content certification.

Entities such as Forbes [4] highlight the relevance of blockchain and some of the solutions include digital content certification as an option for the future after 2021. Our work has also been inspired by the work of Puri et al. [5] on future blockchain applications and that of Foy et al. [6] on digital certificate models for Covid-19. Based on the current state of the art, this work tests the analyzed technology and conducts a prototype of content certification in digital environments.

In view of the above, it is evident that the amount of digital content is expanding and growing at an ever increasing rate and it is, therefore, more important than ever to be able to identify the origin and source of this content. Similarly, it is also important to determine what content is relevant and true, what content has been altered or modified, and to identify any artificial information that is untrue or totally false. It is also necessary to establish which tools are capable of certifying digital content in order to create secure environments within the network. At this moment in time, it is now more than ever necessary to provide entities and users with "undoubted digital trust" so that they can have no reason whatsoever to mistrust the new communication models and the platforms that contain them.

This paper proposes the application of new methods to certify digital content using blockchain in order to secure it. Our objective is that users and entities will have the utmost trust in the digital content managed. The paper is structured as follows: Sect. 2 presents a literature review of successful cases of study in order to contextualize the proposed prototype; Sect. 3 describes our prototype and underlines the evident need for new ways to secure digital content; and, finally, Sect. 4 highlights the usefulness of this prototype using the BESU network on the Alastria platform [7] and outlines our future lines of work.

## 2   Technological Background and Related Work

### 2.1   Designing the Prototype

Our prototype is based on one of the first pieces for building a platform that enables users to detect fake news and content posters in order to certify them at source using a blockchain tool. In order to develop this prototype, a series of steps have been followed to enable us to define both the model and the blockchain technology that will be used:

a)  Analysis of the regulations regarding digital content certification
b)  Detailed study of content likely to be subjected to a certification process
c)  Analysis of current content certification systems
d)  Blockchain technology application: certified content signature
e)  Design and creation of technology testing scenarios

Although a number of solutions already exist to accredit authorship of images and documents and to sign content using a digital signature, certain vulnerabilities can still be exploited to falsify documents. Various institutions add a QR code to their documents to prove that it has not been modified but this code does not provide the security and immutability offered by blockchain technology.

The main task of blockchain in document management is to use distributed networks to validate every document within an organization [8]. Blockchain-based file systems enable a whole wealth of information to be compiled about a given document such as the location, the author, the recipient, when and where it was sent, when it was last modified or the delivery system used. This is possible thanks to the unique fingerprint that is registered in each block of the chain once a transaction has been completed.

It is worth mentioning that of all the different certification modalities, the most difficult one to deal with is audiovisual content because due to the very structure of the blockchain, it is not possible to store much data and in public blockchains, the storage capacity is limited to only a few megabytes. While it can be extremely complicated to compress a video so that it can be registered, it is possible to extract unique features from a video as a mark which makes registration possible and thereby creates an immutable trace of that video. We have used this solution for our prototype at BESU. Although it is not necessary for fourth generation distributed ledger technologies to perform off-chain transactions in federated blockchain models [9], they can store and compress large amounts of data in their tokenized models, as is the case of ARCA [10].

## 2.2   Certification Modalities. A Review of Literature

In this section, we will present the main certification modalities according to the type of data to be certified as referenced in relevant work.

Document Certification:

- **Blockchain and peer-to-peer distributed networks**: In their article [11], the authors highlight the need to certify intellectual property and define when a document has been created or modified without the presence of a third party guarantor. They demonstrate data certification (documents, images, movies and data files) by creating an immutable relationship between owner and data, for which they use Ethereum.
- **CertiDApp**: This is a framework in which documents are certified through smart contracts (university degrees). The user requesting certification communicates with the application by means of the smart contract and the student is sent their certified document. In order to verify that the document belongs to the student, the institutions query the application to verify that the document provided by the student is correct [12].

- **University degrees**: Another example of university degree certification where a federated network is used creates an educational certificate using a smart contract with a digital signature [13].

Image Certification:

- **Copyright in images**: In their article [14], the authors explain how image content attributes are added to the blocks which create copyright and, more specifically, record the loading time of the images and the interactive behavior, generating a hash with data from the application layer.
- **Image Certification**: Another paper [15] offers a method of image privacy protection based on blockchain technology to protect image copyright in networks and to deal with the privacy content in the image in order to protect the rights and privacy of image owners.
- **Registry control**: A further article [16] presents a layered architecture that attempts to control the registry and secure the transfer of medical images with privacy protection. In order to develop this certification, the system accommodates large images with restricted storage blocks, captures image feature vectors and designs a custom transaction structure for image features.

Video certification:

- **Privacy and copyright**: In terms of a blockchain-based data hiding method for digital data protection for certification purposes, the authors [17] combine the blockchain with data encryption techniques and add the hash value to the block with a traceable hash value record. In order to increase data privacy, they design a data-hiding scheme based on data control technology. This interacts with the blockchain and the external chain is used as a controller for the privacy data in the video.
- **BlockSee**: This is a blockchain solution for video manipulation on video IoT devices [18]. It is a blockchain-based video surveillance system that provides validation and immutability to camera settings and surveillance videos, making them readily available to authorized users. The results are quite favorable and opens a new path for new monitoring systems using blockchain.

## 2.3  Current Business Solutions Using Different Approaches

The success of blockchain in terms of data certification on networks is improving and the research lines that are currently being developed are constantly evolving in order to reach the best possible level for prototype development. Accordingly, a large number of companies already use blockchain to certify digital content and some of the currently available commercial solutions are listed as follows:

- **SmartRights**: This is an online platform that certifies any type of document, image, book, music or video using blockchain. Its basic operation attempts to create a digital footprint with a time stamp in the block chain in order to permanently register the authorship and content [19].

- **Acronis Notary**: This product tries to use blockchain technology to prove that a file is authentic and has not been modified. In this case, the concept of fingerprint is used to certify documents, videos, images or emails [20].
- **Blockubox**: This certifies documents in the blockchain as an electronic signature. Using the well-known Alastria network, this solution certifies private evidence of contracts, documents and emails in a public and secure registry. It is available as an Outlook email extension or via web and mobile apps. [21]
- **Stampery**: This generates immutable, independently verifiable records of everything of importance for a user's business. Within its wide range of solutions, stamp.io is an intelligent platform for data certification and the most popular case of application is the certification of blog post content [22].
- **Hashing dna**: This digitally certifies and signs any file or web content in the blockchain. It uses cryptography to certify files from any device and offers users the possibility of signing with their own identity or anonymously [23].
- **Block Tac**: This uses blockchain technology to provide inviolability, immutability and easy verification for every digital certificate and seal, and is supported by IT and business management professionals. One example of the application of this technology is the certification of university degrees, study plans and other university certifications [24].
- **Khipus**: This mobile application was created to simply certify documents and images by registering them in a blockchain. In this way, mobile devices can be converted into a tool which is capable of registering fingerprints that can serve, for example, as evidence for third parties. The application also enables a certified document to be generated with all the information relating to the transaction and this document to be linked with the registry in the blockchain [25].
- **NDL ArcaNet**: This develops the concept of blockchain into a network for the direct transfer of digital assets between end users. This solution provides scalability, efficiency and sustainability to the blockchain ecosystem by eliminating the concepts of mining and consensus and applying blockchain with a totally innovative and disruptive approach in order to store any kind of digital content [10].

## 3   Prototype for Digital Certification Using Hyperledger BESU

On the basis of the previously studied cases, it has been observed that current guidelines for digital online content and the security measures established to guarantee content are not always able to perform the role of ensuring point-to-point content. There are additional problems regarding transparency issues in such processes and this reinforces the need to create new tools to overcome the problems inherent to traditional technologies.

For this reason, an application has been developed as a prototype that aims to provide an extra level of security to secure digital content through the use of blockchain as a possible solution for digital content certification. It has been deployed on an Alastria Hyperledger BESU node, a DLT that serves as an Ethereum client.

We have chosen BESU as an open source client that works under an Apache 2.0 license and is written in Java. It works with various proof-of-work (PoW) consensus algorithms such as Ethash and proof-of-authority (PoA) types such as IBFT, IBFT 2.0, and Click. It is also compatible with extremely prominent blockchain development tools and environments such as Remix, Truffle and web3j, and enables tools such as Prometheus to be used to monitor node performance and tools such as Alethio to be used to monitor network performance. For all of these reasons, we have chosen this network since it enables us to interoperate with other DLTs once this prototype has been deployed on the final platform on which we are working.

Lastly, we would like to point out that BESU also makes it possible to establish a permitting system whereby only certain nodes can operate with the network without other parties being able to access the content.

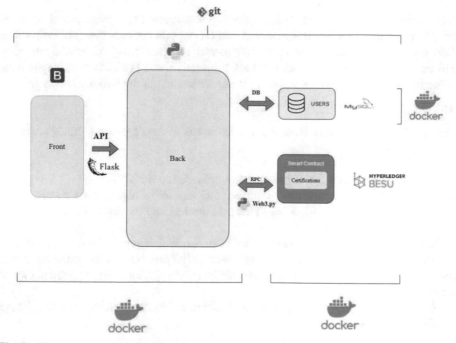

**Fig. 1.** Blockchain platform high level architecture used for the prototype within the Alastria B Network

**Architecture of the prototype:**

The structure of the blockchain platform is shown in the diagram in Fig. 1 which identifies the different parts which interact with each other to achieve overall platform operation. In the following sections, we will briefly review each individual part.

*Front*: This is a visual interface for user interaction with the platform. It is created with Angular and communicates with the back through an API and the Apache Tomcat server.

***Back***: The core of the union between the different parts of the platform will be described in the back part of the architecture. It first receives the requests through an API from the front through the Tomcat server. The back is developed with the Spring Boot language and communicates with the database through its Spring JPA framework due to its usefulness for interacting with SQL-type databases. In turn, the back is connected to the smart contract management application in order to retrieve friendly data from the blockchain. Finally, it communicates with each of the independent projects deployed on the platform through requests to the API and with which it will be possible to view each of the projects that in turn use the blockchain.

***BCP***: This is a relational database created with a *MySQL* workbench and hosted in a docker container. The purpose of this database is to store both the users who will use the platform and their access data in addition to the data of the projects registered on the platform and the independent access URL for each one.

***Smart contract management***: This application was created to provide a tool which is capable of interpreting the data recorded in a blockchain network and processing it as little as possible to make it understandable to users. By accessing this application, they will be able to see every account, contract, transaction, etc. For each of these projects, different contracts are created and all of these are written in the Solidity language.

### How do we certify?

The *certification of images* from the application is a simple process that follows the following steps:

a)  The image to be certified is selected and uploaded to the platform.
b)  Various relevance parameters are hashed and sent to the blockchain network.
c)  One of these hashes is recorded as the primary key and the rest as data.

In order to check whether an image has been certified, it simply needs to be selected and the application will check if the hash used as the primary key has been registered and if the rest matches. If so, a message will be returned indicating the authenticity of the image.

The *document certification* process follows a similar structure to the image certification process:

a)  The chosen document is selected and stored.
b)  A hash of the document content is performed.
c)  The document is then registered in the blockchain network indicating the nature of the file that has been certified.

In the case of documents and images, we upload a file individually, this is processed from the backend, intrinsic data is extracted from the files that makes them unique, and the hash is generated with the *hashlib* library. This hash is the one that is recorded in the blockchain.

When we want to *certificate a mailing list,* we access the certification application and add the sender's email. At this time, the backend processes each of the emails, adding:

subject, description, date, time and attachments, and then generates a hash with these data that is registered in the blockchain [26].

Regarding the query, the files to be consulted are attached or the email data is consulted, then a hash is made again with that data and it is contrasted with the information registered on the network.

In our case we have not implemented a Merkle tree since the approach was designed to generate a code for each file or email sent. It doesn't make sense to add a Merkle tree since we don't have batch tasks like adding email groups or groups of people like in an university certificate, for instance.

In order to check whether a document has been certified on the platform, it is necessary to access the query section and select the file to review (as shown in Fig. 2). The application will then obtain the hash of its content and check its existence. Any change to that content will result in a completely different hash and consequently nothing will be found.

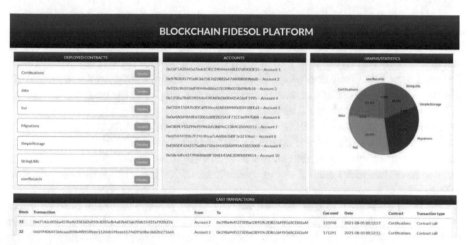

**Fig. 2.** Network data monitoring on the blockchain FIDESOL platform with the records stored in the smart contracts.

This prototype is a simple and scalable solution for digital content certification and this is, in turn, interoperable with other DLTs. The idea is to be able to build a model that can be used for rapid deployment in real environments.

## 4   Conclusions and Future Work

The main objective was to use a public blockchain network that would allow the correct functioning of the developed prototype to be verified in a real environment.

When considering which of the networks on the market to use, we opted for Ethereum since it is the most versatile blockchain and within its ecosystem, we opted for Hyperledger BESU. By analyzing the various commercially available solutions for digital content certification, we were able to choose the fastest and simplest deployment solution. We therefore use the Alastria B Network given its great adaptation to different applications and because it is a public network that provides the system with greater transparency.

In addition to the previously mentioned features, BESU also offers the following technical capabilities that make it ideal for this type of digital certification: it includes a JSON-RPC API which is operable through HTTP or WebSockets so that Ethereum network nodes can be managed through an API; it includes a command-line interface (CLI) to work with network nodes; and it implements Ethereum devp2p network protocols for communication between the different clients.

Finally, BESU also includes a command line interface and a JSON-RPC API that allows the execution of functionalities such as mining, smart contract deployment and dApp development in addition to maintaining, debugging and monitoring nodes on an Ethereum network. The API can also be used through RPC over HTTP or through WebSockets.

This prototype is part of a larger project to develop a platform to fight disinformation and the spreading of fake news from content creators and to prevent the manipulation of their news and associated data. This platform is still in the design stages and one of our first actions was to develop a document certification tool for BESU. We therefore designed a specific Front-End which is easy to use and understand for the management of smart content certification contracts and certificates thanks to the node in the Alastria BESU network deployed by FIDESOL.

In the front-end that we have designed, we show the contracts deployed in the blockchain and statistics of the contracts that generate the most transactions, for the traceability of the system. It also allows us to display a list of the latest transactions in order to have individual control of our interactions within the BESU network, which is public.

By using the Alastria Besu network we have the advantage that each of the transactions carried out does not consume gas, which saves us the work of having to optimize transactions to lower costs. In addition, due to the way our email certification system is designed it presents an inconvenience. In addition, due to the way our email certification system is planned, it presents us with an inconvenience. By validating the veracity of an individual email, it would generate a hash only belonging to one of the leaves of the Merkle tree, so this leaf would not be registered in the blockchain, but would be in the root of the whole tree.

However, in the case of images and documents, we will implement this type of tree in future versions, since it would generate batches of documents and reduce the interaction with the blockchain.

This node is also currently being used at the BlockLab [27] of the MediaLab of the University of Granada for the development of social innovation projects in collaboration with FIDESOL for the innovation of tools for the purposes described in this article in order to empower citizens and to better understand data origin and authenticity.

# References

1. van der Linden, S.: Misinformation: susceptibility, spread, and interventions to immunize the public. Nat. Med. **28**, 460–467 (2022). https://doi.org/10.1038/s41591-022-01713-6
2. World Economic Forum Homepage, Will the world clean up "information pollution" in 2022? https://www.weforum.org/agenda/2022/03/misinformation-cybercrime-covid-pandemic/. Last accessed 12 Apr 2022
3. Russell, S.: Paradigms for verification of authorization at source of electronic documents in an integrated environment. Comput. Secur. **12**(6), 542–549 (1993). https://doi.org/10.1016/0167-4048(93)90049-B
4. Forbes Homepage, "Blockchain 50 2021". https://www.forbes.com/sites/michaeldelcastillo/2021/02/02/blockchain-50/?sh=12c0d3a4231c. Last accessed 14 Apr 2022
5. Puri, N., Garg, V., Agrawal, R.: Blockchain technology applications for next generation. In: Raj, P., Dubey, A.K., Kumar, A., Rathore, P.S. (eds.) Blockchain, Artificial Intelligence, and the Internet of Things. EAI/Springer Innovations in Communication and Computing. Springer, Cham (2022).https://doi.org/10.1007/978-3-030-77637-4_4
6. Foy, M., Martyn, D., Daly, D., Byrne, A., et al.: Blockchain-based governance models for COVID-19 digital healthcare certificates: a legal, technical, ethical and security requirements analysis. Procedia Comput. Sci. **198**, 662–669 (2022). https://doi.org/10.1016/j.procs.2021.12.303
7. Alastria Homepage. https://alastria.io/en/. Last accessed 10 Apr 2022
8. Chang, P., Hwang, M., Yang, C.: A blockchain-based traceable certification system. In: Peng, SL., Wang, SJ., Balas, V., Zhao, M. (eds.) Security with intelligent computing and big-data services. In: SICBS 2017. Advances in Intelligent Systems and Computing, vol. 733. Springer, Cham (2018).https://doi.org/10.1007/978-3-319-76451-1_34
9. Romero-Frías, E., Benítez-Martínez, F.L., Nuñez-Cacho-Utrilla, P.V., et al.: Fundamentals of blockchain and new generations of distributed ledger technologies. Circular economy case uses in Spain. In: Muthu, S.S. (eds.) Blockchain Technologies for Sustainability. Environmental Footprints and Eco-design of Products and Processes, pp. 25–46. Springer, Singapore (2022).https://doi.org/10.1007/978-981-16-6301-7_2
10. NDL ARCA, ByEvolution Homepage. https://byevolution.com/en/. Last accessed 23 Apr 2022
11. Fallucchi, F., Gerardi, M., Petito, M. et al.: Blockchain framework in digital government for the certification of authenticity, timestamping and data property. In: Hawaii International Conference on System Sciences (2021). https://doi.org/10.24251/HICSS.2021.282
12. Ataşen, K., Aslan, B.A.: Blockchain based digital certification platform: CertiDApp. J. Multi. Eng. Sci. Technol. **7**(7), 12252–12255 (2020)
13. Dumepti, N.K., Kavuri, R.: A framework to manage smart educational certificates and thwart forgery on permissioned blockchain. Mater. Today Proc. (2021). https://doi.org/10.1016/j.matpr.2021.01.748
14. Zhao, C., Liu, M., Yang, Y., Zhao, F., Chen, S.: Toward a blockchain based image network copyright transaction protection approach. In: International Conference on Security with Intelligent Computing and Big-data Services, pp. 17–28. Springer, Cham (2018)
15. Dong, X.: A method of image privacy protection based on blockchain technology. In: 2018 International Conference on Cloud Computing, Big Data and Blockchain (ICCBB) pp. 1–4. IEEE (2018). https://doi.org/10.1109/ICCBB.2018.8756447
16. Shen, M., Deng, Y., Zhu, L., Du, X., Guizani, N.: Privacy-preserving image retrieval for medical IoT systems: a blockchain-based approach. IEEE Network **33**(5), 27–33 (2019). https://doi.org/10.1109/MNET.001.1800503

17. Zhao, H., Liu, Y., Wang, Y., Wang, X., Li, J.: A blockchain-based data hiding method for data protection in digital video. In: International Conference on Smart Blockchain, pp. 99–110. Springer, Cham (2018).https://doi.org/10.1007/978-3-030-05764-0_11

18. Gallo, P., Pongnumkul, S., Nguyen, U.Q.: BlockSee: blockchain for IoT video surveillance in smart cities. In: 2018 IEEE International Conference on Environment and Electrical Engineering and 2018 IEEE Industrial and Commercial Power Systems Europe (EEEIC/I&CPS Europe), pp. 1–6. IEEE (2018)

19. SmartRights Homepage. http://smartrights.io. Last accessed 12 Apr 2022

20. Acronis Notary Homepage. http://www.acronis.com. Last accessed 12 Apr 2022

21. Blockubox Homepage. http://blockubox.com. Last accessed 12 Apr 2022

22. Stampery Homepage. https://stampery.com

23. Hashing dna Homepage. https://hashingdna.com. Last accessed 12 Apr 2022

24. BlockTac Homepage. https://blocktac.com. Last accessed 13 Apr 2022

25. Khipus Homepage. https://khipus.io

26. Hinarejos, M.F., Ferrer-Gomila, J.L., Huguet-Rotger, L.: A solution for secure certified electronic mail using blockchain as a secure message board. IEEE Access **7**, 31330–31341 (2019). https://doi.org/10.1109/ACCESS.2019.2902174

27. BlockLab Homepage, MediaLab University of Granada. https://blocklab.ugr.es. Last accessed 03 May 2022

# Prototyping a Smart Contract Application for Fair Reward Distribution in Software Development Projects

Agostino Di Dia[✉], Tim Riebner, Alexander Arntz, and Marc Jansen

University of Applied Sciences Ruhr West, Lützowstraße 5, 46236 Bottrop, Germany
agostino.didia@stud.hs-ruhrwest.de , tim.riebner@stud.hs-ruhrwest.de ,
alexander.arntz@hs-ruhrwest.de , marc.jansen@hs-ruhrwest.de

**Abstract.** The following work describes the development of a project management platform based on the waves blockchain technology. The aim is to use the platform to optimally reward both fairness and the achievement of quality standards in software development. Due to the outsourcing of software projects and the associated processes, situations arise in which different project participants work on the same tasks, while paid differently and in most cases independent of the delivered software quality. This is based on the respective contract conditions an individual software developer has negotiated, without factoring in the actual quality of the code the developer provided. The proposal of this work is a prototypical software application that takes the requirements of a project and measures the corresponding contributions of the developers based on their software quality. Project sponsors can also pay out the enrolled project partners via the platform.

**Keywords:** Software project management · Blockchain application · Smart contract · Reward distribution

## 1 Introduction

During the implementation of technical projects, it is common to set milestones during the planning stage [17]. The milestones of such projects indicate the achieved state of a project and are usually a point in time at which a client pays partial compensation to the contractor.

When developing software, in a team with several system integrators ($SI$ for short), freelancers, or other parties, it may happen that the parties receive different amounts of compensation. The project backer can thereby determine how high the available budget is for a particular project milestone [6]. It is common to observe that several parties contribute to the same milestone. This

J. Prieto et al. (Eds.): BLOCKCHAIN 2022, LNNS 595, pp. 131–141, 2023.
https://doi.org/10.1007/978-3-031-21229-1_13

observation does not mean that the parties will therefore receive the same payment, even if they have performed the same good work or contributed the same software quality. Suppose a software project $P$ has several parties in different geographical locations contributing to a milestone $M$. Each milestone has a predefined budget. Project sponsors, in this case also the customer of the project, tend to keep the schedules, budget, and resources working on the project as low as possible [7]. Therefore, development processes for milestones are often outsourced [5]. Thus, individual parties may feel that they are not being treated fairly during the project, as remuneration is not shared by a certain contributed standard of performance or quality. This causes the motivation of the people involved in the project to drop sharply [13]. Especially in software development, it is sometimes difficult to have an overview of who did what and how much effort a certain implementation required, and whether the financial compensation for it is acceptable. This can lead to conflicts between development teams, especially between multinational teams where negative stereotypes can lead to additional disputes [4]. Since fairness is always an individual perception and not everyone perceives fairness in the same way [26], it raises the research question of how to ensure a fair and transparent method of rewarding software development teams within a project based on their effort in relation to their code quality and complexity.

In addition, when implementing technical projects, especially in software development, it is difficult for organizations to determine or measure whether a defined milestone has been reached. This can have various reasons. For example, the defined milestones may have been insufficiently described during the requirements analysis. This can lead to the fact that no or only superficially described acceptance criteria have been defined [25].

Another reason could be the insufficient technical knowledge of the project sponsors, who do not have the necessary technological expertise [23] to verify the achievement of a milestone, since they cannot, for example, carry out code reviews and thus cannot put the contractor's performance in relation to the compensation. In order for the project owner to accept the milestone, he must either simply believe the contractor that the milestone has been reached or hire an external consultant to review the milestone. Another problem exists in projects where the requirements have changed within the project or the different parties have their own interpretations of the described milestones and thus of the requirements [17]. The transparency of requirements for specific milestones is almost impossible without a clear project structure because even in smaller projects a large number of requirements can be placed on individual milestones. Especially in projects where much of the work is decentralized, there is only a limited level of trust, which decreases over time if no physical meetings take place [18]. The lack of trust among the project partners can thus have a negative impact on the course of the project and its implementation, causing a dispute among the parties.

To obtain an optimal project structure, a work breakdown structure is essential and part of the time schedule [9]. It is common to divide the entire project into further work packages. These packages are then subdivided into further sub-work packages up to the point where exact tasks for implementing a work package can be defined. Ideally, the project parties then define acceptance criteria for specific work packages in order to be able to measure when a work package has been completed. Depending on the chosen project method, the review of the requirements and their acceptance criteria takes place in different time periods. For example, in the "waterfall" methodology, the review of acceptance criteria takes place at the completion of the project [16].

The verification of acceptance criteria and deliverables can be supported in software development projects with the help of platforms such as Gitlab. Other possibilities are the use of Kanban boards, the documentation of software requirements specifications [8] or similar project specifications. Potential manipulation attempts by the contractor on the acceptance criteria are possible by editing such documents. There are also opportunities to change defined requirements on platforms such as Gitlab, where parties can view the generated code. In addition, the branching workflow used by the development teams can make it difficult for the client to keep track. Deliverables may not show the transparency that is needed when implementing a complex technical project, especially when a project has multiple milestones to reach.

Accordingly, another question arises, namely how to support the project management of software projects in order to minimize the problems described above. For this purpose, a blockchain approach was chosen on the basis of which an application was designed and developed to support both project management and the fair distribution of compensation among developers.

## 2   State of the Art

A literature search using the terms "blockchain project management" shows that there has been quite a bit of research in this area in recent years. However, the majority of published articles only deal with the potential opportunities for certain industries to work in a more optimized way, such as the construction industry [10]. It looks at core aspects that improve certain points in the industry, such as the management of subcontractors and their contracts, or the payment of compensation. While there are already some blockchain-based applications, there is not yet an application solution that envisages supporting multi-party project coordination. In addition, a literature search was conducted using the search terms "fair compensation software projects", "fairness software projects" and "fair offshoring software projects" to investigate fair compensation within software projects. The literature review revealed that the topic area of fair distribution of a project budget within software development has not been extensively studied. Therefore, a more detailed investigation on how offshoring affects software quality was conducted. One of the main reasons for offshoring software projects is to reduce the cost of development [2,3,11]. One of the major challenges related to offshoring is that software quality is one of the main issues in offshoring projects [14,15]. This can result from various aspects. While offshoring companies want to reduce their own costs when developing for their clients, they tend to use junior developers who are not as experienced as they perhaps should be in order to meet client expectations regarding quality aspects [15]. Another possibility for not meeting the required quality aspects during development can be the incorrect communication of requirements [20]. This can happen if the requirements have not been described accurately and in detail during handover.

## 2.1   Relevant Work

Blockchain technology is becoming more interesting with the implementation of functions such as smart contracts for other areas of application besides cryptocurrencies. A pilot project investigated how students can form a project group and support project coordination with the help of the blockchain technology waves. After the completion of certain milestones or project progress, the client sent custom tokens to the project group on a waves basis, which they could divide among themselves in the project group in order to be able to reward certain achievements of a project member at a higher rate [19]. The pilot project does not deal with the question of how a possible distribution of remuneration and project coordination could look like in the case of several project parties who do not yet have a basis of trust in any form and the exact work performance is not known. Especially in larger projects that require multiple parties, it is very likely that there will be parties that have to work together for the first time. Also, a related research paper dealt with the development of a blockchain-based smart contract system that ensures fair payment for cloud storage and also ensures payment of penalties based on responsibilities [24]. The paper does not include the viewpoint of software development but focuses mainly on the viewpoint of customers and distributors.

While cost reduction within projects leads to software vendors also trying to reduce the development costs on their side, a fair and transparent division of the budget between the parties should put the development aspect in the foreground to ensure the quality of the software. In order to reduce the share of poor software quality in development projects, this paper describes how fair payment could be ensured by all parties involved in a software project by prototyping a smart contract application.

## 3   Software Architecture

The following section describes the underlying software application architecture. The software components used are shown and the usefulness within the overall context is described. The application was developed fully in the Angular web framework and interacts directly with the blockchain, as shown in Fig. 1. Waves was used as the blockchain, as it promises to be developer-friendly and has low transaction costs [22]. In addition, it offers a test network that provides all the possibilities in the form of the functionalities of waves. The test network was used during development to test the functionalities of the applications within

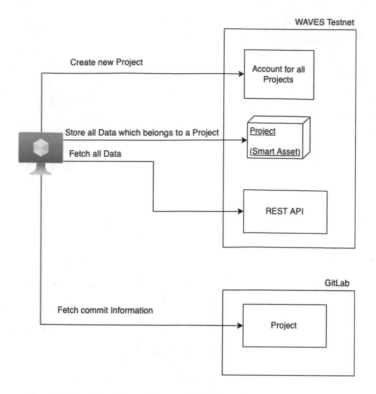

**Fig. 1.** Visualisation of the underlying software architecture.

the blockchain network. The Typescript library provided by waves was used in the application. To sign all transactions, the browser plugin "Waves-Keeper" is required. With the help of waves-Keeper, the provided web services of waves can be used within the browser. Each project created in the application is a so-called "Smart-Asset" in waves. To realize this, the smart contract is compiled via the REST API of waves and the Smart-Asset, which represents a project, is created. In the Smart-Asset, the required functions for saving and validating the data are called. The smart contract was implemented with the functional programming language Ride [21]. The stored data is retrieved via waves' REST API. To verify and make payment after a milestone is completed, the "waves Exchange Widget" is used. To measure the work performance of a project party, for the completion of a milestone, the processed lines of code from the commits of Gitlab are checked and read out via the REST API, see Fig. 1 [12]. In order to provide a distri- bution function for the compensation of the software teams, the corresponding functionality had to be implemented in the built architecture. As shown in Fig. 2, SonarQube is linked to a Gitlab repository. This means that every pull request is automatically checked by SonarQube. Through the Gitlab integration, all infor- mation about the pull request can be retrieved via a REST API, along with the metrics generated by SonarQube. The data is retrieved from the "Commit-

Service". The "CommitService" stores this data in the waves blockchain. Note that the "Data" module shown in Fig. 2 represents a service that reads all general project information from the blockchain associated with a specific project. The "CommitComponent" is the graphical representation where a team adds a pull request to a milestone. When a milestone is completed, it is represented in the "MilestoneSellPageComponent". When the "MilestoneSellPageComponent" is called, the "CommitService" is used to read all stored metrics from the waves blockchain. The "CalculateService" is then used to calculate the distribution of the teams. The distribution is displayed in the "MilestoneSellPageComponent", which is retrieved by the "CommitService" again, and the individual teams can be paid out there.

Blockchain Application

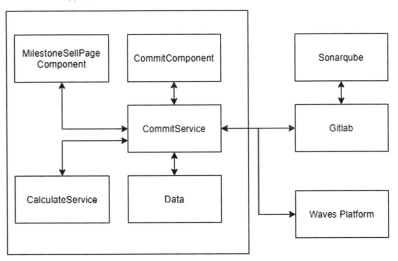

**Fig. 2.** Implemented functional architecture of the distribution of software compensation within the blockchain application.

## 4   Implementation

The following chapter describes the implementation of the blockchain application and how the designed software architecture was implemented. During the implementation, the Single Responsibility Principle was followed, which in turn should reduce the code complexity as much as possible [1]. In waves, all data in the blockchain is stored according to the key/value principle. When information is stored with the same key value, the associated value is overwritten. To ensure that the key for each value remains unique, the transaction ID is also specified at the beginning of the key. This results in the following scheme illustrated using a requirement, which consists of a name and a description:

```
{transactionID Requirement}_requirementName
{transactionID Requirement}_requirementDescription
```

Because both keys have the same transaction ID defined at the beginning, they can be referenced again for a requirement. To create a relation between two object types, the transaction ID of the corresponding object is saved with the attribute name. For example:

```
{transactionID Task}_taskName_
{transactionID Requirement}_requirement
```

In this case, the object Task can be identified on the client, which is in a $N : 1$ relation to the object Requirement. In the waves Explorer, the data can be displayed and thus the storage on the blockchain can be tracked. In Fig. 3 the stored "key/value" pairs from a project are displayed. Each key/value pair is represented by a JSON object. This consists of the attributes "key", "type" and "value". A record can be found again using a key. The attribute type indicates what kind of variable type the value is. The value attribute stores the value that is to be saved. The data view in Fig. 3 shows an example of how a task referenced to a requirement is represented in the blockchain.

```
{
    "key": "4RSRPPoGsC8uFCVCLaRn1dGE8K4Mm3pq5e2rZrWev8aS_requirement_HsmsxGoAPbX4HU8RSBciF91J9yy4CMNNStZCyB5BG
fLJ_taskDescription",
    "type": "string",
    "value": "description"
},
{
    "key": "4RSRPPoGsC8uFCVCLaRn1dGE8K4Mm3pq5e2rZrWev8aS_requirement_HsmsxGoAPbX4HU8RSBciF91J9yy4CMNNStZCyB5BG
fLJ_taskName",
    "type": "string",
    "value": "task 1"
},
{
    "key": "4RSRPPoGsC8uFCVCLaRn1dGE8K4Mm3pq5e2rZrWev8aS_requirement_HsmsxGoAPbX4HU8RSBciF91J9yy4CMNNStZCyB5BG
fLJ_taskStatus",
    "type": "string",
    "value": "OPEN"
},
{
    "key": "HsmsxGoAPbX4HU8RSBciF91J9yy4CMNNStZCyB5BGfLJ_requirementDescription",
    "type": "string",
    "value": "description"
},
{
    "key": "HsmsxGoAPbX4HU8RSBciF91J9yy4CMNNStZCyB5BGfLJ_requirementName",
    "type": "string",
    "value": "req 1"
},
```

**Fig. 3.** Data view on the blockchain with the waves explorer.

The front end is structured in such a way that there is a form for each entity to add corresponding data to a project. The project address must always be specified so that the application saves the data in the correct project. Then all information about the entity is given. When the entity is saved, the address of the newly created entity is displayed. Every time data is saved, waves-Keeper opens and the user has to log in there and sign the transaction.

To create a project, all necessary project information is first stored in an account in which all projects are listed. A new project is then created on the blockchain. For this purpose, the smart contract is compiled via the REST API and then deployed to make it available. The client then receives the project address. The project owner must add the first users to the project, as the Smart-Contract validates whether a user can and may execute the functions. All added users can then use the application and participate in the project accordingly with their work. How new functions can be implemented and the Angular application scaled is explained in the following chapter.

In order to implement the distribution function for the software developers, the criteria first had to be selected on the basis of which the system can make a judgment.

**Table 1.** List of evaluation criteria.

| ID | Description |
|----|-------------|
| I1 | Software code quality |
| I2 | Technical debt |
| I3 | Work performance |
| I4 | Test coverage |
| I5 | Transparency of compensation distribution |

The presented criteria from I1 to I4 from Table 1 are measured within the blockchain application. Secondly, I5 is executed by implementing a graphical user interface (GUI), therefore the implementation is based on a subjective view and there is no guarantee that every user will perceive it as transparent or even clear.

## 4.1   Distribution of Compensation

To distribute the amount of a successful milestone within a project, a mathematical model is defined to show how the distribution will work for the project participants. $m$ is defined as the amount of money contributed to a successful milestone. The number of individual software developers or companies contributing to a milestone is defined as $n$. Each implementation criterion describes a specific factor. This factor also contains a weighting $w$ to allow prioritization within the criteria. It is described as follows:

$$r_n = w_1 \cdot I_1 + w_2 \cdot I_2 + w_3 \cdot I_3 + w_4 \cdot I_4 \text{ with } I_1, I_2, I_3, I_4 \in \mathbb{R} \tag{1}$$

The values for I1 to I4 are numerical scores from the technical measurement of the specific code contributions to a given milestone. The result $r$ is a value that is compared with the $r$ results of all contributors. At the beginning of the implementation, the weights are initialized with a value of 1 and can be adjusted at a later stage. The amount of money a party receives at the end of a milestone is defined as:

$$m_t = \frac{r_n}{\sum_{i=1}^{n} r_n} \cdot m \tag{2}$$

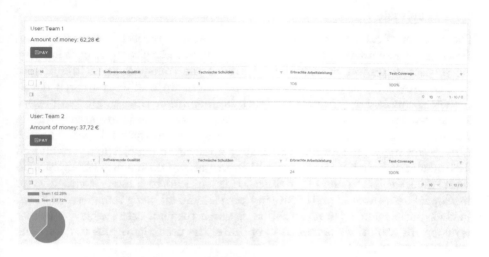

**Fig. 4.** Implemented overview for the distribution of the project budget between the project participants by using the evaluation criteria found in Table 1.

Figure 4 shows the implemented solution for an overview of the distribution of the project budget among the teams. For each team that contributed to the project milestone, a table is shown that contains the measured pull requests. In addition to the pull requests, information about the measured software quality, technical debt, contributed lines of code, and test coverage is displayed. Below the tables, a pie chart visualizes the distribution, while the payment function can be executed for each team within the table of the featured team. This allows the project owner to pay out the amount earned by the team upon successful completion of the milestone.

## 4.2   Limitations

Currently, no requirements or required criteria can be "withdrawn" in the application. This means that as soon as the required criteria are set, they must be fulfilled within the requirement in order to successfully complete a requirement. This also applies to correspondingly created tasks, which must be fulfilled and can no longer be withdrawn. The current blockchain application only works with Gitlab and SonarQube. In addition, each repository must be connected to SonarQube, which is why a seamless connection is not currently supported. In addition, at least the Developer Edition of SonarQube must be used, which will incur licensing costs. Payment is only supported for the waves platform. Measuring individual commits is also not supported. Another limitation is that pull requests can only be contributed by one team. If multiple teams contribute a pull request, distribution between teams will not work.

### 4.3   Conclusion

The solution of a blockchain application for use as a project management tool shows that there are points with which blockchain technology can both support the process of project handling and optimize the transparency of the project services provided. To implement the application, we used current web development technologies to develop a blockchain solution that supports different project parties in the project execution, ensures trust and transparency during the project work, and can also be used to remunerate the work performances fairly accordingly. In addition, the project application enables many small and also independent developers to "enroll" in projects and then also collaborate on them respectively. Among other things, this can lead to a larger community of developers being able to participate in a project and all being remunerated based on their performance. However, future research through user evaluation in real development conditions is needed to optimize the used algorithms to determine the fair distribution.

## References

1. Ampatzoglou, A., Tsintzira, A.A., Arvanitou, E.M., Chatzigeorgiou, A., Stamelos, I., Moga, A., Heb, R., Matei, O., Tsiridis, N., Kehagias, D.: Applying the single responsibility principle in industry: modularity benefits and trade-offs. In: Proceedings of the Evaluation and Assessment on Software Engineering, pp. 347–352 (2019)
2. Aspray, W., Mayadas, F., Vardi, M.Y., et al.: Globalization and offshoring of software. Report of the ACM Job Migration Task Force, Association for Computing Machinery (2006)
3. Barney, S., Mohankumar, V., Chatzipetrou, P., Aurum, A., Wohlin, C., Angelis, L.: Software quality across borders: three case studies on company internal alignment. Inf. Softw. Technol. **56**(1), 20–38 (2014)
4. Barrett, M., Oborn, E.: Boundary object use in cross-cultural software development teams. Human Relat. **63**(8), 1199–1221 (2010)
5. Betz, S., Makio, J., Stephan, R.: Offshoring of software development-methods and tools for risk management. In: International Conference on Global Software Engineering (ICGSE 2007), pp. 280–281. IEEE (2007)
6. Bitterli, P.R.: Informatik-projektentwicklung ohne risiken? Der Schweizer Treuhänder (2000)
7. Boehm, B.W., Ross, R.: Theory-W software project management principles and examples. IEEE Trans. Softw. Eng. **15**(7), 902–916 (1989)
8. Davis, A., Overmyer, S., Jordan, K., Caruso, J., Dandashi, F., Dinh, A., Kincaid, G., Ledeboer, G., Reynolds, P., Sitaram, P., et al.: Identifying and measuring quality in a software requirements specification. In: 1993 Proceedings First International Software Metrics Symposium, pp. 141–152. IEEE (1993)

9. Devi, T.R., Reddy, V.S.: Work breakdown structure of the project. Int. J. Eng. Res. Appl. **2**(2), 683–686 (2012)
10. Hewavitharana, T., Nanayakkara, S., Perera, S.: Blockchain as a project management platform. In: Proceedings of the 8th World Construction Symposium, vol. 1, pp. 137–146 (2019)
11. Huen, W.H.: An enterprise perspective of software offshoring. In: Proceedings. Frontiers in Education. 36th Annual Conference, pp. 17–22. IEEE (2006)
12. GitLab: Gitlab docs (2020). https://docs.gitlab.com/ee/README.html
13. Kadefors, A.: Fairness in interorganizational project relations: norms and strategies. Constr. Manag. Econ. **23**(8), 871–878 (2005)
14. Kannabiran, G., Sankaran, K.: Evaluation of determinants of software quality in offshored software projects: empirical evidences from India. Int. J. Inf. Technol. Proj. Manag. (IJITPM) **11**(1), 32–54 (2020)
15. Matloff, N.: Offshoring: what can go wrong? IT Prof. **7**(4), 39–45 (2005)
16. McCormick, M.: Waterfall vs. agile methodology. MPCS, N/A (2012)
17. Rodrigues, A.G., Williams, T.M.: System dynamics in project management: assessing the impacts of client behaviour on project performance. J. Oper. Res. Soc. **49**(1), 2–15 (1998)
18. Schindler, M.: Wissensmanagement in der Projektabwicklung: Grundlagen, Determinanten und Gestaltungskonzepte eines ganzheitlichen Projektwissensmanagements, vol. 32. BoD-Books on Demand (2002)
19. Schützeneder, P., Lehner, J., Sametinger, J.: Verwendung von blockchain und custom tokens zur projektkoordination-ein pilotversuch. HMD Praxis der Wirtschaftsinformatik **55**(6), 1285–1296 (2018)
20. Serebrenik, A., Mishra, A., Delissen, T., Klabbers, M.: Requirements certification for offshoring using LSPCM. In: 2010 Seventh International Conference on the Quality of Information and Communications Technology, pp. 177–182. IEEE (2010)
21. Waves Technologies: About ride (2020). https://docs.waves.tech/en/ride/
22. Waves Technologies: Waves protocol (2020). https://waves.tech/waves-protocol
23. Tiemeyer, E.: It-projekte erfolgreich managen-handlungsbereiche und prozesse. Handbuch IT-Projektmanagement, pp. 1–37 (2010)
24. Wang, H., Qin, H., Zhao, M., Wei, X., Shen, H., Susilo, W.: Blockchain-based fair payment smart contract for public cloud storage auditing. Inf. Sci. **519**, 348–362 (2020)
25. Wang, X., Zhao, L., Wang, Y., Sun, J.: The role of requirements engineering practices in agile development: an empirical study. In: Requirements Engineering, pp. 195–209. Springer, Berlin (2014)
26. Wiek, U.: Fairness als Führungskompetenz: Strategie und Leitfaden für Führungskräfte und Unternehmen der Zukunft. Springer, Berlin (2018)

# Blockchain in the Public Sector: An Umbrella Review of Literature

Fernando Escobar(✉), Henrique Santos, and Teresa Pereira

Centro ALGORITMI, Universidade do Minho, Guimarães, Portugal
fernando.escobar.br@gmail.com

**Abstract.** Despite being relatively new, blockchain (BC) is already a reality, including in the public sector. Based on a systematic review, this paper reveals the context of BC applied in the public sector, highlighting initiatives and discussing the main concerns that should be considered in these initiatives, particularly related to scalability, lack of laws and regulations, and energy consumption—this one juts the importance of choosing a sustainable consensus mechanism. Also, the potential uses are identified: health, energy, supply chain, food traceability, and e-Gov.

**Keywords:** Blockchain · Public sector · Distributed ledger technology · Smart contract · e-Government · e-Gov

## 1 Introduction

Blockchain (BC) features make it a promising technology to transform many activities [9] since its introduction with Bitcoin—known as the first decentralised digital currency out of the control of any Central Bank [1]. BC can be applied to any online transaction where a trusted authority is required but not desirable [11]. Thus, should be universal and adaptable to different cases [13], potentially altering how applications are developed, creating efficiencies, and driving digital transformation (DT) in perhaps all industries [21].

In this sense, the characteristics of the BC made it an alternative for adoption within public sector (PS) [24] and alter administrative processes, social assistance, regulatory practices, and others [9,18]. In that regard, DT in the PS matures along with a continuum review of policies, processes, and user needs [26]. As an effect, BC in the public sector results in a complete revision of business models [9] and produces benefits, such as reducing the cost of operations, increasing transparency, and promoting innovation and economic growth [24].

The potential impact of the technology depends on the specific context, especially in the public sector [9]. So, systematised knowing the different micro contexts (BC initiatives in the public sector), their concerns and potential applications contribute to a deeper knowledge of the phenomenon and illuminate the path for researchers and practitioners. Thus, the main question of this study is: (MQ) *How are the initiatives of the adoption of blockchain (BC) evolving in the public sector?* For complete coverage, complementary questions are: (CQ1) *What*

J. Prieto et al. (Eds.): BLOCKCHAIN 2022, LNNS 595, pp. 142–152, 2023.
https://doi.org/10.1007/978-3-031-21229-1_14

*are the challenges and concerns of BC application in the public sector?* (CQ2) *What public services can benefit most from BC?* To the best of our knowledge, no studies have yet answered these questions. This work seeks to fill this gap.

Next, Sect. 2 presents the background, and Sect. 3 sets the research method. Then, Sect. 4 presents the findings, and Sect. 5 discusses the results. The paper ends with conclusions, limitations, and suggestions for further work.

## 2 Background

Blockchain is a chain of blocks cryptographically concatenated into an immutable chain as a continuous and incremental list [15]. BC is the integration of core technologies [41,42] as a Distributed Ledger Technology (DLT) that uses broadcasted Consensus Mechanism (CM), with public-key encryption to guarantee transmission and security, and applies Smart Contracts (SC) to execute transactions.

DLT is a multi-party system that works with no central operator or authority [35] and can be classified as permissioned or permissionless [6]. In permissionless ledgers, the process is open to everyone (anonymous or identified by pseudonyms) who is allowed to validate entries with fully devolved authority [7,35]. Also, the transactions are validated by miners through incentives [8], reaching a consensus to maintain integrity [24]. High security is achieved at the cost of speed and scalability [38] serving as a global record that cannot be modified [24]. In opposite, permissioned ledgers preselected the participants [35], who know and accept each other, with no anonymity, tending to show centralisation features [42]. Also, only a specific group of relying parties can validate inputs, so miners are not required [7,8] which makes it much simpler than the consensus used by permissionless [24]—ergo, the permissioned is usually faster than the permissionless [24].

DLT uses public-key encryption [41], an asymmetric scheme for cryptographic digital signatures, to enhance security [7,27]. In BC, each transaction is "hashed" binding the sender to the contents, like a contract's signature; these transaction hashes are collated into a block with a limited number of transactions [27].

Reaching consensus is a fundamental concern in distributed computing [20, 38] to verify whether a transaction is legitimate. The way to ensure the trustworthiness of the records using a predefined cryptographic validation can vary depending on its nature, purpose, and underlying asset [7,27,41]. So, no silver bullet or single CM fits all business needs [38]. Inline, some intrinsic characteristics of BC applications (fast, energy-efficient, easy to scale, censorship-resistant, and tamper-proof) should be considered when designing CM; nevertheless, there will always be unavoidable trade-offs between them [25]. Thus, a vital aspect of designing a CM is the type of DLT: permissioned or not. In permissionless, a CM is necessary that does not rely on the participants' trust or knowledge of their identity. Permissioned does not have to solve this problem and, hence, can process transactions faster and are cheaper than capital/energy-intensive permissionless. On the other hand, permissioned is arguably more open to reducing security, censorship resistance, and reversibility [7,25,38].

SC are another core technology of BC based on reliable and secure protocols [40], implementing automation without human interaction, verification or arbitration [41]. SC are computer programs written on the underlying DLT that execute predefined actions transposing contractual obligations imposed on users into the ledger and are executed automatically when meet specific conditions [6,27,33]. As a result, SC reduce mental and computational transaction costs imposed on parties [40] and reduce the level of trust required in these interactions [35]. Due to their versatility and benefits, SC are being considered widely, especially for regulatory compliance, product traceability, and service management, also combating counterfeit products and fraud [24]. Ergo, the real potential of BC can be fully realised only when combined with SC [2,24]. However, not every BC platform implements SC—e.g. Bitcoin does not use BC.

## 3   Method

The process adopted was derived from PRISMA [22] as a rapid review [44], focused in other reviews [45]—likewise known as an umbrella review.

The articles were obtained from ACM Digital Library, WoS—Web of Science, and Scopus, selected for having research-related works and a filter option that allowed directing to review articles. The search was carried out in Nov 2021, covering the last ten years and limited to the review document type, as a rapid review approach. The search terms were ("government" OR "public sector" OR "public administration" OR "e-gov" OR "public service") AND (application OR use OR adoption OR implementation OR deployment OR implantation) AND (blockchain OR "block chain" OR "block-chain"). Inclusion criteria (IC) were IC1—Review paper; IC2—Written in English; IC3—Published between 2011 and 2021; IC4—Describe initiatives of BC application in the PS; IC5—Describe scenarios or detailed cases of BC application in the PS; IC6—Present concerns, challenges, and/or opportunities of BC application in the PS; IC7—Discuss the potential of smart contracts applied in the PS. To be considered, an article should match at least one IC (from IC4 to IC7). First, was obtained 75 articles among ACM (9), WoS (29), and Scopus (37); removing the duplicates, 57 unique remained. Next, it was carried out a pick based on the title and abstract. Remained 38 being applied to ICs. In the end, remained 26 papers were read in full and analysed.

To guide the analysis, we consider public service and regulated industry sectors covering both the services directly provided by the public sector and the usually-regulated segments (e.g., energy, transport, trade). The BC initiatives in the public sector identified are in Table 1, with the respective inclusion criteria.

Moreover, when organising the information about the challenges and concerns and potential uses referenced in the articles, it is possible to rank them, as shown in Fig. 1—inside the bars, the IDs of analysed papers related.

**Table 1.** Articles analysed and their initiatives.

| ID | Refs. | IC | MQ—initiatives |
|----|-------|----|----|
| AA01 | [3] | IC4; IC6 | Healthcare \| Guardtime \| MedRec |
| AA02 | [13] | IC6 | Not observed |
| AA03 | [29] | IC4; IC6; IC7 | Power ledger |
| AA04 | [15] | IC5 | Not observed |
| AA05 | [17] | IC4; IC6; IC7 | Tradelens \| BDTS CargoX \| ShipChain \| Global shared container |
| AA06 | [2] | IC4; IC6; IC7 | Brooklyn microgrid \| (also, a list of 140 companies and projects surveyed) |
| AA07 | [34] | IC6 | Not observed |
| AA08 | [21] | IC6; IC7 | Smart energy \| Smart cities \| Smart governments |
| AA09 | [4] | IC6; IC7 | Not observed |
| AA10 | [38] | IC4 | Land registry (R3 Corda) \| Canadian government grants information |
| AA11 | [12] | IC6; IC7 | Not observed |
| AA12 | [39] | IC4; IC6 | MiPasa \| Hashlog \| VeChain \| Public health BC consortium \| Hyperchain |
| AA13 | [9] | IC4; IC6 | Dubai's public records system \| Estonian e-residency |
| AA14 | [37] | IC5 | Estonian digital identity \| India's national unique ID (UID) |
| AA15 | [31] | IC4; IC6; IC7 | India's property records system \| Dubai's single blockchain |
| AA16 | [48] | IC4; IC6; IC7 | Oxfam \| Patientory \| Mediledger \| Archangel \| Enerchain \| Energy web chain \| LO3 energy |
| AA17 | [32] | IC4 | Track produce in the US and pork in China (Walmart and IBM) |
| AA18 | [47] | IC6 | Not observed |
| AA19 | [28] | IC4; IC7 | eGov-DAO smart contract framework |
| AA20 | [10] | IC4; IC5 | Civitas \| MiPasa |
| AA21 | [23] | IC4; IC5; IC6; IC7 | Civitas \| VeChain \| User privacy protection \| MiPasa \| Contact tracing \| Public health BC consortium \| Hyperchain \| Acoer coronavirus hashlog \| BC-based data tracking solution |
| AA22 | [30] | IC4 | E-credit confirmation \| E-letter of guarantee (e-LG) \| VAT refunds for tourists \| Tax refund blockchain \| License renewal \| E-bank guarantee |
| AA23 | [19] | IC4; IC6; IC7 | Energy trading \| Smart healthcare \| e-voting \| Supply chain \| Real state |
| AA24 | [36] | IC4; IC6; IC7 | P2P energy trading \| Swiss's electricity market \| Individual P2P interaction \| Smart metering \| Power flow model and smart contracts |
| AA25 | [43] | IC4; IC6; IC7 | Enerchain \| Brooklyn microgrid |
| AA26 | [16] | IC6; IC7 | Not observed |

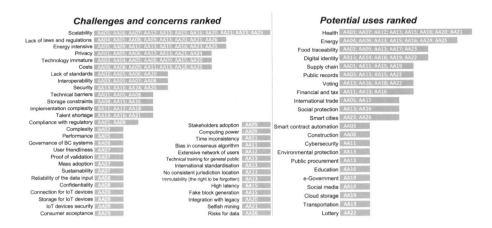

**Fig. 1.** Challenges and concerns ranked/potential uses ranked.

## 4    Under the Umbrella

This section presents BC's characteristics, not in general but in the public sector context, combining state-of-the-art with the findings from the umbrella review.

DLT in the PS has the potential to catalyse innovation to deliver a new kind of trust to a range of government services in terms of data sharing and transparency [24]. A characteristic of these e-Gov services is that only identified and privileged nodes have to write access [24]. Thus, permissioned BC is preferred [46] and better suited for the PS due to its granular control and allowing the owners to enforce rules and limit usage [24,33]. In fact, the decentralisation of e-Gov services through permissioned BC is possible and desirable since it can increase public administration functionality; in opposite, permissionless BC (like Bitcoin) presents serious risks and drawbacks, which offset the benefits [5].

With the belief that authority-based CM with known participants are more suitable for the PS, [38] highlighted 21 (out of 66 reviewed) CM suitable for the PS, based on the argument that government prefers control and authority in the consensus-building process. Furthermore, authority-based CM is more environmentally sustainable due to the lower energy required to validate transactions, producing fewer negative externalities [14], mainly when compared to permissionless CM like PoW (Proof-of-Work) used in Bitcoin.

Applying it, the European Blockchain Services Infrastructure (EBSI), a permissioned BC infrastructure to accelerate the creation of cross-border digital services for public administrations [9], adopts the Proof-of-Authority (PoA), a permissioned CM with a preselected set of identified and trusted parties—this type of CM is best suited for private BC and consortiums [38] like EBSI.

SC application in the PS should substantially improve compliance, competitiveness, cost efficiency, and accountability [33]. Also, SC makes it possible to implement the "absolute law" to process data more efficiently and accurately due they can deal with data with safety and non-falsification [18]. However, [24] emphasises the immaturity of SC solutions.

# 5  Discussion

As presented, BC proves suitable for application in the PS, where the most appropriate approach is a permissioned DLT, with public-key encryption and a consensus mechanism suitable for the public sector.

## 5.1  Initiatives of Blockchain in the Public Sector

Despite still being considered the "next big thing" and technological immaturity, many BC initiatives are a reality in the public sector [18] as shown in Table 1.

Most of the BC initiatives in the PS are related to healthcare. Still, the energy area combined with IoT devices strongly relates to a specific segment and with a relevant number. Moreover, demonstrating a holistic potential that can be interpreted as the possibility of a "blockchain as platform" approach, many government initiatives on Digital ID, government registries, and other e-Government services rank second in the amount identified from the articles reviewed. Furthermore, it is worth noting the initiatives related to international trade (from a global perspective) and smart cities (from a local perspective).

## 5.2  Challenges and Concerns of Blockchain in the Public Sector

Regarding concerns or challenges, according to Fig. 1, items related to scalability were the most mentioned. However, this issue is more linked to the permissionless DLT consensus mechanisms, unsuitable for public sector applications. Likewise, high energy consumption, a frequently cited concern relevant to environmental sustainability, is also related to some CM, particularly Proof-of-Work (PoW)— also not the most recommendable for the public sector.

The second most cited concern examines the lack of laws and regulations. It is hugely relevant in the public sector, guided by regulations and restricted by-laws. However, this concern has a global reach, as it impacts other segments of relevant BC adoption, such as healthcare, supply chain, transportation, and energy. Laws and regulations, in some situations, can be produced at the local level. However, a regional, national or global effort will be needed in other situations. The premise is that with specific regulations, legal certainty is increased; hence the attractiveness of solutions increases too, both for the citizen and for public-private partnerships. Related, the Tech Committee ISO/TC 307 Blockchain and distribution ledger technologies[1] acts for the standardisation [31].

Privacy also makes up the list of top concerns. In times of General Data Protection Regulation (GDPR), in Europe and other globally similar laws, it must be considered when designing BC solutions for the public sector. In this sense, Jun proposes the adoption of additional technologies like zero-knowledge proof, secure multi-party computation, and homomorphic encryption algorithms to accommodate secret and private data in public BC [18].

---

[1] https://www.iso.org/committee/6266604.html.

Technological immaturity is also ranked, with a downward trend as adoption becomes more widespread and the knowledge is generated and shared. In addition, costs complete the list of top 6 concerns. In particular, the initial costs of setting up a BC solution—this issue can make migrating to BC not financially worthwhile and can limit the massification of the technology.

Furthermore, interoperability between blockchains is essential to think about global or local but broad solutions. This necessary data exchange between networks enables greater effectiveness in different solutions, such as trade, health, and energy, between countries, government levels, or powers.

Not at the top, but it is more than a concern, a real bottleneck, the talent shortage needs to be addressed. In this sense, the work by Zou et al. with the situation of BC's labs opened by universities around the world deserves a highlight [48]. This contributes to courses, research, scientific knowledge production, and skilled labour training to deal with this new computational paradigm.

Many other concerns make up the list (41 items in total), and all of them (from the most cited to the least) deserve a pertinence analysis, acting as a checklist for each new BC project planned. This guidance can support the definition of the architecture and the model and demand specific outcomes to respond to concerns. It starts from the premise that, by knowing the concerns and considering them in the planning, with effective answers where they fit, the project's risks are (in part) mitigated, increasing its chances of success.

## 5.3   Public Services Potentially Affected by Blockchain

The potential usage illuminates the future of BC enforcement initiatives in the PS, recording similar experiences that can be benchmarked.

The most cited potential uses are in the health, energy, supply chain (i.e. food traceability), and e-Gov services (digital ID, public records, and voting). Moreover, consolidation of platforms is foreseen as the current initiatives mature. Also, it can help reduce the costs of the initiatives.

Furthermore, related to technical barriers, notably how to fit new BC solutions to legacy applications or existing models, Prajapati et al. listed this as a success factor [31], and Zou et al. proposed a layered approach [48]. This direction can expand blockchain's public sector reach and favour its massification.

Finally, in a propositional approach, Jun presents five principles for the adoption of BC in the public sector: (1) BC statute law—BC enables "absolute law" that cannot be tampered or violated; (2) transparent disclosure of data and source code—an open source strategy; (3) implementing autonomous executing administration—using SC to manage the laws implemented in the BC with the consent of the community; (4) building a governance system—a direct democratic governance system; (5) as a result, making Distributed Autonomous Government—a social operating infrastructure and an information processing apparatus whose rules are decided with the consent of the community [18].

# 6    Conclusions, Limitations and Further Work

This work tries to contribute to the emergence of a body of knowledge about BC in the PS. As sought to demonstrate, BC is considered one of the most significant disruptive technologies in terms to transform governments and society, not just soon but now, as the initiatives listed here, which respond to the MQ. In this sense, the health and energy sectors deserve to be highlighted, in which the BC proves to be a reality in PS in which permissioned BC is most common.

Furthermore, development efforts in the BC in the PS are ongoing and potential adoption of the BC is listed here, in response to the CQ2. In this sense, BC's most cited potential uses are related to the health, energy, supply chain, and e-gov services, with great potential for platform consolidation. These seem to be an avenue of possibility for researchers and practitioners.

Also, every initiative has its degree of uncertainty, translated into risks. The concerns discussed in response to CQ1 bring uncertainty to projects. Among the most mentioned (scalability and energy consumption), they are being mitigated depending on the adoption of proper CM. In addition, lack of laws and regulations, privacy, technological immature, costs, and interoperability—this, is to interlink between BCs (across government, industry and the non-profit sectors) to avoid data silos. Although challenges influence BC initiatives in the public sector, they have not stopped their adoption, needing to be managed as risks.

The rapid review approach is, at the same time, a strength, as it allowed for quick analysis and valuable findings, and also the most significant limitation, regarding the scope of the articles. Additionally, the search terms used in collecting the articles may limit their scope despite being thought out comprehensively. Thus, related concepts may be addressed in future research to build on this. Also, since this work aimed to provide an overview of BC in the PS, it was not carried out with conceptual or philosophical assumptions in detail.

In further work, to specialise proposals of [6] (focused on any use-case) and [21] (on construction), it is foreseen the adaptation of a systematic to assess whether a BC will be suitable for the context of the public sector.

At least, some questions remain open, being central to qualifying the body of knowledge about BC in the PS. These open questions are related to the assessment of the impact of BC adoption in the public sector and the metrics associated with doing this; also, the critical factors that can influence the adoption and use of BC in the PS; and, finally, how public bodies can identify the most suitable governmental process to being supported by BC.

**Acknowledgement.** This work has been supported by FCT—Fundação para a Ciência e Tecnologia within the R&D Units Project Scope: UIDB/00319/2020.

# References

1. Ammous, S.: The Bitcoin Standard: The Decentralized Alternative to Central Banking. Wiley, New York (2018)
2. Andoni, M., Robu, V., Flynn, D., Abram, S., Geach, D., Jenkins, D., McCallum, P., Peacock, A.: Blockchain technology in the energy sector: a systematic review of challenges and opportunities. Renew. Sustain. Energy Rev. **100**, 143–174 (2019). https://doi.org/10.1016/j.rser.2018.10.014
3. Angraal, S., Krumholz, H.M., Schulz, W.L.: Blockchain technology: applications in health care. Circ. Cardiovasc. Qual. Outcomes **10**(9) (2017)
4. Astill, J., Dara, R., Campbell, M., Farber, J., Fraser, E., Sharif, S., Yada, R.: Transparency in food supply chains: a review of enabling technology solutions. Trends Food Sci. Technol. **91**, 240–247 (2019)
5. Atzori, M.: Blockchain technology and decentralized governance: is the state still necessary? J. Governance Regul. **6**(1) (2017)
6. Belotti, M., Božić, N., Pujolle, G., Secci, S.: A vademecum on blockchain technologies: when, which, and how. IEEE Commun. Surv. Tutorials **21**(4) (2019). https://doi.org/10.1109/COMST.2019.2928178
7. Benos, E., Garratt, R., Gurrola-Perez, P.: The economics of distributed ledger technology for securities settlement (2017)
8. Biondi, D., Hetterscheidt, T., Obermeier, B.: Blockchain in the financial services industry. Technical report (2017)
9. Cagigas, D., Clifton, J., Diaz-Fuentes, D., Fernandez-Gutierrez, M.: Blockchain for public services: a systematic literature review. IEEE Access **9**, 13904–13921 (2021). https://doi.org/10.1109/ACCESS.2021.3052019
10. Chamola, V., Hassija, V., Gupta, V., Guizani, M.: A comprehensive review of the COVID-19 pandemic and the role of IoT, drones, AI, blockchain, and 5G in managing its impact. IEEE Access **8**, 90225–90265 (2020)
11. Crosby, M., Pattanayak, P., Verma, S., Kalyanaraman, V.: Blockchain technology: beyond bitcoin. Appl. Innov. **2**(6–10), 71 (2016)
12. Demirkan, S., Demirkan, I., McKee, A.: Blockchain technology in the future of business cyber security and accounting. J. Manag. Analytics **7**(2), 189–208 (2020)
13. Galvez, J., Mejuto, J., Simal-Gandara, J.: Future challenges on the use of blockchain for food traceability analysis. TrAC Trends Anal. Chem. **107**, 222–232 (2018). https://doi.org/10.1016/j.trac.2018.08.011
14. Gola, C., Sedlmeir, J.: Addressing the sustainability of distributed ledger technology. SSRN Electron. J. **2** (2022). https://doi.org/10.2139/SSRN.4032837
15. Hou, J., Wang, H., Liu, P.: Applying the blockchain technology to promote the development of distributed photovoltaic in China. Int. J. Energy Res. **42**(6), 2050–2069 (2018). https://doi.org/10.1002/er.3984
16. Ismagilova, E., Hughes, L., Rana, N.P., Dwivedi, Y.K.: Security, privacy and risks within smart cities: literature review and development of a smart city interaction framework. Inf. Syst. Front. 1–22 (2020)
17. Jovic, M., Filipovic, M., Tijan, E., Jardas, M.: A review of blockchain technology implementation in shipping industry. Pomorstvo Sci. J. Marit. Res. **33**(2), 140–148 (2019). https://doi.org/10.31217/p.33.2.3
18. Jun, M.S.: Blockchain government—a next form of infrastructure for the twenty-first century. J. Open Innov. Technol. Mark. Complex. **4**(1), 1–12 (2018). https://doi.org/10.1186/s40852-018-0086-3

19. Khanna, A., Sah, A., Bolshev, V., Jasinski, M., Vinogradov, A., Leonowicz, Z., Jasiński, M.: Blockchain: future of e-governance in smart cities. Sustainability **13**(21) (2021). https://doi.org/10.3390/su132111840
20. Lamport, L., Shostak, R., Pease, M.: The Byzantine generals problem. ACM Trans. Programm. Lang. Syst. (TOPLAS) **4**(3) (1982)
21. Li, J., Greenwood, D., Kassem, M.: Blockchain in the built environment and construction industry: a systematic review, conceptual models and practical use cases. Autom. Constr. **102**, 288–307 (2019)
22. Liberati, A., Altman, D.G., Tetzlaff, J., Mulrow, C., Gøtzsche, P.C., Ioannidis, J.P., Clarke, M., Devereaux, P.J., Kleijnen, J., Moher, D.: The PRISMA statement for reporting systematic reviews and meta-analyses of studies that evaluate health care interventions. PLoS Med. **6**(7), e1000100 (2009)
23. Marbouh, D., Abbasi, T., Maasmi, F., Omar, I., Debe, M., Salah, K., Jayaraman, R., Ellahham, S.: Blockchain for covid-19: review, opportunities, and a trusted tracking system. Arab. J. Sci. Eng. **45**(12), 9895–9911 (2020)
24. Mark, W.: Distributed ledger technology: beyond block chain. Technical report. UK Government Chief Scientific Adviser, London (2016)
25. Mattila, J.: The blockchain phenomenon—the disruptive potential of distributed consensus architectures. Technical report. ETLA Working Papers (2016)
26. Mergel, I., Edelmann, N., Haug, N.: Defining digital transformation: results from expert interviews. Govern. Inf. Q. **36**(4), 101385 (2019)
27. Natarajan, H., Krause, S., Gradstein, H.: Distributed ledger technology and blockchain. Technical report. World Bank, Washington (2017)
28. Negara, E., Hidayanto, A., Andryani, R., Syaputra, R.: Survey of smart contract framework and its application. Information **12**(7) (2021)
29. Pettit, C., Liu, E., Rennie, E., Goldenfein, J., Glackin, S.: Understanding the disruptive technology ecosystem in Australian urban and housing contexts: a roadmap. Technical report 304. Australian Housing and Urban Research Institute, Melbourne (2018). https://doi.org/10.18408/ahuri-7115101
30. Pongnumkul, S., Bunditlurdruk, T., Chaovalit, P., Tharatipyakul, A.: A cross-sectional review of blockchain in Thailand: research literature, education courses, and industry projects. Appl. Sci. **11**(11) (2021)
31. Prajapati, P., Dave, K., Shah, P.: A review of recent blockchain applications. Int. J. Sci. Technol. Res. **9**(1), 897–903 (2020)
32. Qian, J., Ruiz-Garcia, L., Fan, B., Robla Villalba, J., McCarthy, U., Zhang, B., Yu, Q., Wu, W.: Food traceability system from governmental, corporate, and consumer perspectives in the European Union and China: a comparative review. Trends Food Sci. Technol. **99**, 402–412 (2020)
33. Rab, R.H.C., Gillian, T., Peter, F.: Distributed ledger technologies in public services. Technical report. The Scottish Government, Edinburgh (2018)
34. Raghavendra, M.: Can blockchain technologies help tackle the opioid epidemic: a narrative review. Pain Med. (U.S.) **20**(10), 1884–1889 (2019)
35. Rauchs, M., Glidden, A., Gordon, B., Pieters, G.C., Recanatini, M., Rostand, F., Vagneur, K., Zhang, B.Z.: Distributed Ledger Technology Systems: A Conceptual Framework. Available at SSRN 3230013 (2018)
36. Rosch, T., Treffinger, P., Koch, B.: Regional flexibility markets-solutions to the European energy distribution grid—a systematic review and research agenda. Energies **14**(9) (2021)
37. Samion, N., Mohamed, A.: Innovation of national digital identity: a review. Int. J. Adv. Trends Comput. Sci. Eng. **9**(1.2), 151–159 (2020). https://doi.org/10.30534/IJATCSE/2020/2391.22020

38. Shahaab, A., Lidgey, B., Hewage, C., Khan, I.: Applicability and appropriateness of distributed ledgers consensus protocols in public and private sectors: a systematic review. IEEE Access **7**, 43622–43636 (2019)

39. Sharma, A., Bahl, S., Bagha, A.K., Javaid, M., Shukla, D.K., Haleem, A.: Blockchain technology and its applications to combat COVID-19 pandemic. Res. Biomed. Eng. **38**(1), 173–180 (2020). https://doi.org/10.1007/s42600-020-00106-3

40. Szabo, N.: Formalizing and securing relationships on public networks. First Monday (1997)

41. Tasca, P., Tessone, C.J.: Taxonomy of blockchain technologies: principles of identification and classification. SRN Electron. J. (2017)

42. Tasca, P., Thanabalasingham, T.: Ontology of blockchain technologies. Principles of identification and classification. SSRN Electron. J. (2017). https://doi.org/10.2139/ssrn.2977811

43. Teng, F., Zhang, Q., Wang, G., Liu, J., Li, H.: A comprehensive review of energy blockchain: application scenarios and development trends. Int. J. Energy Res. **45**(12), 17515–17531 (2021). https://doi.org/10.1002/er.7109

44. Vargo, D., Zhu, L., Benwell, B., Yan, Z.: Digital technology use during COVID-19 pandemic: a rapid review. Human Behav. Emerg. Technol. **3**(1), 13–24 (2021). https://doi.org/10.1002/HBE2.242

45. Watt, A., Cameron, A., Sturm, L., Lathlean, T., Babidge, W., Blamey, S., Facey, K., Hailey, D., Norderhaug, I., Maddern, G.: Rapid reviews versus full systematic reviews: an inventory of current methods and practice in health technology assessment. Int. J. Technol. Assess. Health Care **24**(2), 133–139 (2008). https://doi.org/10.1017/S0266462308080185

46. Wüst, K., Gervais, A.: Do you need a blockchain? In: 2018 Crypto Valley Conference on Blockchain Technology (CVCBT), pp. 45–54. IEEE, Zug (2018)

47. Yellamma, P., Anupama, P., Lakshmibhavani, K., Jhansi Siva Priya, U., Ch, K.: Implementation of e-voting system using block chain technology. J. Crit. Rev. **7**(6), 865–870 (2020). https://doi.org/10.31838/jcr.07.06.149

48. Zou, Y., Meng, T., Zhang, P., Zhang, W., Li, H.: Focus on blockchain: a comprehensive survey on academic and application. IEEE Access **8**, 187182–187201 (2020). https://doi.org/10.1109/ACCESS.2020.3030491

# Digital Identity Using Hyperledger Fabric as a Private Blockchain-Based System

Suhail Odeh(✉) , Anas Samara , Ramiz Rizqallah , and Lara Shaheen

Software Engineering Department, Bethlehem University, Bethlehem, Palestine
{sodeh,asamara}@bethlehem.edu

**Abstract.** Recently, with high technology development the need for digital identity as personal identification has become more essential and not limited for the Cybernet world and can also be used every day in the real world. Many popular identity management systems such as uPort, Sovrin and shoCard use the digital identity that will be compared with our method. This extremely personal data must be kept hidden and secure. Moreover, these day-to-day documents are used in their physical form, adding to the risk of being stolen, lost, or forged. In this paper an approach model is proposed for a private blockchain-based software application that converts supported documents into a digital format using Hyperledger fabric, which is one of the popular blockchain technologies. In addition, a verification method using face and text recognition technologies is proposed to ensure the credibility of information access. The main purpose is to keep all the personal documents, such as ID, license, and birth certificate, in one place and accessible at all times, subsequently, reducing the risk of losing or misplacing them during regular activities.

**Keywords:** Hyperledger fabric · Digital wallet · Chain code · Private blockchain

## 1 Introduction

The importance of digital identity nowadays arises from the fact that it provides a complete solution with its ability of providing the identity of the person in both worlds of cyberspace and the real or physical world [1]. Usually, vaccination cards, identification cards, driver licenses, birth certificates, car registration documents and other governmental identification and credentials are used on a day-to-day basis. This requires people to carry many physical documents in their cars, wallets, pockets, or folders to keep them in one place. This extremely personal data must be kept hidden and secure. Moreover, these day-to-day documents are used in their physical form, adding to the risk of being stolen, lost, or forgotten. Many Identity Management Models (IDM) have been proposed in order to resolve the issue of sovereignty that is affected in the identity privacy and security. Self-Sovereign Identity (SSI) users have full sovereignty of their identity and storage-control of their associated personal and confidential data [2]. Naik et al. proposed several specifications to evaluate any SSI solution. Subsequently, they analyzed two emerging SSI solutions uPort and Sovrin [3]. Ferdous et al. examined the properties

© The Author(s), under exclusive license to Springer Nature Switzerland AG 2023
J. Prieto et al. (Eds.): BLOCKCHAIN 2022, LNNS 595, pp. 153–161, 2023.
https://doi.org/10.1007/978-3-031-21229-1_15

of a SSI and explored the impact of SSI over the laws of identity [4]. El Haddouti et al. analyzed and evaluated different IDM Systems: uPort, Sovrin, and ShoCard; under a set of features of digital identity that characterize the success of an Identity [5]. They concluded that the protection of stored data is not clear for some approaches and some privacy challenges may hinder the applications of Blockchain.

Shu Yun Lim et al. made a survey over the IDM blockchain technology; they concluded the main challenges and provided highlights that future work should be directed towards the discovery of new mechanisms to overcome the issues that disrupted the existing IDM by providing a more promising secure platform [6]. Bokkem et al. concluded that blockchain technology is not explicitly required for a SSI solution but it is a good foundation to build up on, due to various technical advantages that the blockchain has to offer [7]. An overall comparison between uPort, Sovrin, and the proposed system is as follows:

**Table 1.** Comparison among uport, sovrin, and the proposed system

| Criteria | Uport | Sovrin | Proposed system |
|---|---|---|---|
| Blockchain, platform | Public, ethereum | Public, hyperledger indy | Private, hyperledger fabric |
| Use of the ledger | Permissionless | Permissioned | Permissioned |
| Consensus protocol | Proof-of-work | Zero knowledge proof | Smart contract (chaincode) |
| Configurability | Less configurable due to ethereum constraints | Fully configurable | Fully configurable |
| Data decentralization | Fully decentralized, raises privacy concerns | Semi-decentralized | Semi-decentralized |
| Fees | Gas fees for transactions | No gas fees | No gas fees |

The approach presented in this paper is a model of digital identity that uses a private blockchain and AI-based mobile application that creates a digital format of people's IDs and government-issued documents and other legal documents to be used for the purpose of citizen identification (Table 1).

In this work, a consistent view of IDM is introduced in order to preserve privacy when Blockchain is used. At the initial phase of the application deployment, users will have full control over their accounts. As a blockchain application, once any user successfully digitizes their documents, they will be immutably added to the blockchain.

The structure of the paper is as follows: The proposed system is outlined in the following section along with the artificial intelligence techniques that have been exploited. Section 3 presents software process methodologies and tools that have been considered in the system development. Section 4 presents a discussion and the quality attributes of the proposed system. And finally, Sect. 5 concludes the proposed identification management system.

## 2   Proposed System

The aim of the proposed system is to provide an identification system that assures privacy and security, which consequently enables trust while using the digital identity between citizens and governmental bodies. The system components are: firstly, blockchain technology that uses a Hyperledger fabric architecture, which is considered an enterprise blockchain platform for building an enterprise-grade system. Secondly Artificial Intelligence/Machine Learning (AI/ML)-Based Software components that enable systems with computer vision and text recognition capabilities, which enhance the user experience and reduce the strain of the effort required to enter legal documents information into the system.

### 2.1   Hyperledger Fabric

Hyperledger Fabric is one of the blockchain projects within Hyperledger. Like other blockchain technologies, it has a ledger that uses smart contracts, whereas it is a system by which participants manage their transactions [8]. Subsequently, Hyperledger Fabric differs from other blockchain systems in that it is private and permissioned rather than an open permissionless system that allows unknown identities to participate in the network [9], which require protocols like "proof of work" to validate transactions and secure the network whereas the members of a Hyperledger Fabric network enroll through a trusted Membership Service Provider [10].

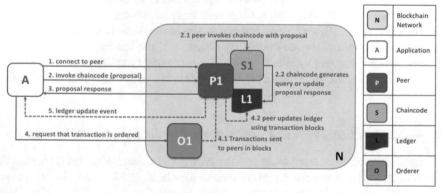

**Fig. 1.**  Hyperledger fabric architecture of an ordered node, one organization with one Peer with one Chain code deployed, and one Ledger of data [11]

Hyperledger Fabric as shown in Fig. 1 is the core of the proposed application's features. It provides a permissioned blockchain system that allows a trusted third party to add the users to the chain. First, a channel is created which is a way of communication between the consortium of organizations. Then organizations are created and added accordingly by a network administrator.

For the sake of simplicity, our proposed system will simulate a consortium of two organizations with a single peer each. Ideally, there would be two peers per organization: one committing peer, and one endorsing peer. The endorsing peer is responsible for running the Chaincode (smart code) and evaluating whether the transaction can be executed on the chain or not. Then a cumulative vote among most endorsing peers will be interpreted and a final decision will be made. If a decision is made to read, create, delete, or update an entity in the ledger, then the committing peer will perform this transaction proposal over all committing peers in the network.

To make this possible, a script written in JavaScript will be used to write a smart contract or what Hyperledger Fabric calls a Chaincode that runs on every endorsing node/peer on the network in order to evaluate a certain transaction request. For example, if a user wants to create an account, if the user account already exists, then the chain code will most likely refuse the transaction with an error code. Every peer will most likely vote on the same output, and the majority decides what to do with the transaction proposal. If the user does not exist, and the user was successfully created on most committing peers, then this will be reflected on all peers in the network.

Once committed, digitized documents will be stored in a CouchDB database that is operated on the committing peer. An "orderer" peer is responsible for sorting transactions and adding them in a certain order in the blockchain blocks. It also determines the best block size and adds blocks to the chain. This operation is done by hashing the data and calculating a total hash of all transactions in a block; each block signs the data and links it to the next block in the chain. This assures that the data is immutably added to the chain, and that any change will only be added to the chain and referring back to previous transactions is possible through a history of transactions stored in the blockchain.

Finally, despite the privacy, immutability, security, and many other features that Hyperledger fabric provides; our approach exploits this technology due to its configurability of its network and organization. Moreover, its open-sourceness allows the configuration of a permissioned, semi-decentralized access to the network via admins and users through a consortium of trusted third-party authorities, and has the option to eliminate gas fees, which subsequently, allows to create a convenient environment for all parties such as users, administrators, and developers to interact with each other in a seamless system (Table 2).

## 2.2 Verification Using Face and Optical Character Recognition Technologies

The availability of computer vision and the robustness in such systems facilitates the process of converting the paper-based documents into the digital format and subsequently makes it automated rather than tedious manual entries for the properties of the legal documents [12]. Moreover, in order to ensure the privacy and security of the digital identity the authentication is required through the user will manually enter his ID number and a chosen password to get his account credentials. This id number will be compared to the one in the second step, where the user takes a picture of the id. Beside that face identification and face verification operations used to verify that a user is who they claim to be. Therefore, comparison of the account id entered by the user and the one in the id will occur. As well as facial features comparison with two selfies taken in the next steps. The user will be prompted to take two selfies as shown in Fig. 2, one alone, and one with a randomly generated code as shown in Fig. 3. This is where Microsoft Cognitive Services [13], particularly the Face API, will be used to detect and extract the data, and

**Table 2.** Comparison among public blockchain, private blockchain, and centralized traditional database

| Features | Traditional database | Public blockchain | Private blockchain |
|---|---|---|---|
| How is governance managed? | Public, ethereum | Public, hyperledger indy | Private, hyperledger fabric |
| Who updates the ledger? | Single party | Unrestricted participants | Restricted participants |
| Are transactions anonymous to the public? | Yes | No | Yes |
| Performance | | Slow transaction speed | Lighter blockchain Fast transaction speed |
| Security | | • Consensus mechanism <br> • Proof of work | • Pre-approved participants <br> • Voting/multi-party consensus |
| Environment | | untrusted | Trusted |
| Typical platforms | | Bitcoin /ethereum | Hyperledger |

compare the two faces together, along with the one present in a user's physical Identity card.

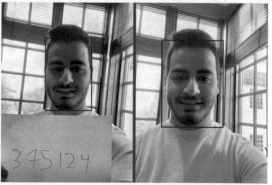

VERIFICATION RESULT: THE TWO FACES BELONG TO THE SAME PERSON. CONFIDENCE IS 0.98186.

**Fig. 2.** Example of face API face recognition and comparison with high confidence using microsoft cognitive services.

CODE: 375124

**Fig. 3.** Example of text extraction using optical character recognition service provided by microsoft cognitive services

## 3   System Development

The development of the proposed system has followed the software process aligned with the typical phases of the software development lifecycle. Specifically, Agile process principles have been followed throughout the construction of the proposed system due to its flexibility as well as quick delivery of the system components considering all the time the policies and requirements [14].

### 3.1   Development and Operations

DevOps was adopted in the introduced system as it combines the development (Dev) and the operations (Ops) together and focuses on end-to-end automation in software development and delivery [15]. DevOps is the combination of practices and tools designed to increase an organization's ability to deliver applications and services faster than traditional software development processes. To speed up development, testing, building, and deployment processes, DevOps technologies were implemented. GitHub as a version of a control system has been used in our approach. GitHub maintains all commits of code pushed to the GitHub cloud architecture and allows going back and forth between system versions as well as accessibility online. Meanwhile, according to Jason Gates. Jenkins, the leading open-source automation server, includes hundreds of plugins to help with project development, deployment, and automation [16]. It was used to automate a pipeline of installation of necessary libraries and services, cloning the code from GitHub repositories, running a sequence of commands, building and testing the code, creating a blockchain network, creating organizations and peers and joining the blockchain network, and then running the REST API on an AWS dedicated virtual machine (Fig. 4).

### 3.2   System Main Features

The backend is the backbone of the proposed system. It is the API of which the frontend will be able to communicate with the blockchain. Furthermore, it will use ExpressJS

**Fig. 4.** Jenkins pipeline for the proposed system

as the server library to create a restful API with the necessary endpoints. Express is a lightweight NodeJS web application framework that offers a comprehensive set of features for web and mobile applications [17]. Using these endpoints, a user will be able to communicate with the most important endpoints, which are responsible for user Create-Read-Update-Delete (CRUD) operations, exchanging the data, and digitizing supported documents operations. Moreover, to maintain authenticity and authorization, JSON Web Tokens, also known as JWT's, will be used as a form of validating a user's login request [18]. JWTs are an authentication mechanism that does not require the use of a database [19]. User's credentials will be used to authenticate and create a valid token signed by the server's secret key which is stored in a ".env" file. This will mean that the user is authenticated and will then be permitted to use the endpoints that they are authorized to use. Finally, the server will have the necessary NodeJS SDK packages of the Hyperledger fabric in order to communicate with the chaincode deployed on the blockchain peers.

Another important feature of the proposed system is the wallet. According to the Hyperledger fabric documentation, "A wallet contains a set of user identities. An application run by a user selects one of these identities when it connects to a channel. Access rights to channel resources, such as the ledger, are determined using this identity in combination with an MSP" [20]. These identities are created using X.509 cryptosystem by creating private and certificate keys for every user. This certificate will have been signed by the admin of the organization who created the user as illustrated in Fig. 5, which in our case Org1 will be the one who creates the users, and the admin of Org1 will be the responsible party to sign the certificate. According to [20], There are different types of wallets according to where they store their identities: CouchDB - database based wallet storage system, In Memory - a volatile in-memory storage, and File System wallet storage system which we will be using in our approach. In our approach, a file system will be used to store these wallet identities of each user in the format of [ID_NUMBER.id] files in our backend. Private and certificate keys will be used from the wallet to provide an identity of the user to the blockchain in order to perform transactions on the blockchain. Therefore, every permanently immutable transaction will be referenced to a user.

## 4 Discussion

The proposed system achieves a number of quality attributes and provides its customers with an easy-to-use platform, decreasing the quantity of documents they must carry on a daily basis. First, this approach guarantees a secure method of transmitting hashed data by immutably preserving all transactions in the chain and requiring a set of organizations to endorse or commit transactions made in the system via a smart contract that works as a consensus protocol between organizations. Each government body has its own smart

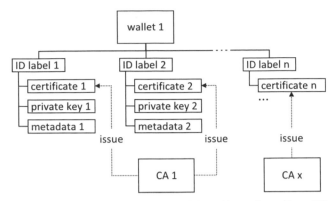

**Fig. 5.** A Fabric wallet can hold multiple identities with certificates issued by a different certificate authority. Identities comprise certificate, private key and fabric metadata [20].

contract that allows it to accept or reject any document for its users. Moreover, as data is being transmitted, it will be sent over HTTPS protocol, applying a sense of security for authentication processes, and peers-to-servers communications. Second, the proposed system considers a user's capacity to recover their account and data. Third, the system will reject any false documents which raise safety concerns, in order to protect those who may be represented by a fake ID from being harassed or mistaken for a real person in a variety of scenarios when identification is required. The program also protects the user's physical ID from being stolen and put at risk. Finally, data will be copied across the blockchain network's peers/nodes; this means that data is not monopolized. Also, in the private blockchain the data will be stored on the ledgers (peers) in each organization. This ensures data loss prevention. Through correct blockchain network configurations, a history of immutably endorsed transactions including digitizing documents and requests to read or update them will be viewable by all government organizations and kept hidden from users who are only accessing their accounts, which increase privacy as reidentification through transactions can be disabled through these configurations.

## 5  Conclusion

The main purpose that the blockchain-based application serves is to reduce the dependency on trusting a physical, government-issued document when exchanging it with any party. The application allows the user to digitize their documents, once and only once, and use it multiple times in exchange transactions. This independence is achieved by communicating with a consortium of ministries within a country, then validating the existence of a user, using AI services to extract the necessary data out of the documents, and finally adding them immutably to the blockchain.

The proposed approach represents a private, blockchain-based application that will stand out at its configurable, secure, and free-of-use features. Its independence of Blockchain or Ethereum based services will allow transactions to be exempt from gas fees, and therefore, free to use by our users. Ministries, who are acting as the organization party in the blockchain, will be able to maintain their administrative configurations, and number of endorsing and committing peers to use to maintain the integrity of data. Finally, our approach stands out at its applicability on a small scale, permissioned, private network of organizations, and its extensibility to include other governments in the future.

# References

1. Sullivan, C., Burger, E.: Blockchain, digital identity, e-government. In: Treiblmaier, H., Beck, R. (eds.) Business Transformation through Blockchain: vol. II, pp. 233–258. Springer International Publishing, Cham (2019).https://doi.org/10.1007/978-3-319-99058-3_9
2. Mühle, A., Grüner, A., Gayvoronskaya, T., Meinel, C.: A survey on essential components of a self-sovereign identity. Comput. Sci. Rev. **30**, 80–86 (2018)
3. Naik, N., Jenkins, P.: Self-Sovereign identity specifications: govern your identity through your digital wallet using blockchain technology. In: 2020 8th IEEE International Conference on Mobile Cloud Computing, Services, and Engineering (MobileCloud), pp. 90–95 (2020)
4. Ferdous, M.S., Chowdhury, F., Alassafi, M.O.: In search of self-sovereign identity leveraging blockchain technology. IEEE Access **7**, 103059–103079 (2019)
5. El Haddouti, S., El Kettani M.D.E.-C.: Analysis of identity management systems using blockchain technology. In: 2019 International Conference on Advanced Communication Technologies and Networking (CommNet), pp. 1–7 (2019)
6. Lim, S.Y.: Blockchain technology the identity management and authentication service disruptor: a survey. Int. J. Adv. Sci. Eng. Inf. Technol. **8**(4–2), 1735–1745 (2018)
7. Van Bokkem, D., Hageman, R., Koning, G., Nguyen, L., Zarin N.: Self-sovereign identity solutions: the necessity of blockchain technology. ArXiv preprint arXiv:1904.12816 (2019)
8. Li, D., Wong, W.E., Guo, J.: A survey on blockchain for enterprise using hyperledger fabric and composer. In: 2019 6th International Conference on Dependable Systems and Their Applications (DSA), pp. 71–80 (2020)
9. Androulaki, E.: Hyperledger fabric: a distributed operating system for permissioned blockchains. In: Proceedings of the thirteenth EuroSys conference, pp. 1–15 (2018)
10. Uddin, M.: Hyperledger fabric blockchain: Secure and efficient solution for electronic health records. Computer. Mater. Continua **68**(2), 2377–2397 (2021)
11. Peers—hyperledger-fabricdocs master documentation. https://hyperledger-fabric.readth edocs.io/en/release-2.2/peers/peers.html. Accessed 28 Apr 2022
12. Shapiro, L.G., Stockman, G.C.: Computer Vision. vol. 3. Prentice hall New Jersey, (2001)
13. Del Sole, A.: Introducing Microsoft Cognitive Services, pp. 1–4. Springer, In Microsoft Computer Vision APIs Distilled (2018)
14. Flora, H.K., Chande, S.: A systematic study on agile software development methodologies and practices. Int. J. Comput. Sci. Inf. Technol. **5**(3), 3626–3637 (2014)
15. Ebert, C., Gallardo, G., Hernantes, J., Serrano, N., DevOps. IEEE Softw. **33**(3), 94–100 (2016).https://doi.org/10.1365/s40702-017-0296-3
16. Gates, J.M., Collins, D.I., Braun, J.R.: CI tools as Lego Blocks: Build Your Ideal Custom Solution. (2021)
17. Mardan, A.: The Comprehensive Book on Express. js. (2014)
18. Rahmatulloh, A., Gunawan, R., Nursuwars, F.M.S.: Performance comparison of signed algorithms on JSON Web token. IOP Conf. Ser. Mater. Sci. Eng. **550**(1), 12023 (2019)
19. Ahmed, S., Mahmood, Q.: An authentication-based scheme for applications using JSON web token. In: 2019 22nd International Multitopic Conference (INMIC), pp. 1–6 (2019)
20. Wallet—hyperledger-fabricdocs master documentation. https://hyperledger-fabric.readth edocs.io/fa/latest/developapps/wallet.html. Accessed 28 Apr 2022

# Cryptocurrencies, Systematic Literature Review on Their Current Context and Challenges

Yeray Mezquita[(✉)], Marta Plaza-Hernández, Mahmoud Abbasi, and Javier Prieto

BISITE Research Group, University of Salamanca, Edificio Multiusos I+D+i, Calle Espejo 2, 37007 Salamanca, Spain

yeraymm@usal.es, martaplaza@usal.es, Mahmoud.abbasi@ieee.org, javierp@usal.es

**Abstract.** The financial crisis of 2008 breached society's trust in banks and centralized financial institutions. Bitcoin appeared as a response to those transgressions, proving that the blockchain technology it implements is reliable. Since then, this technology has become widespread in the form of new cryptocurrencies and an increasing number of people have begun to invest in them. In the absence of a body to regulate the cryptocurrency market, users play with and encourage the fluctuations in their price, causing them to be treated as investment assets rather than transaction assets. This work carries out and then discusses a systematic study of the state of the art and of the general challenges faced by cryptocurrencies.

**Keywords:** Cryptocurrencies · Systematic review · Challenges

## 1 Introduction

The 2008 financial crisis breached the society's trust in banks and centralized financial institutions; these entities led to the crisis by lending large sums of money while keeping very little in reserve [11,18]. Bitcoin emerged in 2009, in response to those transgressions. Its solution was a currency that could operate without a central authority [41]. Bitcoin comprises a series of cryptographic protocols that completely transform the process of making transactions. Thus, these protocols have brought the financial system one step closer to a truly democratic economy constructed by the community [8,22,26,36,51].

Due to the relative immaturity of blockchain technology, cryptocurrencies face technical issues such as network latency, network governance, transaction load, etc. Besides, while the appearance of a truly democratic economy is one of the strengths of using cryptocurrencies, it can also become the biggest problem for their adoption. In the absence of a body to regulate cryptocurrencies, users play with and encourage the fluctuations in their price, causing them to be treated as investment assets rather than transaction ones [14,19,29,43,49,55].

J. Prieto et al. (Eds.): BLOCKCHAIN 2022, LNNS 595, pp. 162–172, 2023.
https://doi.org/10.1007/978-3-031-21229-1_16

This article provides a systematic review of the general challenges faced by cryptocurrencies and their underlying technologies, which helps in organizing the works previously done in the literature. Everyone interested in using cryptocurrencies, in any kind of way, will find this work useful, because it will help them to understand the challenges, at any level, that cryptocurrencies are facing.

Following the introduction, Sect. 2 presents the methodology used for the systematic literature review. In each of its subsections, it is presented the discussion on each of the research questions that have arise in this study. Finally, Sect. 3 concludes this work.

## 2   Proposed Framework

In this work, a systematic review of the literature [2,42] is carried out regarding the use of cryptocurrencies, the general vulnerabilities they face, and the cryptocurrency-related crimes that it is possible to commit with them. To achieve our goals, we have proposed the following questions to be answered during our study:

- **Q1**: What research has been carried out previously in the literature regarding the use of cryptocurrencies?
- **Q2**: What are the vulnerabilities that affect the adoption of cryptocurrencies?
- **Q3**: How are crimes committed with crypto affecting the global socio-economic ecosystem?

For the purpose of this study, we have used the following scientific databases to search for articles: ScienceDirect, IEEE Explore, ACMDigital Library, and Springer-Link. Given that the Google Scholar results consisted mostly of articles that had not been peer-reviewed and that were not related to our study, this source has not been considered.

We have identified 4 keywords that are relevant to the topic of our study: Blockchain, Cryptocurrency, Challenge, and Adoption; each of these keywords is represented by the following search strings:

- **Blockchain**: ("distributed ledger technology" OR blockchain*)
- **Cryptocurrency**: (crypto* OR cryptocurrency*)
- **Challenge**: challenge*
- **Adoption**: adopt*

To obtain relevant works stored in each database, four queries have been carried out with the combination of the above keywords and search strings:

- **SS1**: Blockchain AND Cryptocurrency AND Challenge
- **SS2**: Blockchain AND Cryptocurrency AND Challenge AND Adoption
- **SS3**: Cryptocurrency AND Challenge
- **SS4**: Cryptocurrency AND Challenge AND Adoption

All publications available in the databases that matched the search criteria and had been published no later than May 2020, were included in this study.

## 2.1  Q1: What Research Has Been Carried Out Previously in the Literature Regarding the Use of Cryptocurrencies?

This question has been asked to motivate the study of this work. We propose two new questions associated:

- RQ1.1: Has the possibility of mass adoption of cryptocurrencies been sufficiently studied?
- RQ1.2: Is it justified to carry out this study despite what has been done previously?

To find an answer to RQ1, a study has been made from papers that have addressed the topic of the use of cryptocurrencies from different perspectives. All the works that do not make a review of the challenges of the cryptocurrencies or a study of the value that their features provide have been excluded from this section.

In [10] and [53] a review of the fundamentals of cryptocurrencies is shown. These reviews indicate that the risks of buying cryptocurrencies lie in the absence of a legal basis, which prevents the excessive price fluctuations to which they are subjected daily. It also shows how the possibility of wallet theft is real, and that cryptocurrencies are often used for money laundering and illegal activities. In [35] it has been studied the frameworks proposed in the literature for the prediction of cryptocurrency prices. The conclusion reached by the study is that the prices are still dependent on the opinion of influencers that can manipulate the prices and the regulation of the governments.

In [30] a review has been made, although only on Bitcoin, of the most relevant aspects that affect the confidence of the users in that cryptocurrency. It is stated that cryptocurrencies rely on three technological elements: blockchain networks, cryptocurrency wallets, and exchange platforms. From those three constructs, 11 attributes have been detected as capable of building trust in investors, from which there are three that are the most relevant: the number of transfers made, the immutability of the blockchain, and the openness of the platform. The transfers made with a cryptocurrency are related to the usability given in its economic ecosystem. The other two attributes are related only to the technology used: openness creates transparency and immutability creates accountability.

Eyal [13] has studied the challenges that cryptocurrencies face in terms of resources spent, user privacy, and the risks of using smart contracts. To optimize the number of resources spent, a small review of the alternatives in the consensus algorithms is made, also addressed in works such as [49]. In addition, it has been studied the options of working off-chain and using a layer of software that works over a blockchain network. Regarding the improvement of the users' privacy, it is proposed the mixing of transactions to avoid their traceability (e.g. through token ring signatures), and the use of cryptographic keys called Zero-Knowledge Proofs (ZKP) [45]. In [15] the alternatives that exist when dealing with the privacy of the identity of users and the transactions carried out have been studied also. This study concludes that the methods have limitations, for example, ZKPs require a lot of computing power. Besides, the possibility of complete privacy carries the risk of users abusing it and committing crimes that states cannot punish.

From this sub-section, it can be summarized that the biggest disadvantage cryptocurrencies face is the fluctuation in their price, which prevents them from being used daily. However, they are still being used in illegal activities, such as money laundering and drug dealing, a problem that the legislation must tackle through regulation [5,50], as has been stated in numerous works of the literature [4,46]. Besides, it has been studied how distributed networks challenge existing legal mechanisms of allocating responsibility, concluding that collaboration between legal scholars and technological developers is necessary [5], and answering RQ1.

Finally, regarding RQ1.1 it has been shown that there are works that study the cryptocurrencies' flaws and aspects that influence the mass adoption of this market. But, it is no work in the literature that put them all together, being the justification of this paper's study asked in RQ1.2.

## 2.2   Q2: What Are the Vulnerabilities that Affect the Adoption of Cryptocurrencies?

In the Sect. 2.1, it has been found the economic implications of the fluctuating nature of cryptocurrency prices. By answering Q2, we seek to identify what are the factors that allow those fluctuations and are harmful to the mass adoption of cryptocurrencies [52]. In this subsection, we will focus on the works whose studies point out types of vulnerabilities and problems that affect the use of cryptocurrencies as transactional assets.

Because of the fluctuating nature of the cryptocurrencies' price, there have been numerous articles that have tried to find a pattern in it [35]. In [44] it is mapped the change in features on users and network activities to understand the dynamics of the Bitcoin cryptocurrency. To analyze the results it has been proposed a machine learning model to predict the changes in the prices. This model is based on the computed correlation between features such as hash rate, number of users, transaction rate, the total number of bitcoins, and price. Following that trend, in [1] has done an analysis of the factors that impact the adoption decision of a cryptocurrency, along with the impact of those factors based on the quantification of users' judgments. That manuscript suggests that the users of cryptocurrencies are influenced by economic, technical, social, and personal factors. From the defined factors, it was found that the top criteria used by the investors were the investment opportunity, subjective norms, business acceptance, privacy, and global attention. Being the economic and social factors as the general rules.

However, the problem observed from the review carried out is the little information that society has about how to use cryptocurrencies and the technology on which they rely. Most of the attacks are due because users are not careful enough with their passwords for their wallets, and how to store them, not because of the technology [16]. E.g. some of the recommendations on how to anticipate this kind of problem are not to reuse wallet addresses, to introduce new address types using stronger hashing and signature schemes, etc.

There have been cases of coin theft via smart contracts, like in the case of the DAO attack [32]. Due to a vulnerability in the code of the smart contract, it was stolen millions of dollars in Ether. As for this kind of problem, users must make sure that these smart contracts had been previously audited correctly, before being deployed in the corresponding blockchain network [7,31,33–35,37–40,40].

Exchanges are the most widely used gateway to bring users closer to the main cryptocurrencies and the great risks users are facing when investing in them. Attacks on exchanges created a strong alarm in the community, as they have been the main gateway for users (although mostly speculative investors). Because of that, [23] has proposed an economic model to capture the short-term incentives of cryptocurrency exchanges concerning making security investments and establishing transaction fees, which would help to build the trust of the users.

While blockchain technology has been created with the intention of replacing the current, centralized financial system, authors of [21] claim that blockchain technology is capable of replacing intermediaries while ensuring the security of other kinds of platforms. Although this technology offers resistance to traditional cyberattacks, hackers have developed new forms of cyberattacks, specifically targeted at blockchain technology [6,17,25,27,36].

There also exist vulnerabilities regarding the anonymity of the users that use cryptocurrencies. In [28] the issue is developed in terms of the regulation advantages and drawbacks that pseudo-anonymity currently has in the Bitcoin ecosystem. While studying also the risk that genuinely anonymous transactions, for example in terms of illegal activities, like money laundering or drug dealing, would entail.

From this subsection, it can be concluded that cryptocurrencies are seen by the users as investment instruments, and not as transactional ones. It has been shown that by using artificial intelligence models, patterns in price fluctuation can be predicted over a week, which helps speculation and avoid its mainstream adoption as a transactional asset. Another economic aspect is that investors tend to keep the social aspect in mind, what others think of a cryptocurrency, more than the technology behind it. On the other hand, it can be concluded that one of the ways to resist high speculation is a healthy economic ecosystem underlying a robust network of nodes [1,44].

Many of the studied papers talk about the technology behind a cryptocurrency as one of the most important features to determine its price. The highlights in this respect are the network of nodes that ensures the immutability of the blockchain; the transparency that is obtained from a network of nodes governed in a distributed manner; and the cost of keeping the network operational [39]. However, the price also depends on the economic ecosystem surrounding a cryptocurrency. However, it has been found that most of the users are speculative investors and not people with real intentions to participate actively in those ecosystems. Therefore, factors such as visibility in social networks and the use given to a cryptocurrency are strongly related to its market capitalization [1,35,44].

## 2.3  Q3: How Are Crimes with Crypto Affecting the Socio-economic Ecosystem in the World?

Due to the pseudo-anonymous character of crypto, many illegal activities take advantage of it in the dark web [3,48]: money laundering, purchase of weapons and drugs. This is a characteristic of cryptocurrencies that makes them be targeted by states and has caused the prohibition of their use in some of them.

In the literature, studies have been carried out on the social ecosystem of the dark web in the market for illegal substances. In [50], it has been made an exploratory study on how New psychoactive substances (NPS) are being sold with the help of cryptocurrencies. Another legal problem arises when financial scams appear in the cryptocurrency market. It is demonstrated in [24], with a proposed detection system, that the pump-and-dump schemes happen and may be detected by examining real-world use cases. Given that this kind of scams should be penalized by law to help protect investors.

To tackle the previously mentioned problems, cryptocurrencies as structures that are designed to circumvent the regulation can always be prohibited. To avoid prohibitionary solutions, a combination of established and novel concepts for regulating new technological phenomena could be implemented [12]. That's the case of Gibraltar, which is regulating DLTs also as a competitive tool and a means for creating new public value. The study carried out in [46] contributes to the emerging view of smart regulation as an enabler and protector rather than an inhibitor and obstacle in areas of rapid innovation.

Right now it is unclear whether or not regulatory approaches based on principles rather than hard rules, like in the case of Gibraltar, would be practical only for small and highly agile jurisdictions. In [54] it is stated that a mix of soft law and hard law tools is needed for a successful regulatory framework. Those tools will help install sandbox regimes for a transitional period and at the same time inform about the legal consequences of participating in distributed ledger technologies that lack clear structures of responsibility. As another example, in [20] is stated that virtual currencies pose a serious threat to be used for money laundering, and by so, weakening the European Union's financial system. That's why directive (EU) 2018/843 (the fifth anti-money laundering Directive) intends to mitigate these risks by introducing a definition of virtual currencies within Union law.

Regulating, not only helps protect investors from fraudulent token issuers but also helps to fulfill other worthwhile goals, such as providing additional funding for small to medium-sized enterprises and financial inclusion [47]. In the Initial Coin Offering (ICO) area, [9] has stated that ICOs must be regulated, because a ban in that area, may stifle the innovation and is not conducive to the development of FinTech and blockchain technology.

According to [5], it is stated that current General Data Protection Laws are a challenge to distributed networks. Nodes of a blockchain network, according to article 26 of the General Data Protection Regulation (GDPR), are obliged to determine their respective responsibilities. Therefore, a solution to allocating responsibility for data protection in distributed networks lies outside the GDPR.

Using the Bitcoin network as an example, the transfers stored constitute personal data within the meaning of Article 4(1) GDPR, because the users behind these transfers are identifiable via their Bitcoin address(es). Also, full nodes qualify as establishments under Article 3(1) GDPR and the Bitcoin network provides a service within the meaning of Article 3(2)(a) GDPR to data subjects in the EU. Therefore, it is concluded that the data processing carried out within the Bitcoin blockchain, and therefore in the rest of the cryptocurrency blockchains, falls within the material and the territorial scope of the GDPR.

By answering Q3, it has been found that the most effective way to tackle the challenges posed by crypto and DLTs is by applying a sandbox of a mix of soft laws and hard laws to the legal framework of a country. This way it is possible to regulate the activities carried out, protecting investors and impeding crimes like laundering fraud, while allowing innovation in the area. Which could lead to a potential replacement of the current financial system.

## 3   Conclusion

The capitalization boom of the cryptocurrency market has contributed to the increase in the number of them. Each one of the cryptocurrencies is supported by a DLT platform, whose technology builds user trust in the cryptocurrency and therefore contributes greatly to its value. This work provides a systematic review of the general challenges faced by cryptocurrencies and their underlying technologies. Focusing on answering three research questions: (Q1) What research has been carried out previously in the literature regarding the use of cryptocurrencies? (Q2) What are the vulnerabilities that affect the adoption of cryptocurrencies? (Q3) What crimes are being carried out with crypto that poses a threat to its adoption?

From the studies related to Q1, it can be summarized that it has been stated in numerous works that cryptocurrencies need some kind of regulation. Because although they face great fluctuations in their price, they are still being used in illegal activities, such as money laundering and drug dealing. Besides, it has pointed out how distributed networks challenge existing legal mechanisms of allocating responsibility, concluding that collaboration between legal scholars and technological developers is necessary.

Works related to Q2 point out how the technology behind a cryptocurrency is one of the most important features to determine a cryptocurrency price. The highlights in this respect are the network of nodes that ensures the immutability of the blockchain; the transparency that is obtained from a network of nodes governed in a distributed manner; and the cost of keeping the network operational. However, the price also depends on the economic ecosystem surrounding a cryptocurrency. Therefore, factors such as visibility in social networks and the use given to a cryptocurrency are strongly related to its market capitalization. In addition, through this study it has been found that the illegal activities contracted with cryptocurrencies are not only related to money laundering: procurement of illicit services such as distributed denial of service attacks; malware binaries; botnets; and the purchase of illegal products including weapons, drugs, and falsified or stolen documents.

By answering Q3, it has been found that the most effective way to tackle the challenges posed by crypto and DLTs is by applying a sandbox of a mix of soft laws and hard laws to the legal framework of a country. This way it is possible to regulate the activities carried out, protecting investors and impeding crimes like laundering fraud, while allowing innovation in the area. Which could lead to a potential replacement of the current financial system.

**Acknowledgements.** The research of Yeray Mezquita is supported by the predoctoral fellowship from the University of Salamanca and co-funded by Banco Santander. This research was also partially supported by the project "Technological Consortium TO develop sustainability of underwater Cultural heritage (TECTONIC)". Authors declare no conflicts of interest.

# References

1. Alzahrani, S., Daim, T.U.: Evaluation of the cryptocurrency adoption decision using hierarchical decision modeling (HDM). In: 2019 Portland International Conference on Management of Engineering and Technology (PICMET), pp. 1–7. IEEE (2019)
2. Barn, B., Barat, S., Clark, T.: Conducting systematic literature reviews and systematic mapping studies. In: Proceedings of the 10th Innovations in Software Engineering Conference, pp. 212–213 (2017)
3. Barone, R., Masciandaro, D.: Cryptocurrency or usury? Crime and alternative money laundering techniques. Eur. J. Law Econ. **47**(2), 233–254 (2019)
4. Bortnikov, S.: The state sovereignty in questions of issue of cryptocurrency. In: International Scientific Conference Digital Transformation of the Economy: Challenges, Trends, New Opportunities, pp. 564–573. Springer, Berlin (2019)
5. Buocz, T., Ehrke-Rabel, T., Hödl, E., Eisenberger, I.: Bitcoin and the GDPR: allocating responsibility in distributed networks. Comput. Law Secur. Rev. **35**(2), 182–198 (2019)
6. Casado-Vara, R., Novais, P., Gil, A.B., Prieto, J., Corchado, J.M.: Distributed continuous-time fault estimation control for multiple devices in IoT networks. IEEE Access **7**, 11972–11984 (2019)
7. Castellanos-Garzón, J.A., Mezquita Martín, Y., José Luis Jaimes, S., López, S.M.: A data mining approach applied to wireless sensor networks in greenhouses. In: International Symposium on Distributed Computing and Artificial Intelligence, pp. 431–436. Springer, Berlin (2018)
8. Crosby, M., Pattanayak, P., Verma, S., Kalyanaraman, V., et al.: Blockchain technology: beyond bitcoin. Appl. Innov. **2**(6–10), 71 (2016)
9. Deng, H., Huang, R.H., Wu, Q.: The regulation of initial coin offerings in China: problems, prognoses and prospects. Eur. Bus. Organ. Law Rev. **19**(3), 465–502 (2018)
10. Dilek, S.: Cryptocurrencies in the digital era: the role of technological trust and its international effects. In: Blockchain Economics and Financial Market Innovation, pp. 453–474. Springer, Berlin (2019)
11. Earle, T.C.: Trust, confidence, and the 2008 global financial crisis. Risk Anal. Int. J. **29**(6), 785–792 (2009)

12. Eisenberger, I.: Innovation im recht (2016)
13. Eyal, I.: Blockchain technology: transforming libertarian cryptocurrency dreams to finance and banking realities. Computer **50**(9), 38–49 (2017)
14. Fatima, N.: Enhancing performance of a deep neural network by comparing optimizers experimentally. ADCAIJ Adv. Distrib. Comput. Artif. Intell. J. Salamanca **9**(2), 79–90 (2020). ISSN: 2255-2863
15. Feng, Q., He, D., Zeadally, S., Khan, M.K., Kumar, N.: A survey on privacy protection in blockchain system. J. Netw. Comput. Appl. **126**, 45–58 (2019)
16. Giechaskiel, I., Cremers, C., Rasmussen, K.B.: When the crypto in cryptocurrencies breaks: bitcoin security under broken primitives. IEEE Secur. Priv. **16**(4), 46–56 (2018)
17. González-Briones, A., Castellanos-Garzón, J.A., Mezquita Martín, Y., Prieto, J., Corchado, J.M.: A framework for knowledge discovery from wireless sensor networks in rural environments: a crop irrigation systems case study. Wirel. Commun. Mobile Comput. **2018** (2018)
18. Gros, D., Roth, F.: The financial crisis and citizen trust in the European central bank. CEPs working document (334) (2010)
19. Gupta, S., Meena, J., Gupta, O., et al.: Neural network based epileptic EEG detection and classification (2020)
20. Haffke, L., Fromberger, M., Zimmermann, P.: Cryptocurrencies and anti-money laundering: the shortcomings of the fifth AML directive (EU) and how to address them. J. Bank. Regul. 1–14 (2019)
21. Hawlitschek, F., Notheisen, B., Teubner, T.: The limits of trust-free systems: a literature review on blockchain technology and trust in the sharing economy. Electron. Commer. Res. Appl. **29**, 50–63 (2018)
22. Hussain, A., Hussain, T., Ali, I., Khan, M.R., et al.: Impact of sparse and dense deployment of nodes under different propagation models in manets (2020)
23. Johnson, B., Laszka, A., Grossklags, J., Moore, T.: Economic analyses of security investments on cryptocurrency exchanges. In: 2018 IEEE International Conference on Internet of Things (iThings) and IEEE Green Computing and Communications (GreenCom) and IEEE Cyber, Physical and Social Computing (CPSCom) and IEEE Smart Data (SmartData), pp. 1253–1262. IEEE (2018)
24. Kamps, J., Kleinberg, B.: To the moon: defining and detecting cryptocurrency pump-and-dumps. Crime Sci. **7**(1), 1–18 (2018). https://doi.org/10.1186/s40163-018-0093-5
25. Lazarenko, A., Avdoshin, S.: Financial risks of the blockchain industry: a survey of cyberattacks. In: Proceedings of the Future Technologies Conference, pp. 368–384. Springer, Berlin (2018)
26. Li, T., Chen, H., Sun, S., Corchado, J.M.: Joint smoothing and tracking based on continuous-time target trajectory function fitting. IEEE Trans. Autom. Sci. Eng. **16**(3), 1476–1483 (2018)
27. Li, T., Corchado, J.M., Sun, S.: Partial consensus and conservative fusion of Gaussian mixtures for distributed PHD fusion. IEEE Trans. Aerosp. Electron. Syst. **55**(5), 2150–2163 (2018)
28. Li, Y., Susilo, W., Yang, G., Yu, Y., Du, X., Liu, D., Guizani, N.: Toward privacy and regulation in blockchain-based cryptocurrencies. IEEE Netw. **33**(5), 111–117 (2019)
29. López, A.B.: Deep learning in biometrics: a survey. ADCAIJ Adv. Distrib. Comput. Artif. Intell. J. **8**(4), 19–32 (2019)
30. Marella, V., Upreti, B., Merikivi, J., Tuunainen, V.K.: Understanding the creation of trust in cryptocurrencies: the case of bitcoin. Electron. Markets 1–13 (2020)

31. Martín, Y.M., Parra, J., Pérez, E., Prieto, J., Corchado, J.M.: Blockchain-based systems in land registry, a survey of their use and economic implications. CISIS **2020**, 13–22 (2020)
32. Mehar, M.I., Shier, C.L., Giambattista, A., Gong, E., Fletcher, G., Sanayhie, R., Kim, H.M., Laskowski, M.: Understanding a revolutionary and flawed grand experiment in blockchain: the dao attack. J. Cases Inf. Technol. (JCIT) **21**(1), 19–32 (2019)
33. Mezquita, Y., Alonso, R.S., Casado-Vara, R., Prieto, J., Corchado, J.M.: A review of k-NN algorithm based on classical and quantum machine learning. In: International Symposium on Distributed Computing and Artificial Intelligence, pp. 189–198. Springer, Berlin (2020)
34. Mezquita, Y., Casado-Vara, R., González Briones, A., Prieto, J., Corchado, J.M.: Blockchain-based architecture for the control of logistics activities: pharmaceutical utilities case study. Logic J. IGPL **29**(6), 974–985 (2021)
35. Mezquita, Y., Gil-González, A.B., Prieto, J., Corchado, J.M.: Cryptocurrencies and price prediction: a survey. In: International Congress on Blockchain and Applications, pp. 339–346. Springer, Berlin (2021)
36. Mezquita, Y., Gil-González, A.B., Martíndel Rey, A., Prieto, J., Corchado, J.M.: Towards a blockchain-based peer-to-peer energy marketplace. Energies **15**(9), 3046 (2022)
37. Mezquita, Y., González-Briones, A., Casado-Vara, R., Chamoso, P., Prieto, J., Corchado, J.M.: Blockchain-based architecture: a mas proposal for efficient agri-food supply chains. In: International Symposium on Ambient Intelligence, pp. 89–96. Springer, Berlin (2019)
38. Mezquita, Y., González-Briones, A., Casado-Vara, R., Wolf, P., Prieta, F.D.I., Gil-González, A.B.: Review of privacy preservation with blockchain technology in the context of smart cities. In: Sustainable Smart Cities and Territories International Conference, pp. 68–77. Springer, Berlin (2021)
39. Mezquita, Y., Parra, J., Perez, E., Prieto, J., Corchado, J.M.: Blockchain-based systems in land registry, a survey of their use and economic implications. In: Computational Intelligence in Security for Information Systems Conference, pp. 13–22. Springer, Berlin (2019)
40. Mezquita, Y., Valdeolmillos, D., González-Briones, A., Prieto, J., Corchado, J.M.: Legal aspects and emerging risks in the use of smart contracts based on blockchain. In: International Conference on Knowledge Management in Organizations, pp. 525–535. Springer, Berlin (2019)
41. Nakamoto, S., et al.: Bitcoin: a peer-to-peer electronic cash system (2008)
42. Petersen, K., Feldt, R., Mujtaba, S., Mattsson, M.: Systematic mapping studies in software engineering. In: 12th International Conference on Evaluation and Assessment in Software Engineering (EASE), vol. 12, pp. 1–10 (2008)
43. Pimpalkar, A.P., Raj, R.J.R.: Influence of pre-processing strategies on the performance of ml classifiers exploiting TF-IDF and bow features. ADCAIJ Adv. Distrib. Comput. Artif. Intell. J. **9**(2), 49 (2020)
44. Saad, M., Choi, J., Nyang, D., Kim, J., Mohaisen, A.: Toward characterizing blockchain-based cryptocurrencies for highly accurate predictions. IEEE Syst. J. **14**(1), 321–332 (2019)
45. Sasson, E.B., Chiesa, A., Garman, C., Green, M., Miers, I., Tromer, E., Virza, M.: Zerocash: decentralized anonymous payments from bitcoin. In: 2014 IEEE Symposium on Security and Privacy, pp. 459–474. IEEE (2014)
46. Scholl, H.J., Bolívar, M.P.R.: Regulation as both enabler of technology use and global competitive tool: the Gibraltar case. Gov. Inf. Q. **36**(3), 601–613 (2019)

47. Tjio, H., Hu, Y.: Collective investment: land, crypto and coin schemes: regulatory 'property'. Eur. Bus. Organ. Law Rev. 1–28 (2020)
48. Todorof, M.: Fintech on the dark web: the rise of cryptos. In: Era Forum, vol. 20, pp. 1–20. Springer, Berlin (2019)
49. Valdeolmillos, D., Mezquita, Y., González-Briones, A., Prieto, J., Corchado, J.M.: Blockchain technology: a review of the current challenges of cryptocurrency. In: International Congress on Blockchain and Applications, pp. 153–160. Springer, Berlin (2019)
50. Van Hout, M.C., Hearne, E.: New psychoactive substances (NPS) on cryptomarket fora: an exploratory study of characteristics of forum activity between NPS buyers and vendors. Int. J. Drug Policy **40**, 102–110 (2017)
51. Vergara, D., Extremera, J., Rubio, M.P., Dávila, L.P.: The proliferation of virtual laboratories in educational fields. ADCAIJ Adv. Distrib. Comput. Artif. Intell. J. **9**(1), 85 (2020)
52. Weaver, N.: Risks of cryptocurrencies. Commun. ACM **61**(6), 20–24 (2018)
53. Wilson, C.: Cryptocurrencies: the future of finance? In: Contemporary Issues in International Political Economy, pp. 359–394. Springer, Berlin (2019)
54. Zetzsche, D.A., Buckley, R.P., Barberis, J.N., Arner, D.W.: Regulating a revolution: from regulatory sandboxes to smart regulation. Fordham J. Corp. Fin. L. **23**, 31 (2017)
55. Zubair, S., Al Sabri, M.A.: Hybrid measurement of the similarity value based on a genetic algorithm to improve prediction in a collaborative filtering recommendation system. ADCAIJ Adv. Distrib. Comput. Artif. Intell. J. **10**(2), 165–182 (2021)

# Blockchain Assisted Voting in Academic Councils

João Alves[1(✉)] and António Pinto[2]

[1] ESTG, Politécnico do Porto, Porto, Portugal
8000055@estg.ipp.pt
[2] CIICESI, ESTG, Politécnico do Porto, CRACS and INESC TEC, Porto, Portugal
apinto@inesctec.pt

**Abstract.** Councils are a common organisational structure of Portuguese Universities and Polytechnic Institutes. They make the key decisions, in these organisations, by nominal voting at assembly meetings. The COVID pandemic forced the remote work upon most organisations, including universities and polytechnic institutes. Assuming that a remote assembly requires additional efforts in order to guarantee the integrity of the majority decisions taken by votes expressed by its members, opportunity arises for the use of a blockchain-assisted voting system. Benefits of blockchain, such as verifiability, immutability, tamper resistant, and its distributed nature appear to be a good fit. We propose a novel blockchain-assisted system to support the decision making of academic councils that operate by nominal voting in assemblies, gathering remotely and online.

**Keywords:** Blockchain · Academic councils · Electronic voting system

## 1 Introduction

In Portugal, a common organisation structure present in higher education institutions is the council, examples being the scientific council or the pedagogical council. A significant part of the activities within Faculties of Portuguese Universities and Polytechnic Institutes are governed by these academic councils. Usually, each University comprises multiple Faculties, each one having multiple academic councils. Their decision making process is based on majority decisions by nominal member voting without anonymity. They operate in a way much similar to the nominal group technique firstly developed in 1971 [7]. The councils are usually composed of members, a president and a secretary. The president will present one or more topics to be discussed and voted, each member will manifest

This work is financed by National Funds through the Portuguese funding agency, FCT-Fundação para a Ciência e a Tecnologia, within project LA/P/0063/2020.

himself either for or against the decision, and the votes are counted. Decisions are then written in a minutes book that accounts the votes of all members on all issues but without a formal assurance that the vote of each member was considered or rightly considered. Moreover, if no additional measures of integrity are considered, a section of a minutes book may, in principle, be altered without its members being aware of such change. Such alteration, if performed by someone with bad intentions, may even occur long after the end of the mandates of the related members.

Electronic Voting System (EVS) are considered as a way to perform a more effective act of voting in democratic societies [8]. It reduces the cost of electoral acts by requiring less human resources per act. While EVS have yet to become the *de facto* way of voting of modern days, the forced push to remote working due to the COVID pandemic [5] may increase the pace of its adoption. That lead the authors to consider a blockchain-assisted EVS to support the decision making process of academic councils that are gathering remotely or online. In [1], the same authors already analysed the general adequacy of blockchain technologies when applied to EVS. The conclusion was that EVS can benefit from the integration of blockchain, but requires a case-by-case analysis. No solution that fulfils the common requirements of EVS, plus voter anonymity, while being scalable and without requiring the spending of digital currency, was identified. Key shortcomings where related to the anonymity requirement that, while not present in academic council voting, is a common requirement of general EVS.

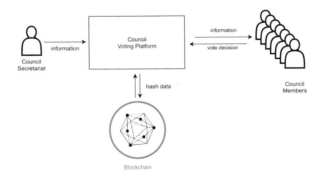

**Fig. 1.** Reference scenario.

The adopted reference scenario, depicted in Fig. 1, comprises an EVS for academic councils based on a web platform, made available to the members, where it is possible to access a topic under discussion and to record a member's vote. The President, or Secretary, of the academic council starts a meeting and puts topics up for discussion, one at a time. Each topic, after being discussed, is then subjected to a nominal vote by each member. Members cast their vote on their device when the President closes each topic. The votes are then collected and counted, and a decision is made. After the decision is computed, it is also stored in a ledger or blockchain. The decision will consist of the result of a secure hash function over all collected votes.

The authors, in the work herein, propose a novel platform that will automate vote counting and decision recording in situations that have integrity requirements but do not require voter anonymity, as is the case of the adopted reference scenario.

This paper is organised in sections. Section 2 describes EVS and presents their main principles. Section 3 describes and compares existing solutions that are similar to ours or that implement some of the features also included in the proposed solution. In Sect. 4 we enumerate the identified requirements and present the proposed solution, including its architecture and the description of the implemented prototype. Section 5 concludes our work.

## 2   Background

The adoption of EVS is a step that must be taken on the path to complete dematerialisation. Nonetheless, EVS must maintain some of the key characteristics of the traditional ballot systems. Vote secrecy is such a key characteristic, in most cases. In [8], multiple generic principals for EVS are identified. These being: Generality, Freedom, Equality, Secrecy, Directness, and Democracy. Generality means that every voter has the right to participate in an electoral process. Freedom means that the election process must ensure that it occurs without any violence or coercion over the voters. Equality means that all candidates are equally qualified to be elected and that voting rights are equal among all voters. Secrecy means that no one should be able to know the vote of another person. Directness means that there shouldn't be intermediaries in the voting process. Lastly, Democracy means that an EVS must respect the requirements of traditional electoral systems. EVS may assume a relevant role in guaranteeing that voters are not coerced [14].

However, additional requirements must also be met due to the nature of electronic voting. New features are being added and new requirements are being imposed [8]. These requirements relate to the preservation of attributes and properties, such as transparency, accountability, security, accuracy and legitimacy of the system. Voters should be able to understand how elections are conducted. The overall security of the voting process must be assured and fraud proof. From the voter's point of view, the EVS must confirm his participation in the electoral act. The indiscriminate access to the voting platform must be guaranteed, as well as a record of the correct, independent and separate counting of votes.

Multiple EVS proposals can be found in literature, namely those described in [2,3,10,13,15]; all supporting voter anonymity. Vote anonymity is not a requirement of all types of elections. The adopted reference scenario does not require vote anonymity and, therefore, none of these solutions can be directly used in the reference scenario. Other more suited solutions exist and were classified as related work, these will be analysed in the next section.

# 3   Related Work

Verify-Your-Vote (VYV), described in [6], is presented as a secure online blockchain based e-voting protocol. Authors use blockchain as a bulletin board, supporting election acts that are verifiable. Voters can only vote after being authenticated. Requirements such as vote privacy, receipt-freeness, fairness, individual and universal verifiability are also supported. Authors identified, as a main goal and as a requirements, that their solution must be able to be formally verified at the protocol level, using a ProVerif tool.

Hardwick et al. in [9], proposed an EVS based on blockchain technology that addresses some requirements that are less frequent, such as the capability of a voter to change its vote after being cast. Additionally, the authors strive for a maximum degree of decentralisation where the voters control the network by being part of it, as its peers.

Bistarelli, et al., proposed an EVS based on Bitcoin [4]. It comprises 3 phases: pre-voting, voting and post-voting. The resulting implementation is completely decentralised as you can vote without any intermediaries. All votes can be verified by anyone who reads this public ledger. This solution can only be applied when voter anonymity is not a requirement. Uncoercibility, confidentiality and neutrality are also not satisfied. Once a vote is cast, it is broadcasted to the peer-to-peer network and, by not being confidential, it can influence future voters.

The Secure Electronic Voting System (SecEVS), proposed in [12], is focused in secure e-voting in university campus elections. They make use of Merkle root hashes, like Bitcoin, guaranteeing the privacy of the transmitted data, voter confidentiality, and the uniqueness of each vote, avoiding duplication. Anonymity is not considered a requirement. The scenario of SecEVS appears similar to the one proposed herein, but the key difference resides on the fact that the envisioned reference scenario is not compatible with voter anonymity.

**Table 1.** Comparison of related work.

| Related EVS | VYV | Hardwick | Bistarelli | SecEVS |
|---|---|---|---|---|
| Confirmation | Y | Y | Y | N |
| Eligibility | Y | Y | Y | Y |
| Verifiability | Y | N | Y | Y |
| Unreusability | – | N | Y | Y |
| Uncoercibility | N | Y | N | N |
| Integrity | Y | Y | N | Y |
| Uniqueness | Y | Y | N | Y |
| Validation | Y | Y | Y | Y |
| Ballot counting | Y | – | N | Y |

Table 1 compares the related work with respect to multiple characteristics. Namely confirmation, eligibility, verifiability, unreusability, uncoercibility, integrity, uniqueness, validation, and ballot counting. Confirmation requires a process to confirm the participation in the voting process. Eligibility requires a process to verify if a voter is eligible to vote. Verifiability requires processes to allow each voter verify is own vote. Unreusability requires that each voter can only vote once, preventing future changes in votes already made. Uncoercibility means that the system must prevent any attempt to coerce the vote of others. Integrity refers to maintaining each vote as it was cast, making it difficult or impossible to tamper with. Uniqueness requires the guarantee that a voter can only cast one vote, per voting act. Validation requires a process that enables the verification of the overall voting process. Ballot counting requires a vote counting process that can be validated.

Requirements such as the confirmation of the participation in the voting process, or only allowing eligible voters to vote, or permitting voters to verify that their vote is considered in the result, or that votes are not reused are almost satisfied by all solutions. For instance, the solution proposed by Bistarelli [4] supports vote confirmation, voter eligibility, unreusability, integrity, separation, validation and requires the use of currency to cast votes. Uniqueness, ballot counting, and uncoercibility are not satisfied by this solution. The SecEVS solution considers a reference scenario that is closer to the one adopted in our work, in the sense that both address voting in academic institutions and do not require the expenditure of digital currency for its operation. Nonetheless, they differ because SecEVS is geared towards an large electoral vote, while, in our case, multiple voting processes are to be expected per plenary meeting.

The analysis of the existing literature and related work, complementary to previous work [1] on this subject, lead the authors to conclude that the solution proposed herein is novel and not previously addressed by our peers.

## 4    Proposed Solution

The proposed solution, named blockchain-assisted Academic Council EVS (bACEVS), aims to specifically address the recording of decisions made in academic councils. With the proposed solution, a faithful record of voting acts is maintained as well as a significant increase in usability and verifiability, of the entire decision making process, is expected. Please recall that, while voter anonymity and voting privacy are not required within the envisioned reference scenario, characteristics such as voter eligibility, ballot counting, integrity, verifiability and uniqueness of votes still need to be addressed. Such considerations lead to the identification of the following list of key requirements:

1. Support an operation based on the fact that members are elected for a mandate (period of time).
2. Support multiple Academic Councils of multiple Faculties within one organisation.
3. Support the management of Academic Councils and their members.

4. Schedule meetings and record the list of members participating in each one.
5. Insert, modify and remove topics subject to decision by voting.
6. Support the association of external digital documents that are relevant to the topics to be voted on.
7. Register the participation of all members in a voting act, including their vote and the overall decision.
8. Guaranty the immutability of records pertaining voting acts.
9. Support overall availability by using a redundant, decentralised system.
10. Restrict access to the information of each council, allowing only access to the current members of each council.

The crossed analysis of the reference scenario with these key requirements resulted in the definition of a set of use cases. The main use cases that were identified comprise two actors: President and Member. The Member will be able to cast a vote or issue an opinion when a topic is open for voting. Members will also be able to view the information recorded on the blockchain. The President can start a meeting, manage topics that will be voted on, and close meetings. The President, at the topic level, can open a topic for discussion within a meeting, collect its votes and generate the decision, closing the topic. The decision is forwarded to the blockchain, as soon as the President closes the topic (Fig. 2).

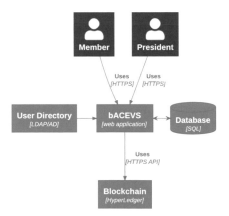

**Fig. 2.** Architecture of the proposed solution.

## 4.1   Architecture

The proposed solution comprises a web application, named bACEVS. The web application is expected to be accessed by members with their smartphone, tablet or personal computers, using Transport Layer Security (TLS) encrypted web connections (HTTPS). The web application will read and write information to a database where information about Meetings, Topics, Users, Faculties and the

Institution is stored. The web application will also interact with an user directory, such as an Active Directory (AD) database, for user authentication. The use of an AD database is expected to simplify the adoption of the proposed solution because it is commonly used in academic institutions. The web application will also connect to the blockchain, in order to read and write decisions onto the Blockchain, when appropriate. The setup will be replicated on each Faculty of an University or Polytechnic Institute while using a common Blockchain with, at least, one peer per Faculty. ions voted in each topic.

Table 2 presents the adopted notation. Letters P and M represent system users, P for president and M for normal members. $ID_m$ represents the unique identifier of member $m$. $NM_i$ represents a number to be used only once (nonce) securely generated for the meeting $i$. $NT_i$ represents a nonce securely generated for the topic $i$. $Ts_i$ represents a timestamp. $SEK$ represents a session encryption key. $PriK_m$ represents the private key of member $m$. $PubK_m$ represents the public key of member $m$. $D_{T,M}$ represents the decision generated from the discussion of topic $T$ in meeting $M$. $HMAC(D_{T,M}, SEK)$ represents the result of a secure Hash-based Message Authentication Code (HMAC) function, such as HMAC-SHA256, having $D_{T,M}$ and the $SEK$ as inputs.

**Table 2.** Adopted notation.

| P, M | Members |
|---|---|
| $ID_m$ | ID of member $m$ |
| $NM_i$ | Nonce for meeting $i$ |
| $NT_i$ | Nonce for topic $i$ |
| $Ts_i$ | Timestamp $i$ |
| $SEK$ | Session encryption key |
| $PriK_m$ | Private key of member $m$ |
| $PubK_m$ | Public key of member $m$ |
| $D_{T,M}$ | Decision on the topic $T$ of meeting $M$ |
| $HMAC(D_{T,M}, SEK)$ | HMAC of decision $D_{T,M}$ and key SEK |

Figure 3 describes the communications exchanged in the decision making process between the all entities. This exchange assumes that all members are online and authenticated in the bACEVS. The first step is the opening of a meeting, performed by the council's President (or Secretary), and consists of selecting an option in the bACEVS web-based user interface. In Step 2, the bACEVS will securely generate both a session key (SEK) and a nonce for the meeting ($NM_i$). These values, in Step 3, will be transmitted to all members actively participating in the meeting. Steps 4 up to 9 are repeated per topic that is included in the meeting. In Step 4, the President opens a meeting topic for discussion, by using the web-based user interface. The bACEVS system will securely generate a topic nonce, in Step 5, and communicate it to all active members, in Step 6. Next, the bACEVS will grant access to the documentation related to the topic being discussed to all active members. A period of discussion on the topic is expected at this time and, after this discussion, the President can request a vote collection (Step 7) which will trigger the bACEVS to issue vote request messages (Step 8) to all active members. In reply, all active members will return their vote (Step 9) to the bACEVS. All votes will be joined in a list. The list of votes and the current SEK will be the inputs to a secure HMAC function that, at last, will return the overall decision on topic $T$ of meeting $M$.

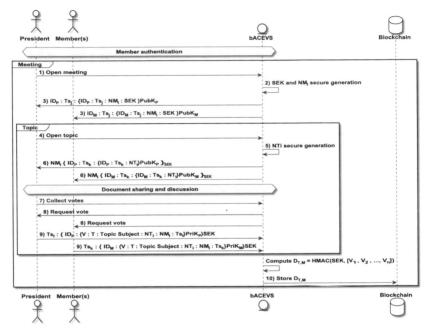

**Fig. 3.** Decision token (D$i$) generation.

The decision $D_{T,M}$ is the result of a HMAC function having as input a session key $(SEK)$ and a list of the votes $([V_1, V_2, \ldots, V_n]$, with $n$ being the number of members participating on the decision) of all members on a specific topic $(T)$ of a specific meeting $(M)$. Each one of these votes consists of multiple concatenated fields, all encrypted with the member's private key and structured as follows: $V_i = \{V : T : M : Topic\ Subject : NM_i : NT_i : T_s\}_{PriK_i}$ with $V$ being the vote cast by the member $i$ on topic $T$ of meeting $M$. $V$ can either be 1 (yea), 2 (nay) or 3 (abstention).

## 4.2   Prototype Implementation

A prototype was implemented to validate the proposed solution and comprises four distinct equipment: (1) a bACEVS server; (2) an authentication server; (3) a personal computer, simulating the President's workstation; and (4) a smartphone, simulating a voting member. Both servers were setup as Virtual Machines

(VM) on a host machine with 16 GigaBytes (GB) of Random Access Memory (RAM) memory, an Intel I5 processor and a 240 GB Solid State Drive (SSD) storage, running the MS Windows 10 operating system. Oracle Virtual Box version 6.0.14 was the adopted virtualisation platform. The bACEVS server VM was deployed with 2 cores and 8 GB of RAM, running Community ENTerprise OS (CENTOS) version 7, the Linux, APACHE, MySQL, and PHP (LAMP) stack version 7.9.2009, and PHP version 7.3.31. The authentication server VM, was deployed with 2 cores and 4 GB of RAM, running Windows Server 2019 with AD services. The President's workstation consists of a personal computer with 16 GB of RAM, an Intel I5 processor and a 240 GB SSD, running the MS Windows 10 operating system. The member's smartphone is an iPhone 6s with 4 GB of RAM, and an Apple A9 processor, running the iOS version 15.2.1.

The selected blockchain technology was the Hyperledger Fabric. The key reasoning behind this decision was based on the application of the framework proposed in [11] to the reference scenario, resulting in the decision to adopt a Distributed Ledger Technology (DLT) that imposes permissions on both read and write accesses, but without requiring properties such transparency or equal rights. To implement the interface between the Hyperledger Fabric and the bACEVS server, the relevant smart-contracts were implemented using IBM Blockchain Platform for VSCode version 2.0 and its Local Test network.

## 5    Conclusions

The COVID pandemic forced the remote work upon universities and polytechnic institutes and their academic councils. Academic councils take decisions based on nominal voting, without anonymity. A blockchain-assisted voting system can fully support the decision making process of academic councils in remote, online meetings. The proposed bACEVS system implements such an approach while offering additional guaranties such as vote integrity and verification. The adopted blockchain allows for having a node per faculty in the university, supporting redundancy but still enforcing council confidentiality. A prototype of the proposed solution was implemented and tested. A real world test is expected to be pursued next.

## References

1. Alves, J., Pinto, A.: On the use of the blockchain technology in electronic voting systems. In: International Symposium on Ambient Intelligence, pp. 323–330. Springer, Berlin (2018)
2. de Balthasar, T., Hernandez-Castro, J.: An analysis of bitcoin laundry services. In: Nordic Conference on Secure IT Systems, pp. 297–312. Springer, Berlin (2017)
3. Ben Ayed, A.: A conceptual secure blockchain based electronic voting system. Int. J. Netw. Secur. Appl. 9(3), 01–09 (2017). https://aircconline.com/ijnsa/V9N3/9317ijnsa01.pdf

4. Bistarelli, S., Mantilacci, M., Santancini, P., Santini, F.: An end-to-end voting-system based on bitcoin. In: Proceedings of the Symposium on Applied Computing—SAC '17, pp. 1836–1841 (2017)
5. Brynjolfsson, E., Horton, J.J., Ozimek, A., Rock, D., Sharma, G., TuYe, H.Y.: Covid-19 and remote work: an early look at us data. Technical report. National Bureau of Economic Research (2020)
6. Chaieb, M., Yousfi, S., Lafourcade, P., Robbana, R.: Verify-your-vote: a verifiable blockchain-based online voting protocol. Lecture Notes in Business Information Processing, vol. 341, pp. 16–30 (2019)
7. Delbecq, A.L., Van de Ven, A.H.: A group process model for problem identification and program planning. J. Appl. Behav. Sci. **7**(4), 466–492 (1971)
8. Gritzalis, D.A.: Principles and requirements for a secure e-voting system. Comput. Secur. **21**(6), 539–556 (2002)
9. Hardwick, F.S., Gioulis, A., Akram, R.N., Markantonakis, K.: E-voting with blockchain: an e-voting protocol with decentralisation and voter privacy. In: Proceedings—IEEE 2018 International Congress on Cybermatics: 2018 IEEE Conferences on Internet of Things, Green Computing and Communications, Cyber, Physical and Social Computing, Smart Data, Blockchain, Computer and Information Technology, iThings/Gree, pp. 1561–1567 (2018)
10. Hopwood, D., Bowe, S., Hornby, T., Wilcox, N.: Zcash protocol specification. Technical report, 1.10. Zerocoin Electric Coin Company (2016)
11. Hunhevicz, J.J., Hall, D.M.: Do you need a blockchain in construction? Use case categories and decision framework for DLT design options. Adv. Eng. Inf. **45**, 101094 (2020)
12. Singh, A., Chatterjee, K.: Secevs: secure electronic voting system using blockchain technology. In: 2018 International Conference on Computing, Power and Communication Technologies (GUCON), pp. 863–867 (2018)
13. Tarasov, P., Tewari, H.: Internet voting using zcash. Cryptology ePrint Archive, Report 2017/585 (2017). https://eprint.iacr.org/2017/585.pdf. Accessed: 2017–06-29
14. Wang, K.H., Mondal, S.K., Chan, K., Xie, X.: A review of contemporary e-voting: requirements, technology, systems and usability. Data Sci. Pattern Recogn. **1**(1), 31–47 (2017)
15. Wu, Y.: An e-voting system based on blockchain and ring signature. University of Birmingham, Master (2017)

# Scalable and Transparent Blockchain Multi-layer Approach for Smart Energy Communities

Marta Chinnici[1]($\boxtimes$), Luigi Telesca[2], Mahfuzul Islam[2], and Jean-Philippe Georges[3]

[1] ENEA-C.R Casaccia, Via Anguillarese 301, 00123 Rome, Italy
marta.chinnici@enea.it
[2] Trakti Ltd, Trento, Italy
{luigi,mahfuzul.islam}@trakti.com
[3] Université de Lorraine, CNRS, CRAN, 54000 Nancy, France
jean-philippe.georges@univ-lorraine.fr

**Abstract.** The Blockchain paradigm applied to energy communities represents both a challenge and an opportunity. Indeed, it introduces a new distributed and transparent framework to manage the communities' services. In the meantime, it guarantees the traceability of the energy produced and consumed, informing consumers of the origins and costs of their supply, making tariffs more transparent with the result of promoting energy flexibility. Blockchain enables a real digitalization of the energy sector, maintaining a high level of trust and compliance. The security of the transaction and the guarantee that the platform on which an energy network is based cannot be tampered with has prompted energy communities to start a blockchain-based system. The paper proposed a blockchain-based open and transparent ecosystem for smart energy communities within different and several actors interfacing thanks to the blockchain system and its services interact in a natural and disintermediate way. It allows, in a peer-to-peer (p2p) manner, to implement participatory logic aimed at reducing and self-consumption of energy and direct participation of citizens in the electricity system. A proof of concept is created and tested using an innovative smart legal contract framework created by Trakti and a scalable blockchain ecosystem thanks to the Trakti platform through an end-to-end approach. Indeed, this platform implements a new way of automatically managing and incentivizing the energy community via a smart contract connecting actual data and services that the smart energy community can consume using community energy tokens. The approach demonstrates how the new smart legal contract makes it possible to define an energy token, which remunerates the energy exchanges in communities between producers and consumers within a smart grid and converts the energy into community services.

**Keywords:** Blockchain · Energy communities · Energy market · Energy flexibility · P2P · Smart legal contracts

J. Prieto et al. (Eds.): BLOCKCHAIN 2022, LNNS 595, pp. 183–197, 2023.
https://doi.org/10.1007/978-3-031-21229-1_18

# 1  Introduction

With the current energy and environmental crisis, more sustainable energy production and consumption approaches are gaining momentum. More participatory and bottom-up approaches are becoming mainstream thanks to the proliferation of information system technologies and off-grid, on-grid microgeneration systems. Citizens are also discovering the power of clustering their energy assets in pools of micro-producers to create Energy communities closer to their values and local communities. They see this as a substantial effort to change the world, and they ask regulators for more incentives and to reduce the administrative burden of setting up energy communities. Energy communities provide flexibility for the electricity system and energy efficiency for the citizens by organizing the energy actions. It mainly sets out new regulations that help consumers produce, use, and exchange or trade electricity [1, 6]. With micro-generation, we refer to the small-scale generation of electric power by energy communities composed of citizens, small businesses, and public administration to meet their own needs as alternatives or supplements to traditional centralized grid-connected control. The energy community concept is a pivotal point in the design of the future European energy infrastructure. It implies the strict collaboration of market players (utilities), "energy designers", policymakers and citizens, all aiming to develop intelligent energy delivery, fostering renewable sources and technology innovation in distributed generation. This benefits the economy, sustainability, and energy security [2]. The success of the energy community paradigm relies on several factors like renewable sources and generation system availability, innovative technological solutions, normative regulation, and political, psycho-social, and cultural dimensions. Those conditions need to be sustainable for all community members. According to this novel approach and the security aspect, using blockchain infrastructures and smart contracts for energy representation and exchange provides several advantages. Blockchain offers a distributed and trusted technology infrastructure that can be used as an Information and Communication Technology (ICT) backbone for an open energy market [6, 7]. The security of the transaction and the guarantee that the platform on which an energy network is based cannot be tampered with has prompted energy communities to start a blockchain-based system. The paper proposed a blockchain-based open and transparent ecosystem for smart energy communities within, different actors interfacing thanks to the blockchain system, and its services interact in a natural and disintermediate way. It allows, in a peer-to-peer (P2P) manner, to implement participatory logic aimed at reducing energy waste while promoting self-consumption of energy and direct participation of citizens in the electricity system. Trakti smart legal contract framework and its blockchain platform (scalable) [5] that, through an end-to-end approach, implement a new way of automatically managing and incentivise via smart contract the energy community by connecting real data and services that can be consumed by the smart energy community using community energy tokens. The results show that the new smart legal contract makes it possible to define an energy token, which remunerates the energy exchanges in communities between producers and consumers within a smart grid and converts the energy into community services.

The paper is organised as follows: Sect. 1 – Introduction; Sect. 2 – Related Works: Blockchain and Energy Communities; Sect. 3 – Methodology: Blockchain multi-layer approach for Smart Energy Communities; Sect. 4 – Results Discussion; Sect. 5 – Conclusions and Future Works.

## 2 Related Works: Blockchain and Energy Communities

This section presents our proposal's essential points; we discuss fundamental concepts like blockchain & energy communities to blend these ingredients to create a blockchain-based open, transparent and scalable ecosystem for smart energy communities. Sustainable energy communities offer the opportunity to move from energy consumption to more sustainable patterns. Reducing energy consumption is a priority and a significant challenge for all. It must be tackled from a wide range of angles, squeezing every opportunity for optimization and savings. At the same time, distributed systems and tech for energy generation are gaining traction in our society. Blockchain, an emerging technology for realizing distributed ledgers, has recently attracted extensive research attention. Such a ledger intends to achieve decentralized transaction management, which means that any node joining the ledger can initiate transactions equally according to rules, and the transaction does not need to be managed by any third party. All transactions in the system are stored in blocks, which are then linked as a chain and organized in chronological order. Moreover, transactions written in blocks are immutable and transparent to all peers. With all these attractive characteristics, blockchain drastically differs from traditional centralized trust entities and has become a significant enabler to future financial systems [8]. In recent years, the blockchain has developed rapidly, from Bitcoin [9], the first decentralized cryptocurrency, to Ethereum [10] with smart contracts, followed by the emerging permissioned blockchain (e.g. Hyperledger fabric [11]). Because of the wide adoption of Blockchain, blockchain-based applications have been getting involved in our daily lives. When blockchain system users increase extensively, the scalability issues of major public-chain [12] platforms (e.g. Bitcoin and Ethereum) have arisen and greatly affected blockchain development. Transaction throughput and transaction confirmation latency are the two most talked-about performance metrics of blockchain, and both of them have not reached a satisfactory level in recent popular blockchain systems [13], which leads to a lousy user quality of experience. However, compared with the centralized payment system like banks, these two metrics cannot be easily improved in a blockchain, a self-regulating system that needs more consideration to maintain decentralization. After numerous studies on the particularities of blockchain, some researchers raise the view of the Blockchain Trilemma [14]. Similar to the CAP theory [15] in the traditional field of the distributed system, the Blockchain Trilemma points out that three essential principles of a blockchain framework, decentralization, security, and scalability, cannot perfectly co-exist. For instance, considering a simplified circumstance, adding a centralized coordinator into the system can reduce the consumption (e.g. computational resources consumed by proof-of-work [16]) for all users in the system to reach a consensus on a set of transactions. Therefore, balancing or achieving these three aspects of the blockchain system well is essential for future blockchain development suitable for more complex and larger-scale scenarios in our daily lives. To improve the scalability of the blockchain, many companies and researchers have proposed many different solutions; in [8], the authors classify them according to the hierarchical structure of the blockchain. Even if the current state-of-the-art on this issue proposes a plethora of solutions to become a scalable blockchain system, the token allocation models remain static. In [17], the authors proposed an advance beyond state of the art to introduce a dynamic allocation of the token to become the scalable whole of the blockchain ecosystem thanks to the introduction to the Trakti platform. In this work, the same authors continued the previous work to show the end-to-end approach to allocating the token dynamically with the end to provide the scalability of the ecosystem for the energy community. This dynamic approach also introduces new market concept elements and novelty in the token economy.

Micro-generation is the capacity for consumers to produce electrical energy in-house or in a local community. The concept of "market" indicates the possibility of trading the electricity that has been micro-generated among producers and consumers. User acting both as a producer and consumer is called a "prosumer". Traditionally, this market has been served by pre-defined bilateral agreements between prosumers and retail energy suppliers. This means that electricity-generating prosumers have not had real access to the energy market, which remains a privileged field for institutionalised energy suppliers. So far, this fact has heavily impacted the actual diffusion at a large scale of micro-generation due to the limited economic advantages this energy generation approach would bring to the prosumers. Indeed, the main options considered so far by the technical literature were completely centralised, and their viability (under a prosumer perspective) was in general challenged as they introduced additional management fees and costs and assumed the intervention of a trusted third party, reducing once again the potential gains of end-users [2]. New approaches should be developed, enabling end-users to have free access to the energy market. In this context, the advent of distributed ledgers, i.e., blockchains, can be considered beneficial. Blockchain technologies offer the opportunity to capture value produced, exchanged and consumed fully automatedly. Their shared and transparent ledger of transactions and operations can support smart communities in their day by day operations, increasing transparency, simplifying energy accounting and automating value-sharing among community members. In particular, using a blockchain for energy representation and exchange provides several advantages. First, it gives the possibility to have a trusted and decentralised direct exchange between two parties. No intermediaries or third parties are needed to fulfil transactions. The data on the blockchain is public, easily verifiable by interested parties, consistent, and always available. Even if the transactional data are available, the user's real identities can be shielded by specific services and wallets and remain pseudonymous in the blockchain; as for the transactions, blockchain addresses and not personal data are used. Moreover, blockchains are resistant to denial-of-service attacks due to their decentralised nature and lack of a central point of failure. Data on the blockchain are immutable, meaning that it cannot be altered once inserted in the blockchain, providing a reliable point of reference. Finally, the contractual relations among community members, intermediaries, distributors, and third-party services can be fully automated thanks to the usage of smart contracts. With these features, blockchain provides a trusted technology that can be used as an ICT backbone for an open energy market. According to this approach, self-generated electricity could be consumed within the house, accumulated in batteries for later use, or returned to the grid. Thanks to the distributed and pervasive nature of the blockchain, the produced energy could be redeemed elsewhere. For example, when charging an electric vehicle abroad or selling through the blockchain to the best buyer, according to a mechanism like that of a stock-exchange market. To sum, for its nature, blockchains present some attractive advantages for smart energy communities:

- Disintermediation and trustless model: exchanges (or transactions) do not require intermediaries or trusted third parties; moreover, the parties fully guarantee that the transactions will be executed as expected.
- User empowerment: transactions and data are in control of the users' community.
- Resilience: Blockchains do not have a central point of failure due to their decentralised nature.
- Transparency and immutability: every modification in public blockchains is visible to everybody; moreover, the transactions stored in a blockchain cannot be altered or deleted.
- Automation and control over a heterogeneous and fragmented number of players with many different types of contractual relations and IT/Energy infrastructure.

For these reasons, blockchain and especially smart contracts can be used to implement other decentralized services apart from currency transactions in which trust is built-in based on blockchain intrinsic properties. A smart contract is a distributed computer program running on a blockchain network capable of executing or enforcing a predefined agreement using a blockchain, when and if specific conditions are met. Its main goal is to enable two or more parties to perform a trusted transaction without needing intermediaries. Moreover, smart contracts inherit the characteristics of blockchains and thus have no downtime, censorship or third part interference. Exploiting and applying the potentialities of blockchains to energy communities is a step forward to providing a novel ecosystem for smart energy communities.

In the following, we underline those elements as critical factors for the success of every technological paradigm [2]:

1) Community aspects: sense of community, energy consumes, community renewables project;
2) Digital competencies: digital skills, knowledge, digital divide, niche market;
3) Energy citizenship: social, civic and politic participation, social capital, engagement;
4) Attitudes: conservatives, enthusiastic, techno scepticism, innovators, trust in progress;
5) Motivations: costs reduction, environment, grassroots' actions, mainstream energy policy, hedonic.

In this perspective, microgeneration through innovative digital paradigms should consider the stratification of users' digital competencies. This aims to open Digital Single Market and Society, enhance collaborative and participatory practices to different skilled citizens, and minimize the digital divide. Consequently, the European Commission in [2] argues that tailor-made micro-generation resources addressing other citizen skills can be more effective for supporting and spreading the communities that called energy communities. However, the microgeneration implies tight coordination among its members and has often been approached from a top-down perspective, explicitly trying to design optimal solutions for a few participants simultaneously. While it can, in principle, achieve high efficiency, its adoption and structural implementation are necessarily very slow. Conversely, market approaches are more dynamic but fail to consider possible bottom-up cooperation effects among users, which can significantly enhance the manageability of the distribution network by creating aggregated consumption profiles. Moreover, pure price bidding mechanisms ignore social factors, such as the users' willingness to adapt and their personal preferences or goals (green energy, cheap energy, etc.). Therefore, since the energy community building as an incentive for citizen participation in the micro-generation market can be enumerated among the intrinsic nature of the blockchain paradigm – e.g., the use of smart contracts for energy exchange and the services exchange – we define this community as smart energy community. This paper is a real-case implementation, demonstrating the proper functioning of the entire chain, up to the actual generation of rewards and bonuses, thus validating the opportunity of smart energy communities as a proof of concept.

## 3   Methodology: Blockchain-Based Smart Energy Communities

Although blockchain and smart contracts offer an enormous opportunity for scaling smart energy communities, both infrastructure and smart contract flexibility and usability remain an issue. Privacy-related aspects connect to the blockchain, plus the programmability and flexibility of smart contracts and their connection to the real world of legal provisions and regulatory requirements, remain an issue. Most of the current blockchain infrastructure for the industrial application they do fail to offer a complete end-to-end and compliant approach. Using Trakti's mature contract technology platform providing a comprehensive end-to-end framework [5] to programme and execute not only smart contracts but smart (legal) contracts on an open source blockchain infrastructure (Ethereum) and real data coming from ENEA, we create an open and transparent relationship ecosystem for smart energy communities. Within this ecosystem, we can find different actors that interfacing through the blockchain, and its services interact in a peer-to-peer way (P2P). The ecosystem envisions the share of the energy services more securely and transparently between different end users that adhere to a standard community contract and aim to encourage better use of electrical resources, distributed production, and self-consumption using blockchain infrastructure to execute smart legal contracts (see Fig. 1). The ecosystem consists of a social fabric made of individuals, families, small companies, and small energy producers, which together form the prosumer layer. A mix of selfish and altruistic interests such as environmental concerns or cost reduction, possibly further motivated by specific governmental incentives is driving the behaviour of the prosumers. These motivations are captured in our model by the incentive layer automated via the smart legal contract that the communities will implement to thrive sustainable consumption, tailor-made micro-generation adoption and resources sharing.

This goal will be achieved by creating a Local Energy Market on smart legal contracts using "Smart energy tokens" as a participation incentive for the actors of the designed

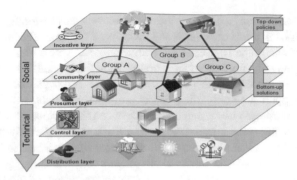

**Fig. 1.** A sketch of pivotal elements in smart energy community.

ecosystem to trade community services, making all the players in the supply chain active actors. As mentioned, some platforms already offer the possibility to generate ERC20 tokens correlated to sustainable energy behaviour or energy production from small energy producers. In most cases, their model is static and does not allow any innovation or community-based mechanisms to be embedded on the fly in the model. Everything is pre-determined, and communities can use the model, and they can't create new incentive structures in the model. As we know, in most cases, DApps and Smart contract implementation do consider a specific vertical logic and do not allow adaptive model covering the long tail of possibility and agreement among community members.

### 3.1   Trakti Smart Legal Contract and Blockchain Framework

Trakti [5] is a unified platform for smart, self-executing and compliant contracts running on a private and public blockchain, streamlining the contracting process of medium-large enterprises, financial, insurance and regulated market operators. Trakti offers a full end to end framework and all the tools needed to securely streamline the acquisition, negotiation, and signature flows, in full compliance, and manage all corporate contracts in a unique trusted infrastructure. As of today, Trakti is the only platform in the market to offer a modelling and execution environment for smart legal contracts automating the life-cycle of all sort of contracts. Trakti no-code modelling environment allows the possibility to connect legal prose to smart contract execution on blockchain networks and offers the possibility to integrate additional services like KYC/AML, credit checks and IoT integrations to streamline the entire end to end procedure in a unique compliant infrastructure. Since Trakti infrastructure can support both traditional contracts and smart contracts with blockchain implementations it is very appealing for companies that do not have to manage an additional complexity layer to monitor and maintain their contracts on multiple blockchain infrastructures. Via a combination of contract management expertise and targeted use of blockchain technology, Trakti offers smart contracts functionality in an end-to-end CLM system it has applied across enterprise use cases, at real customers, at supply chain scale. Most CLM vendors excel at digitizing contracts and obligations or streamlining negotiation, but they rarely venture into true automation powered by smart contracts. Trakti does live up to the standard of actually automating CLM processes. By adopting Trakti smart legal contract framework, we empower an extensive range of possibilities since users can customise the community smart legal contract in terms of rewards and incentive structures. The negotiation among the shareholders and orchestration of the community services in the smart community is facilitated by implementing smart legal contracts in the blockchain system. The smart legal contracts are modelled and implemented in the Trakti platform using dynamic blockchain to support the onboarding procedures and negotiations among the actors. In this way, the intra-ecosystem agreements between the various stakeholders will be guaranteed and facilitate the initial validation of the Local Market Place of Community Services. The energy token will be used in the Local Market Place to buy community services to involve and make all supply chain players active. The energy tokens are used to remunerate the energy exchanges among prosumers, producers and consumers and reward the users based on their provided services.

## 3.2 Blockchain-Based Smart Energy Communities High Level Architecture

The originality of our approach is to consider user motivation in all its social relevance. Our main objective is to enable users to self-organize into groups of various sizes with broad freedom in determining on how far they are willing to commit to sustainable energy production in microgrids and eventually reduce green-house gas emission. Instead of top-down approaches implemented through discounts or top down monetary incentive or infrastructures, the community will independently promote collaboration among users, the adoption of decentralised and innovative solution with the objective to fulfil greater common objectives. With our community-based approach, we want to address energy efficiency at community level (voluntary groups of people and small-scale companies) which, at a neighbourhood, city or regional level, organize themselves around proclaimed high level objectives such as environmental considerations, cost saving or social involvement implemented into a trusted, transparent and distributed IT framework governed by community mechanisms implemented in the blockchain. For achieving their objectives communities can independently define their concrete and measurable targets on blockchain via smart legal contract that can be easily subscribed to evaluate and autonomously transfer the benefit of their sustainable energy production capacity or joint electrical consumption. To some extent, they will need the support of a community manager, eg. Micro-community coordinator, imagined by the writer in the figure of the condominium administrator, initialising the smart legal contract used for community governance and interaction. The community will also have to cooperate with the present owners and managers of the distribution (and to some extent production) systems. Based on the previous consideration, the following high level architecture (Fig. 2) to implement the Local Market Place for energy community services is proposed.

Figure 3 instead explain the technological stack used specifying the blockchain, the smart contract technology and the protocols, the interfaces for users, contract management and marketplace. The open blockchain infrastructure used it has been the Quorum blockchain, a JP Morgan open-source initiative, already in use by the ENEA team.

The layer for the modelling and orchestration of smart contracts, as well as the definition of an initial prototype of a marketplace connected to it for the consumption of community services it has been provided by Trakti. Trakti environment and framework it has been composing and deploying smart contracts coded in Solidity to deploy (see the

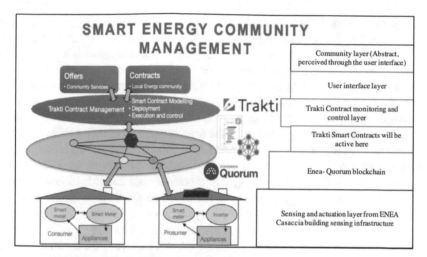

**Fig. 2.** Smart Energy Community management.

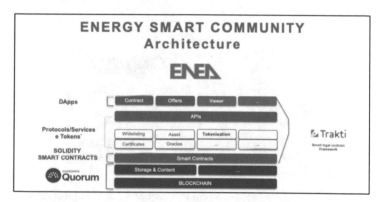

**Fig. 3.** The architecture layers of ENEA/Trakti platform.

orange layer in Fig. 3) compatible with Quorum blockchain used in ENEA. Using real-world data pulled from ENEA building at the ENEA R.C Casaccia premises (Rome) – the ENEA sensing infrastructure reads and analyses the data from energy sensors making consumption and production data accessible – we can easily model the communities services exchange (e.g. babysitting, car sharing, etc…) in smart energy communities that can be investigated using Trakti smart legal contract blockchain framework. Data are sent to the blockchain through a software layer equipped with APIs that can be called up from the outside, which it sends to the blockchain and records copies of them in the relational DMBS for analysis. Considering the Energy Tokens produced by the blockchain and the related readings, a further group reward logic has been implemented instead of the individual. This is because the tokens issued in the various cycles could be used as a proxy for defining the contribution in terms of energy efficiency that the energy micro-energy community, on the whole, could get based on collective behaviour. Based on the tokens assigned, the facilitator stakeholder, in the event of signing a contract made through the Trakti platform, recognize the members of the micro-community, identified through the wallet list acquired during an onboarding procedure, an additional bonus equal to the bonus received divided the number of wallets listed in the contract (whitelisting). Indeed, the micro-community coordinator, imagined by the writer in the figure of the condominium administrator, should have obtained from the condominiums, with the onboarding and whitelisting procedure, the authorization to sign a connected agreement that allows the distribution of the premiums. The energy profile is obtained and distributed equally among the subscribers' wallets. Therefore, in this scenario, it is planned to distribute, in a billing cycle to be defined, equally among all participants, an additional number of Energy Tokens equal to those obtained individually by the various members of the micro energy community identified in the wallets. Subscriber ID of the whitelisting agreement. In the Fig. 4, both the contract between the administrator of the micro community and the condominiums, and the contract signed between the administrator and the facilitator are present. At each billing cycle, the contract (Platform Agreement) will invoke the contract (whitelist) to obtain the ID of the wallets for the calculation of the rewards and the distribution of the additional bonus.

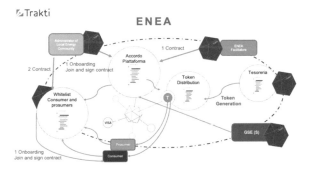

**Fig. 4.** The implemented scenario in Trakti-platform using real data.

## 4   Technical Requirement and Results Discussion

This section provides a further detail on the smart legal contract approach before showing the achieved results.

### 4.1   Smart Communities: Reward Smart Contract with Trakti Platform

The developed smart contract is consisting of several modules: Campaign, Perk, Subscriber and Reward. In detail, in Trakti platform a user can define and use two different type of rewarding system using smart contract:

1. **Individual Reward Smart Contract**: A rewarding contract where each user earns rewards depends on their own activities.

2. **Community Reward Smart Contract**: A rewarding contract where each user earns rewards depends on any members of their group. All the rewards are distributed equally among each member of the group.

The first one works by distributing the rewards for the activities of each user. In Trakti, individual reward smart contract has three distinct phases: campaign creation, campaign subscription, reward negotiation and distribution. Meanwhile, in the Community Reward Smart Contract, there are mainly three parties involved. Community creator, subscriber and reward provider. In addition to that, there can be any number of other reward beneficiaries. For example, a condo manager can create a community and ask all the condo members to join in that community. He can then negotiate different reward rules with a service provider. Whenever a community member earns rewards from the service provider, it will be equally distributed according to the rules negotiated with the provider. Said that, in the platform there are two pathways to create a fully functional reward system. A campaign creator can first do negotiation with a reward provider. This will generate a master contract in Trakti. Under that master contract, the campaign creators can then onboard subscribers to his campaign. The second way is campaign creator will first create an onboarding contract in Trakti. He will create two onboarding templates under that master contract. First onboarding template is to onboard subscriber to the campaign and second onboarding template is to negotiate and onboard reward provider to the campaign. The process works like (Fig. 5):

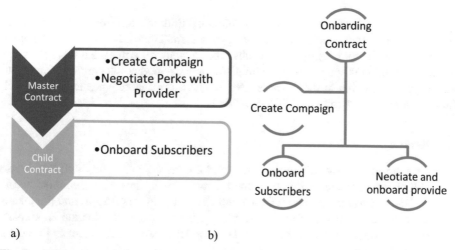

**Fig. 5.** a) Reward campaign with first negotiating rewards, then onboarding subscribers; b) Reward campaign with first onboarding subscribers, then negotiating rewards.

ENEA has a community reward system that follows the second workflow (onboarding subscribers and then negotiating rewards). The community manager is the condominium (building) administrator who creates an onboarding contract in Trakti. In the master template, he initiates the smart contract, which will enable him to add subscribers and negotiate rewards later in a separate child onboarding contract. After creating the onboarding contract, the condominium (building) administrator adds an onboarding template for all the condominium users to join the campaign. ENEA called it as whitelisting of users. When enough users get whitelisted, the condominium manager adds another onboarding template and negotiates rewards with a provider (in ENEA use case). All the rewards generated are then distributed equally among all whitelisted users. Detailed use case and step by step process of the workflow is given below. To create a community reward system, the condominium administrator first creates an onboarding contract in Trakti which launches a new campaign in the smart contract. There is a marketplace template to create a community reward campaign. In the following the steps:

1. **Step 1:** Go to add new contract model, select community reward campaign from the marketplace template
2. **Step 2:** The condominium administrator edits it accordingly and publish the template.
3. **Step 3:** After publishing the template, the condominium manager goes to his contract page and creates a new onboarding contract
4. **Step 4:** He then selects his newly published community reward campaign contract model as master agreement template. This allows him to do the configuration of his campaign like start and end date, maximum number of subscribers allowed etc..
5. **Step 5:** After submission, the admin waits for the blockchain transactions to be completed. It can take a couple of minutes to finish the transaction. The smart contract will be initialized according to the configuration and a decentralized application (dApp) interface will be loaded on his contract detail page.

After creating the campaign, the condominium (building) admin can add an onboarding template to start onboarding (whitelisting) members. There is a marketplace template to create a simple community reward subscription contract. After whitelisting is completed, the condominium (building) administrator adds another onboarding template to start negotiating terms of the rewards ENEA. There is a marketplace template which allows configuring up to two additional rewards.

## 4.2   Results

Based on the previous details about the smart contract creation for ENEA use case on Trakti platform, tokens assignment and the reward mechanism are presented in this section. Token distribution contract, the control and monitoring dApps, and the tokens assigned contract under the defined scenario have been implemented on the Trakti platform (Fig. 6). The test with real data (Fig. 7) showed the tokens assignment on the active experimental campaign.

In the Fig. 8, several tokens are received in the system, and one pending is calculated based on the readings acquired by the Quorum blockchain and the due sums that need to be distributed. It is possible to notice as there could be a series of signed contract campaigns capable of generating a distribution of tokens over time.

In Fig. 9, details of bonus received are shown. Considering such results, it demonstrates the proper functioning of the entire chain, up to the actual generation of rewards and bonuses, thus validating as a proof of concept the opportunity of smart energy

# ENERGY SMART COMMUNITY
## Flows

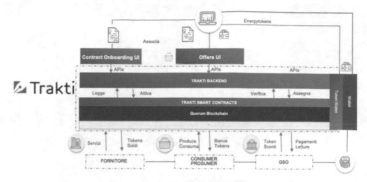

**Fig. 6.** Smart energy community flow.

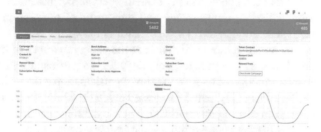

**Fig. 7.** Detail of token assignment in experimental contract campaign.

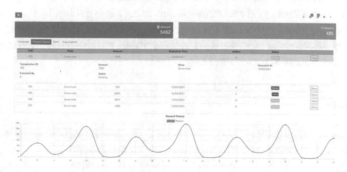

**Fig. 8.** Detail of amount of the tokens' system with pending and distributed.

communities. Energy tokens can be hence defined to support the renumeration in the communities and lead also to a conversion of the energy into communities services.

In Fig. 9, the offers, dashboard and ENEA-Marketplace are presented.

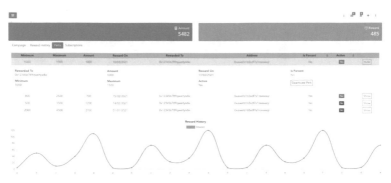

**Fig. 9.** Detail of bonus received.

## 5    Conclusions and Future Works

To conclude, blockchain can benefit energy system operations, markets and consumers. They offer disintermediation, transparency and tamper-proof transactions. Still, most importantly, blockchains provide novel solutions for empowering consumers and small renewable generators to play a more active role in the energy market and monetise their assets. This paper shows, as in a peer to peer (P2P) way, a participatory logic aimed at reducing and self-consumption of energy and direct participation citizens to the electricity system, a proof of concept is created and tested using the Trakti blockchain platform (scalable) [5]; through an end-to-end approach, a new dynamic smart legal contract with real data for the local marketplace within a smart energy community is implemented. The results show that the new smart legal contract makes it possible to define an energy token, which remunerates the energy exchanges in communities between producers and consumers within a smart grid and converts the energy into community services. In this way, a blockchain-based system for the smart energy communities is possible and realise their potential. Blockchain represents, in the meantime, the challenges and opportunities to create a smart urban environment [17].

## References

1. Energy communities [online] Energy-European Commission. Available at: https://energy.ec.europa.eu/topics/markets-and-consumers/energy-communities_en
2. Kounelis, I., Giuliani, R., Geneiatakis, D., Di Gioia, R., Karopoulos, G., Steri, G., Neisse, R., Nai Fovino, I.: Blockchain in energy communities, a proof of concept. EUR 29074 EN, Publications Office of the European Union (2017). ISBN 978-92-79-77773-8. https://doi.org/10.2760/121912, JRC110298
3. Merlinda, A., et al.: Blockchain technology in the energy sector: a systematic review of challenges and opportunities. Renew. Sustain. Energy Rev. **100**, 143–174 (2019)
4. Mannaro, K., Pinna, A., Marchesi, M.: Crypto-trading: blockchain-oriented energy market. AEIT International Annual Conference. IEEE (2017)
5. www.trakti.com/
6. Rapuano, A., Iovane, G., Chinnici, M.: A scalable Blockchain based system for super resolution images manipulation. In: 2020 IEEE 6th International Conference on Dependability

in Sensor, Cloud and Big Data Systems and Application, DependSys 2020, pp. 8–15 (2020). ISBN: 978-172817651-2

7. Iovane, G., Nappi, Chinnici, M., et al.: A novel blockchain scheme combining prime numbers and iris for encrypting coding. In: IEEE 17th International Conference on Dependable, Autonomic and Secure Computing, IEEE 17th International Conference on Pervasive Intelligence and Computing, IEEE 5th International Conference on Cloud and Big Data Computing, 4th Cyber Science and Technology Congress, DASC-PiCom-CBDCom-CyberSciTech 2019, pp. 609–618 (2019), ISBN: 978-172813024-8

8. Zhou, Q., Huang, H., Zheng, Z.: Solutions to scalability of blockchain: a survey. Gen (2020). https://doi.org/10.1109/ACCESS.2020.2967218

9. Nakamoto, S.: Bitcoin: a peer-to-peer electronic cash system. (2008)

10. Wood, G.: Ethereum: a secure decentralised generalised transaction ledger. Ethereum Project Yellow Paper **151**, 1–32 (2014)

11. Cachin, C.: Architecture of the hyperledger blockchain fabric. In: Proceedings of the 3rd Workshop on Distributed Cryptocurrencies Consensus Ledgers, vol. 310 (2016)

12. Wang, H., Zheng, Z., Xie, S., Dai, H.N., Chen, X.: Blockchain challenges and opportunities: a survey. Int. J. Web Grid Services **14**, 352 (2018)

13. Zheng, P., Zheng, Z., Luo, X., Chen, X., Liu, X.: A detailed and real-time performance monitoring framework for blockchain systems. In: Proceedings of 40th International Conference on Software Engineering Software Engineering: Practice-ICSE-SEIP, pp. 134–143 (2018)

14. *The Scalability Trilemma in Blockchain*, Sep. 2019, [online] Available: https://medium.com/@aakash_13214/the-scalability-trilemma-in-blockchain-75fb57f646df

15. Gilbert, S., Lynch, N.: Brewer's conjecture and the feasibility of consistent available partition-tolerant web services. SIGACT News **33**(2), 51 (2002)

16. Li, X., Jiang, P., Chen, T., Luo, X., Wen, Q.: A survey on the security of blockchain systems. Future Gener. Comput. Syst.

17. Chinnici, M., Telesca, L., Islam, M., Georges, J.-P.: Blockchain-Based smart energy communities: operation of smart legal contract. In: Proceeding: 24th HCI International Conference, HCI'22 Conference, Thematic area Universal Access in Human-Machine Cooperation and Intelligent Environments, LNCS 2022, pp. 324–336, vol 13520. https://doi.org/10.1007/978-3-031-18158-0_24

18. Iannis, K. et al.: Blockchain in Energy Communities, JRC Technical reports (2017)

# Blockchain Consensus Algorithms: A Survey

Pooja Khobragade[✉] and Ashok Kumar Turuk

NIT Rourkela, Rourkela, Odisha, India
520cs1004@nitrkl.ac.in, akturuk@nitrkl.ac.in

**Abstract.** Blockchain is one of the emerging technologies based on Distributed peer-to-peer networking. Blockchain gained popularity in 2009 with the launch of Bitcoin. Since, then blockchain has found applications in various domains such as finance, supply chain, health care, agriculture, pharmacy, IOT, automobile, energy, and many more. Blockchain is decentralized in nature which providing an immutable and tamper-proof ledger of transactions that includes data integrity and security. The consensus algorithm is a common agreement between block nodes to become a part or publish a new block in the blockchain. In this study, we survey different consensus algorithms reported in the literature.

**Keywords:** Blockchain · Consensus algorithm · PoW · PoS · DPoS · PoET · PBFT · PoC · PoB · PoA · VR · Raft · Paxos

## 1 Introduction

Blockchain is an emerging technology that gained popularity with its first application in Bitcoin. It is a decentralized, distributed, tamper-proof technology. The problem of the centralize system such as single point of failure and data integrity, can be overcome by Blockchain. It provides a trusted environment for participants. Peers in blockchain share information over the network [1]. Blockchain is distributed digital ledger of cryptographically signed transaction groups together to make a chain like structure called blockchain [2]. Transactions are stored in a chronological order with a time stamp assigned to each block. Transactions are continuously growing, and new blocks are added to the blockchain, with the consensus of other block nodes. According to the application scenario Blockchain is broadly classified into two categories as follows: permission-less blockchain is when participants can join the network without permission or prior authority. However in a permissioned blockchain only authorized users can join the network [3]. Blocks are created and added to the existing chain of blocks through a consensus protocol, which is agreed upon among its peers. In this work, we have surveyed different blockchain consensus algorithms reported in the literature.

© The Author(s): under exclusive license to Springer Nature Switzerland AG 2023
J. Prieto et al. (Eds.): BLOCKCHAIN 2022, LNNS 595, pp. 198–210, 2023.
https://doi.org/10.1007/978-3-031-21229-1_19

The consensus protocol can be broadly classified into two types: (i) Byzantine fault-tolerant (BFT) and (ii) Crash fault-tolerant (CFT) as shown in Fig. 1. CFT consensus builds a degree of resiliency, and they are mainly used in an environment where nodes with a certain degree of closure and credibility. It solves consistency problems such as process crashes, network failure, machine downtime, etc. The consensus is reached even if some network components does not work properly. CFT does not guarantee to provide security under malicious activity. However, the BFT consensus algorithm provides a solution to the consistency problem as well as deals with the malicious node and assures security in the network. The fault tolerance capability of nodes is higher than the CFT algorithm [4].

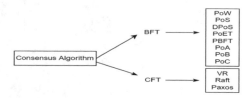

**Fig. 1.** Classification of consensus algorithm

The rest of the paper is organized as follows: Section 2 explains the need for consensus among the peers in blockchain. Section 3 describes the different consensus algorithms and analyses the working of the consensus algorithm. Section 4 concludes the paper.

## 2 Consensus Protocol

Blockchain is a distributed decentralized network. The consensus protocol is a general agreement among all nodes to synchronize the network transaction [5]. In blockchain, all nodes perform a series of operations, and it is difficult to get consistency between nodes.

Six properties of the consensus algorithm (i) **Cooperation**: Nodes should work as a whole group for the common benefit of the group, rather than individual benefit. (ii) **Collaboration**: All nodes in the system should have a shared goal and work together to achieve the same objective. (iii) **Inclusive**: Maximum number of participants from the group should be done. (iv) **Agreement seeking**: Obtain as much as agreement from the individual nodes in the system. (v) **Participatory**: Active group members participation is required to get success and (vi) **Democratic**: Each node in the system casts an equal-weighted vote. Figure 2 shows the property of the consensus algorithm.

No central authority is present in the network, but all the transactions are secure and valid because of the consensus algorithm. The primary purpose of the consensus algorithm is to provide security to the blockchain network [6].

The goal of the consensus algorithm is to reach a joint agreement in terms of a network transaction [7]. A Consensus algorithm also solves: (i) The consistency problem is that multiple nodes try to perform a series of operations. It isn't easy to obtain the results saved at each node to reach consistency. (ii) Solves Byzantine's general problem of reaching the common agreement in a distributed network with possible malicious nodes. (iii) Activeness refers to the nodes in the blockchain network that are active and participate in the consensus algorithm and provide adequate computing power to the blockchain network. (iv) Prevent the double-spending problem.

**Fig. 2.** Properties of consensus algorithm [8]

### 2.1  Goals of Consensus Algorithm

The goals of the consensus algorithms are: (i) **Termination**: For every transaction, there are only two states, accept or reject. Honest nodes either accept or reject the transaction. (ii) **Agreement**: At every new transaction, honest nodes accept or reject the transaction. If all honest node accepts transactions, it stored in the same sequence in all nodes in the blockchain network. (iii) **Validity**: Only valid transactions are accepted by the nodes that become part of the network. (iv) **Integrity**: All accepted transactions generate valid hash chains, so the nodes are consistence with each other to provide integrity to the network.

### 2.2  Components of Consensus Algorithm

There are five key components of the consensus algorithm [4]. They are: (i) **Block Proposal**: Where the validators or miners select the next proposal for a block, (ii) **Information Propagation**: The process of verifying the proposed block, which is verified by all the selected nodes, (iii) **Block validation**: Full nodes distribute all the transaction information across the whole network, (iv) **Block Finalization**: Validators reaches to common agreement either accept or reject the block. (v) **Incentive Mechanism**: Miner who follows all the rules get rewards, and those who break the rule get penalties. The above prevents malicious activity in the network.

# 3   Types of Consensus Protocol

Different consensus algorithms provide different capabilities like storage, computing power, and other configurations [9].

## 3.1   Proof-of-Work (PoW)

Proof-of-Work (PoW) was proposed by Dwork and Moni in 1993 [10]. However, it is used by "Satoshi Nakamoto" in his application for Bitcoin uses PoW consensus algorithm [11]. PoW is required to solve a complex problem. The node that can solve the problem obtains the right to add a new block into the blockchain.

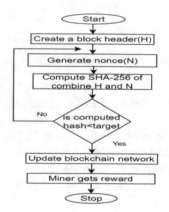

**Fig. 3.** PoW consensus process

Figure 3 shows the flowchart of the PoW consensus process. A miner compute the SHA256 of a block header which contains a fixed value and a variable value (nonce). The fixed value is computed apriori from the transaction information in all blocks. The miner obtains all rights to add a block to the blockchain network, if the computed value is less than the target value. For computed value greater than the target value, the value of nonce is changed, and the hash of the header is computed. The above process continues until the header's computed hash value is less than the target value. Solving the problem is an intensive task. Nodes adjust the nonce value and compute the hash of the header until it is less than the target value. To modify a block, an attacker must redo the block's PoW and all the blocks after it [12].

## 3.2   Proof-of-Stake (PoS)

In PoW, nodes invest their resources and computation power in solving a complex problem. PoW algorithm requires a high computation of power for mining, which leads to increased energy usage. Moreover, the transaction rate of PoW

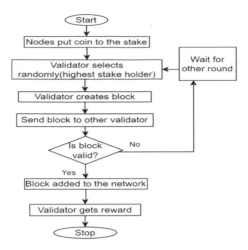

**Fig. 4.** PoS consensus process

is low. To overcome the limitations of PoW, King and Nadal proposed Proof-of-Stake (PoS). In PoS, nodes put a certain coin at stake to become a part of the validation process. The more a node has a stake, the higher the chance of becoming a validator [12]. The validator is chosen pseudo-randomly and becomes a part of the consensus algorithm [13]. A node having the highest stake can monopolize the validation process. Figure 4 shows the flowchart of the PoS consensus process.

### 3.3    Delegated Proof-of-Stake (DPoS)

Delegated Proof-of-Stake (DPoS) is more energy-efficient compared to PoW and PoS. In PoS, the highest stakeholder can control all validation processes. To overcome the chance of centralization in PoS and enhance security, DPoS is proposed. In DPoS, the validator is voted by stakeholders for producing a new block. The number of votes allocated to a participatory node depends upon the number of currencies held by the node. Nodes participating in the voting process are the decision-makers in the consensus algorithm [14]. The time limit is decided for each delegated node. If the delegated node cannot generate a block within the allocated time limit, then the delegated node is dismissed. The stakeholders choose a new delegate node, and the next round of block creation begins. Figure 5 shows the flowchart of the DPoS consensus process.

### 3.4    Practical Byzantine Fault Tolerance (PBFT)

Practical Byzantine Fault Tolerance (PBFT) was proposed by Liskov and Castro [15] to reduce the algorithm's complexity to a polynomial-time of Byzantine general problems. PBFT has different phases: pre-preparation, preparation, submission, and reply. In the pre-preparation phase, the master node sends information

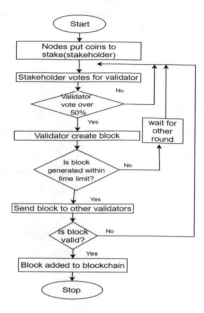

**Fig. 5.** DPoS consensus process

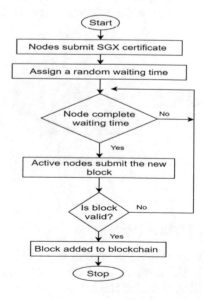

**Fig. 6.** PoET consensus process

to all the nodes. In the preparation phase, nodes receive information and send it to other nodes except themselves. In the submission phase, all nodes receive 2f + 1 information, where 2f is the number of honest nodes, and one is the master node. In the reply phase master gets a reply only from an honest node. Figure 7 shows the PBFT consensus process.

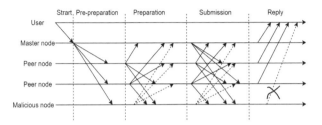

**Fig. 7.** PBFT consensus process

### 3.5 Raft

In the Raft algorithm, a node uses the log replication of other nodes to maintain a unified transaction. Nodes are divided into three categories: leader, follower, and candidate. The leader is responsible for interactive communication. Followers become voters in the voting activity, and candidates are transformed from followers and can be part of the leader selection process [16]. Figure 8 shows the Raft consensus process.

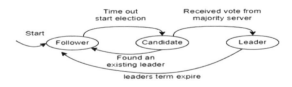

**Fig. 8.** Raft consensus algorithm

### 3.6 Proof of Elapsed Time (PoET)

The Proof-of-Elapsed-Time (PoET) is similar to PoW. It requires fewer computational resources. In PoET a separate timer is attached to each node, which allocates a random waiting time for the node. Every miner gets a random and fair waiting time and decides which miner to publish a block. The miner whose waiting time finished first gets a chance to publish the next block and broadcast it to the network. Figure 6 shows the flowchart of the PoET consensus process.

### 3.7   Proof of Activity (PoA)

Proof-of-activity (PoA) combines the PoW and PoS consensus algorithm. PoA works similar to PoW but with less complexity. In PoA, the miner solves a cryptographic puzzle-like PoW and claims for reward, then shifts to PoS [17]. The block does not contain transactions, instead contains templates that include header information and reward address. The header information is used to select the validator randomly. The more the stake value a validator has, more is its chance to selected and sign the block. Once block is signed by all validator, it become a part of the blockchain. If a chosen validator does not sign the block, then the block is discarded, and the next block with a high stake will be use. Figure 9 shows the flowchart of the PoA consensus process.

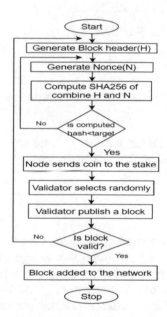

**Fig. 9.** PoA consensus process

### 3.8   Proof of Burn (PoB)

The Proof-of-burn (PoB) consensus algorithm is a mechanism to burn the coin in a verifiable manner to generate a new coin [18]. The burned coin is destroyed forever. It requires less energy than PoW because it burns its coin instead of burning energy. Miners put their coin in burn address or unspendable address. The coin in this address is not immediately destroyed, but the miner can not spend this coin anywhere else. The higher the amount spent by a miner, greater the chance of being chosen to mine a block. If the miner can publish a valid block, it gets a reward. If not, the miner only wastes the coin. The unspendable

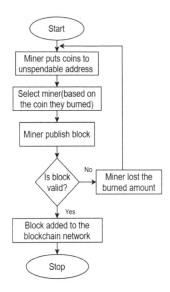

**Fig. 10.** PoB consensus process

address is the blockchain address with no private key, and burn coins are locked in this address and lost forever. Figure 10 shows the flowchart of PoB consensus process.

### 3.9  Proof of Capacity (PoC)

Proof-of-capacity (PoC) overcomes the limitation of PoW, PoS, and PoB. It requires less storage space and consumes less energy as compared to PoW, PoS, and PoB. PoC works in two phases: plotting and mining [19]. In PoC, before mining a block, the miner stores the list of possible solutions for the correct block. Solutions are stored at free disk storage. Nodes use these solutions to plot a graph. The more solutions a node stores, the higher its chance of becoming a block's next publisher .The higher capacity holder wins the consensus. After plotting miner start the mining process. Figure 11 shows the process flowchart of the PoC consensus process.

### 3.10  Viewstamped Replication (VR)

Viewstamped replication (VR) is a replication protocol rather than a consensus protocol. It performs log replication of nodes to get consistency in the network. In VR, one node is selected as a primary node, and the others are backup nodes [20]. The Primary node decides the sequence of commands. The Backup node follows the sequence and executes the command in order. If the primary node fails, the VR algorithm performs a view change operation, and selects the next primary node from the voting process. The next node becomes the primary node

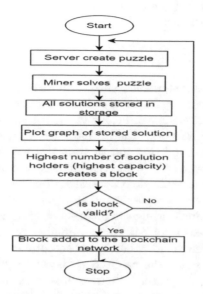

**Fig. 11.** PoC consensus process

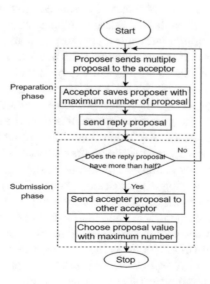

**Fig. 12.** Paxos consensus process

if it gets f + 1 votes from a total of 2f nodes. The new primary node starts a new consensus process and the old primary node goes to a recovery state and cannot perform any operation.

### 3.11  Paxos

Paxos solves network-related problem such as message delay, machine down-time, network failure, data loss etc. Nodes are divided into three categories: proposer, acceptor, and learner. A node can belong to more than one category. Proposer sent a proposal to the network. The acceptor votes on specified proposal and learner collects all the accepted proposals from the acceptor. Paxos operates in two phase: The preparation phase and the submission phase. In the preparation phase, the proposer sends multiple proposals with proposal number to acceptor. The acceptor compares all proposals and stores the proposal, and returns the maximum proposals value sent by the proposer, and discards the proposer who has proposal number less than a maximum number of proposal. In submission phase, if a proposer receives more than 1/2 of acceptance out of the total number of acceptor, proposer sends the received proposal to all other acceptors and continues the preparation phase. A consensus is reached when more than half of the recipients have the same proposal value [21]. Figure 12 shows the process flow of Paxos consensus Process.

## 4  Methodology

The following metrics are constructed for the comparison of the blockchain consensus algorithm: Throughput, Energy consumption, scalability and cost. Throughput is the maximum agreement rate to verify the transaction expressed as the number of transactions per second (TPS). In a decentralized system, if any new block is arrived there is some time gap, for all nodes to agree on it. If there is no time gap, many blocks arrive at some fraction of delay and do not get any optimization benefits, and it just becomes a chain of transactions. Maximum throughput means the maximum rate at which the blockchain confirms a transaction. Blockchain scalability is another essential factor in consensus algorithms, and higher scalability means the blockchain can achieve more transaction per second. Blockchain can achieve high throughput by modifying existing consensus algorithms or changing other parameters [3]. Energy consumption is the next important factor in the consensus algorithm.The amount of energy consume by the blockchain network is depends on it's consensus mechanism. Two factors contributing to the cost of consensus algorithm are: transaction fees and transaction verification time. Nodes pay transaction fees to the miner to verify their transaction. Not all cryptocurrencies require transaction fees. Blockchain uses some methods to eliminate transaction fees. Table 1 shows a comparison of different consensus algorithms.

Consensus is a set of rules to reach a decision in a decentralized environment. Without consensus, the blockchain is reduced to a platform for storing encrypted

**Table 1.** Comparative analysis of consensus algorithm

| Blockchain type | Consensus algorithm | Energy consumption | Throughput (transaction per second) | Scalability | Cost | Example of representative blockchain |
|---|---|---|---|---|---|---|
| Permission less | PoW | High | Low | Low | High | Bitcoin, Ethereum |
| | PoS | Medium | Low | Medium | Medium | Peercoin, Ethereum |
| | DPoS | Medium | High | Medium | Low | Bitshares, Steem and Steemit |
| | PoA | High | High | Low | High | Decred, Espers |
| | PoB | Low | Medium | Low | Medium | Bitcoin, Slimcoin |
| | PoC | Low | High | Medium | High | Burstcoin |
| Permissioned | PBFT | Low | High | Low | Low | Hyperledger Fabric, Zilliqa |
| | PoET | Low | Medium | High | High | Hyperledger, Sawtooth |
| | VR | Low | High | Low | Low | – |
| | Paxos | Low | Medium | Low | High | etcd, Kubernetes |
| | Raft | Low | High | High | Medium | Quorum etcd, Kubernetes |

data. From the Table 1, we can conclude that energy consumption is always low for permissioned blockchain. Permissionless blockchain chains mostly faced scalability issues. The blockchain cost depends upon the transaction fee and transaction storage. If energy consumption and storage demand are high, the cost is also high.

## 5 Conclusion

Consensus is one of the vital technology in blockchain. Over the period, many consensus algorithms have been developed to solve security, scalability, energy consumption, and fault tolerance. The consensus algorithms have their limitation in terms of throughput. Improvement in the consensus algorithm can significantly improve the overall performance of the blockchain network. The consensus algorithm applies on blockchain depending on the network type and application scenario. This paper summarizes the popular blockchain consensus algorithm. Choosing a suitable consensus algorithm is essential for making an efficient blockchain network.

## References

1. Siva Sankar, L., Sindhu, M., Sethumadhavan, M.: Survey of consensus protocols on blockchain applications. In: 2017 4th International Conference on Advanced Computing and Communication Systems (ICACCS), pp. 1–5. IEEE (2017)
2. Yaga, D., Mell, P., Roby, N., Scarfone, K.: Blockchain technology overview (2019). arXiv preprint arXiv:1906.11078
3. Sanka, A.I., Irfan, M., Huang, I., Cheung, R.C.C.: A survey of breakthrough in blockchain technology: adoptions, applications, challenges and future research. Comput. Commun. **169**, 179–201 (2021)

4. Xiao, Y., Zhang, N., Lou, W., Thomas Hou, Y.: A survey of distributed consensus protocols for blockchain networks. IEEE Commun. Surv. Tutorials **22**(2), 1432–1465 (2020)
5. Gu, W., Li, J., Tang, Z.: A survey on consensus mechanisms for blockchain technology. In: 2021 International Conference on Artificial Intelligence, Big Data and Algorithms (CAIBDA), pp. 46–49. IEEE (2021)
6. Nguyen, G.-T., Kim, K.: A survey about consensus algorithms used in blockchain. J. Inf. Process. Syst. **14**(1), 101–128 (2018)
7. Xiao, Y., Zhang, N., Li, J., Lou, W., Hou, T.: Distributed consensus protocols and algorithms. Blockchain Distrib. Syst. Secur. **25**, 40 (2019)
8. Jayabalan, J., Jeyanthi, N.: A study on distributed consensus protocols and algorithms: the backbone of blockchain networks. In: 2021 International Conference on Computer Communication and Informatics (ICCCI), pp. 1–10. IEEE (2021)
9. Sadek Ferdous, Md., Chowdhury, M.J.M., Hoque, M.A., Colman, A.: Blockchain consensus algorithms: a survey (2020). arXiv preprint arXiv:2001.07091
10. Yadav, A.K., Singh, K.: Comparative analysis of consensus algorithms of blockchain technology. In: Ambient Communications and Computer Systems, pp. 205–218. Springer, Berlin (2020)
11. Nakamoto, S.: Bitcoin: a peer-to-peer electronic cash system. Decentralized Bus. Rev. 21260 (2008)
12. Rama, R.K.: Overview of blockchain technology: consensus algorithms, applications. https://www.researchgate.net/publication/350727529_Overview_of_Blockchain_Technology_Consensus_Algorithms_Applications. https://doi.org/10.13140/RG.2.2.17450.75205
13. Zhang, S., Lee, J.-H.: Analysis of the main consensus protocols of blockchain. ICT Express **6**(2), 93–97 (2020)
14. Yang, F., Zhou, W., Wu, Q.Q., Long, R., Xiong, N.N., Zhou, M.: Delegated proof of stake with downgrade: a secure and efficient blockchain consensus algorithm with downgrade mechanism. IEEE Access **7**, 118541–118555 (2019)
15. Castro, M., Liskov, B., et al.: Practical byzantine fault tolerance. In: OsDI, vol. 99, pp. 173–186 (1999)
16. Wu, Y., Song, P., Wang, F.: Hybrid consensus algorithm optimization: a mathematical method based on POS and PBFT and its application in blockchain. Math. Probl. Eng. **2020** (2020)
17. Bentov, I., Lee, C., Mizrahi, A., Rosenfeld, M.: Proof of activity: extending bitcoin's proof of work via proof of stake [extended abstract] y. ACM SIGMETRICS Perform. Eval. Rev. **42**(3), 34–37 (2014)
18. Karantias, K., Kiayias, A., Zindros, D.: Proof-of-burn. In: International Conference on Financial Cryptography and Data Security, pp. 523–540. Springer, Berlin (2020)
19. Aggarwal, S., Kumar, N.: Cryptographic consensus mechanisms. In: Advances in Computers, vol. 121, pp. 211–226. Elsevier, Amsterdam (2021)
20. Oki, B.M., Liskov, B.H.: Viewstamped replication: a new primary copy method to support highly-available distributed systems. In: Proceedings of the Seventh Annual ACM Symposium on Principles of Distributed Computing, pp. 8–17 (1988)
21. Xiong, H., Chen, M., Canghai, W., Zhao, Y., Yi, W.: Research on progress of blockchain consensus algorithm: a review on recent progress of blockchain consensus algorithms. Future Internet **14**(2), 47 (2022)

# Decentralised Argumentation for Data Vetting in Blockchains

Subhasis Thakur[✉][iD] and John Breslin[iD]

National University of Ireland, Galway, Ireland
subhasis.thakur@nuigalway.ie, john.breslin@nuigalway.ie

**Abstract.** Although it is difficult to overwrite the data kept in blockchains, there are numerous incidents of false data insertion in blockchains. Current blockchain technologies can not prevent such false insertion into blockchains. In this paper, we present a blockchain model that can prevent such immutable lies. We have several contributions in this paper: we developed a decentralised argumentation protocol that allows auditors to decide the validity of a claim, we developed an incentive system for the auditors not to withheld any evidence for or against a claim, we developed methods to execute the decentralised argumentation protocol in blockchain offline channels for high scale execution of the proposed data vetting method. We prove that the proposed data vetting method executed in the blockchain offline channel network is correct.

**Keywords:** Data vetting · False data · Blockchain offline channels · Data audit

## 1 Introduction

Blockchain can prevent data overwriting by acting as a decentralised database. However, such data security is often misrepresented as data correctness. There are numerous examples of false data insertion into blockchains. Supply chain data is one of the most vulnerable for such activity. It is estimated that current supermarkets have on average 33 thousand items including items procedure from 2400 km away.[1] Such an FSC network usually consists of numerous actors. Blockchain can prevent data overwriting but it cannot prevent the inclusion of incorrect information. There are frequent incidents of product mislabelling in supply chains. Studies have revealed a product mislabelling rate of 78% in meat products [1,12,14] as these products contains unspecified meat from unknown sources. A study revealed that 16 out of 23 companies have mislabelled products which include traces of donkey meat and GMO organism [8]. The research found chicken and duck were present in red meats and red meat in chicken food items [5]. There are mislabelled organic products [9], wrong third-party certification [13]. Often the mislabelling is caused by products procured outside the EU [4].

---

[1] https://www.fmi.org/our-research/supermarket-facts.

J. Prieto et al. (Eds.): BLOCKCHAIN 2022, LNNS 595, pp. 211–224, 2023.
https://doi.org/10.1007/978-3-031-21229-1_20

Data vetting for supply chains by a centralised entity such as a supermarket chains may not be trustworthy.

In this paper, we propose a data vetting method for blockchains. We used multi-agent argumentation to develop the data vetting method. Argumentation allows autonomous agents to engage in a turn-taking protocol to prove or disprove a statement. We used IPFS blockchain as a data store for arguments which are the evidence for or against a claim. Our solution allows a group of auditors to engage in argumentation as they access this argument data store and argument attack relation distributed Hash table. We have developed payment schemes for the auditors as they get paid for their auditing work in fair and there is economic incentive to prevent malicious behaviour of the auditors. We implement the decentralised argumentation protocol for data vetting in a blockchain offline channel environment. Our main results are as follows: (1) We developed a decentralised argumentation protocol for data vetting with economic incentives for the auditors to deter them from data withholding or biased argumentation. (2) We have implemented the decentralised argumentation protocol in a blockchain offline channel environment. This allows high scale execution of the proposed data vetting method. The paper is organised as follows: in Sect. 2 we discuss related literature, in Sect. 3 we present the decentralised argumentation method for data vetting, and we conclude the paper in Sect. 4.

## 2    Related Literature

In [6] argumentation framework was proposed. It included various semantics of valid sets of arguments. Bench-Capon [2] extended Dung's argumentation framework to a value-based argumentation framework. Leite et al. [11] proposed method to merge multiple argumentation frameworks for a combined decision-making platform. Panisson and Bordini [15] have explored computational models of argumentation in agent-oriented programming languages. Rahwan and Larson [17] explored Pareto optimality in argumentation frameworks. Kakas and Moraitis [10] proposed decision-making methods using multi-agent argumentation. Roth et al. [18] provided game-theoretic modeling of strategic argumentation. Dziuda [7] explored strategic argumentation methods where an expert withheld arguments. IPFS blockchains was proposed in [3]. We advance the state of the art in argumentation theory and data vetting in the following directions: (1) We have developed a decentralised argumentation protocol that improves current state-of-the-art centralised argumentation protocols. (2) We have developed economic incentives for the auditors to prevent them data withholding and biased argumentation. The state-of-the-art argumentation protocols lack such economic incentives. (3) We have developed a high scale and secure execution model for such decentralised argumentation in a blockchain offline channel environment.

# 3   Decentralised Argumentation for Data Vetting

An overview of the proposed data vetting solution is shown in Fig. 1. It is as follows: (1) There is an offline channel network created from a proof of work or state based public blockchains. Our solution will also work with any permissioned blockchain network where Hashed Timed Locked contracts or similar scripts can be executed. (2) The auditors and the principles (who issue data vetting jobs) are part of this blockchain offline channel network. These entities are also participant of a peer to peer data network. (3) The peer to peer data network will host the argument graph data. (4) Argument graph data will be stored in this peer to peer data network by a verified set of data contributors. The principles (who issue the data vetting jobs) will verify the identity of such data contributors. (5) Data contributors will create labelled graph data with various parameters of an application scenario. For example, in case of a perishable supply chain, such parameters include RFID data, weather data, IoT data on products and trucks moving the products, etc. The data contributors will create argument graphs (by identifying the attack relations) and upload it to thee peer to peer data network.

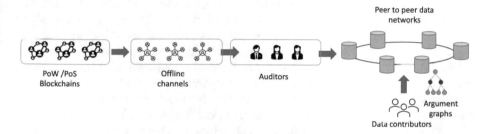

**Fig. 1.** Overview

    In this proposed data vetting solution, our scope for this paper are as follows: (1) We present a formulation of multi-agent argumentation framework to produce argument graphs to be used for data vetting. (2) We present a decentralised argumentation protocol for data vetting using the blockchain offline channel network. (3) We present an incentive model for the auditors (who are vetting the data using this decentralised argumentation protocol).

    We assume the following: (1) We assume that the data contributors are either honest or there exists incentives to encourage them to create correct argument graph. (2) We assume that the peer to peer data network can efficiently store and retrieve graph data. (3) In future, we will develop additional solutions to overcome these assumptions.

## 3.1   Argumentation

Argumentation Framework (AF) allows a set of autonomous agents to make decisions from a set of conflicting data or evidence. An AF is represented as a tuple $< A, \succ >$ where $A$ is a set of arguments and $\succ$ is a total order among the arguments. An argument is an abstract information that may be represented as data or evidence. $A_1 \succ A_2$ ($A_1, A_2 \in A$) means argument $A_1$ defeats argument $A_2$. Given a set of arguments, we can use Dung's argumentation semantics to define the subset of arguments that are valid.

**Definition 1.** *An argument $A_1 \in A$ is acceptable with respect to a subset of arguments $E \subseteq A$ if $E$ defends $A_1$, i.e., for any $A_2 \in A$ if $A_2 \succ A_1$ then there exists $A_3 \in E$ such that $A_3 \succ A_2$. A set of arguments is conflict free if it do not contain a pair of arguments $A_1, A_2$ such that $A_1 \succ A_2$. A set of arguments $E \subseteq A$ is admissible if and only if $E$ is conflict free and all arguments in $E$ are acceptable with respect to $E$.*

We will use graphs to represent admissible arguments. Given an argumentation framework $(A, \succ)$ and an argument $a^* \in A$ (will be referred to as the root argument), we will sequentially form the admissible set of arguments that includes $a^*$ or prove that such set does not exist. We will sequentially form trees with root as $a^*$. $G^0 = (a^*, \emptyset)$ will represent the root argument tree with only one argument $a^* \in A$ as the vertex and there is no edge in this graph. A sequence of augmented graphs from $G^0$ will represent the sequential formation of an admissible set of arguments that includes $a^*$ or represent the failure to do so. $G^1 = (a^* \cup A^1, \succ^1)$ where $A^1 \in A$ is set of arguments that attacks $a^*$ and $\succ^1$ represents edges for such attack relations. $G^2 = (a^* \cup A^1 \cup A^2, \succ^1 \cup \succ^2)$ will represent the augmented argument graph from $G^1$ such that $A^2 \in A$ attacks any argument in $A^1$ and $\succ^2 \in \succ$ represents such attack relations. $G^3 = (a^* \cup A^1 \cup A^2 \cup A^3, \succ^1 \cup \succ^2 \cup \succ^3)$ will represent the augmented argument graph from $G^2$ such that $A^3 \in A$ attacks any argument in $a^* \cup A^2$ and $\succ^3 \in \succ$ represents such attack relations. Hence the argument graph to be created during an even (odd) numbered step can produce arguments to attack arguments presented by graph augmentation during odd-numbered previous steps. In any augmented graph $G^i = (a^* \cup A^1 \cup \cdots \cup A^{i-1}, \succ^1 \cup \succ^2 \cdots \cup \succ)$, $a^*$ is part of an admissible set of arguments if there is no leaf node the argument graph at an odd-numbered distance from the root node.

We will use this notion of a sequence of argument graphs to represent the process of decentralised data vetting where each auditor will construct an argument graph to argue for or against the validity of the root argument.

## 3.2   IPFS as Argument Store

IPFS works as a decentralised database where a peer can access content using the name or identification information of the content rather than the location of the content. IPFS is built on a peer-to-peer network where distributed hash tables are used to store location information of contents. Content will be stored at multiple peers of the network and all peers maintain and synchronize its Hash tables as the location information of the contents. In order to add new content to the IPFS, a peer will generate a unique ID of the content and update its local hash table. Next, it informs its neighbors in the peer-to-peer network

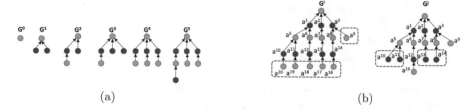

(a)                                                (b)

**Fig. 2.** (a) Sequence of argument graphs. Note that the graph $G^5$ is not admissible, (b) in the argument graph $G^i$ (left) 'all winners' payment scheme can be applied and auditors who provided the leaf nodes will get paid evenly. In the argument graph $G^j$, 'strict winner' payment scheme can be applied as the auditors who provided $a^{10}, a^{11}, a^{13}, a^{14}$ will be rewarded. In $G^j$ 'chosen strict winner' payment scheme can be used as both 'all winners' and 'strict winners' payment scheme are applicable.

about the updated hash tables who will update their hash tables and inform their respective neighbors. This process continues until all local hash tables are synchronized. If a peer wants to access the new content then it will look into the hash table and access the content from the peer who is storing it. Upon accessing it, the peer will have the option to keep a local copy of the content if the peer is acting as a host. Hence multiple copies of the document are stored across the peer-to-peer network. The content file can be split into multiple small files and the content identification information may have links to these small chunks of the content. As the content is stored in multiple locations, it is difficult to censor and overwrite content kept in IPFS.

### 3.3   Decentralised Argumentation

In the previous section, we describe a sequence of argument graph creation. We will present a protocol for decentralised argumentation protocol for the data vetting which uses such a sequence of argument graphs. It is as follows: Let there are $n$ auditors $(V_1, \ldots, V_n)$ who want to check the validity of data kept in a blockchain. $A$ be the set of all arguments and $\succ$ be the attack relations among the arguments. The sets $A, \succ$ will be kept in an IPFS blockchain as mentioned before. The auditors are arranged in a total order $>$ and they will argue about the validity of an argument in that order. The argumentation protocol is as follows:

1. The auditors will argue about the validity of an argument $a^* \in A$. An argument will be regarded as a valid argument if it is part of an admissible subset of arguments $S \subseteq A$.
2. The auditors will gradually construct the admissible set or fail to do as each of them may construct a new argument graph. The auditors will take turns to create the argument graph using the ordering $>$ among the auditors. Let $>$ is such that $V_1 > V_2 > \cdots > V_n$.

3. $V_1$ will have the first opportunity to create the first argument graph $G^1$ from $G^0 = (a^*, \emptyset)$. $V_1$ will have a fixed finite time to create such an argument graph. If $V_1$ creates $G^1$ (as shown in Fig. 2(a)) then, it will inform all other auditors. If $V_1$ fails to do so then $V_2$ will have to opportunity to create $G^1$ and so on.
4. This process will continue until either the argument graph reaches a critical height or a fixed time duration has passed.
5. Outcome of this protocol is proof of the validity of the root argument as the existence of an argument graph that is admissible and it is not a subgraph of any other argument graph.

The auditors will be paid for their data vetting service. We will evaluate the following payment schemes:
(1) All winners: In this payment scheme auditors who have created any of the leaf nodes of the final argument graph will get paid evenly. (2) Strict winners: In this payment scheme auditors who have created the leaf node(s) of the final argument graph which has an odd-numbered distance from the root node will get paid evenly. (3) Chosen strict winners: In this payment scheme, the payment scheme 'strict winners' will be chosen over 'all winners' in an argument graph where both payment schemes can be applied.

The 'all winner' payment scheme rewards any auditor who has developed an argument that can not be proved wrong. This payment scheme can be used for both cases where there is an admissible argument graph and where there is no admissible argument graph that includes $a^*$ (in this case, there may be leaf nodes both at an odd and even-numbered distance from $a^*$). Let $\delta$ be the total reward for vetting data. It will be distributed evenly among the winners according to these payment schemes.

Let $\Delta$ be the total reward for vetting the data with root argument $a^*$. The amount of reward will gradually increase as the argument graph is built by the auditors. Given an argument graph $G^i$ with maximum distance $p$ from the root $a^*$, the total reward will be $\delta^p$ ($p$ and $\delta$ are positive integers and $p, \delta > 1$). $\delta^p$ will be evenly distributed among all auditors who have constructed at least one leaf node of $G^i$. If the auditing for the argument $a^*$ is permitted to be executed for $p$ iterations then $\delta^p \leq \Delta$.

We assume that there is a finite cost to create arguments. Hence auditors may strategically reveal their arguments to minimise their expense. We assume that all auditors are equally likely to construct an argument, find an appropriate argument. Any auditor will choose one of the following strategy when its turn to present arguments:
(1) Reveal: It may choose to create a new set of arguments by constructing a query to the IPFS data store for arguments and by doing so it will incur a certain cost. This strategy will be useful for an auditor if it is not a winner according to the payment scheme and latest argument graph. (2) Withhold: It may choose not to create a new set of arguments to avoid the cost of querying the IPFS data store. It can do so if it is currently getting paid as per the most recent

argument graph and payment scheme. (3) Biased: It may choose either to attack or support the root argument.

**Lemma 1.** *Any auditor is better off with the strategy to reveal arguments for all payment schemes if $\delta > d^p$ where total reward at iteration $p$ is $\delta^p$ and $d$ is the average degree of the argument graph.*

*Proof.* Consider the argument graph formation shown in Fig. 3(c). Let the argument graph $(G^{p-1})$ till iteration $p-1$ is the most recent and the argument graph $(G^p)$ for the iteration $p$ can be constructed from it. Assume that an auditor is a winner at the current argument graph $G^{p-1}$ however it has an argument to create $G^p$ (at least subset of $G^p$ which is a strict superset of $G^{p-1}$). The maximum payoff of the auditor for $G^{p-1}$ is $\delta^{p-1}$ (in the best-case scenario, the auditor has constructed all arguments in the leaf nodes of $G^{p-1}$) and the minimum payoff of the same auditor for $G^p$ is $\delta^p/d^p$ (in the worst-case scenario, exactly one auditor has constructed argument corresponding to one leaf node of $G^p$) where $d$ is the average degree of the entire argument graph stored in the IPFS data store. Hence, the auditor will always reveal arguments if the following holds:

$$\delta^{p-1} < \frac{\delta^p}{d^p}, \frac{1}{\delta} < \frac{1}{d^p}, \delta > d^p. \tag{1}$$

This condition is sufficient for the auditors to always reveal for all payment schemes. Case 1: If the 'all winner' payment scheme is chosen then the auditor will get at least $\delta^p/d^p$. Case 2: If 'strict winner' or 'chosen strict winner' scheme is chosen and the auditor can augment $G^{p-1}$ then it will always do so as it will get at least $\delta^p/d^p$.

**Lemma 2.** *An auditor is better off by not choosing 'bias' strategy.*

*Proof.* Let the 'bias' is against the root argument. In this case, (a) if the maximally augmented argument graph does not correspond to an admissible set then the arguments against the root argument may pay off. However, any non-biased auditor may also present the same argument against the root argument. Thus 'bias' strategy does not bring more payoff than the 'non-bias' strategy. (b) if the maximally augmented argument graph does correspond to an admissible set then, the 'bias' strategy will have less payoff than the non-bias strategy.

Let the 'bias' is for the root argument. In this case, (a) if the maximally augmented argument graph does not correspond to an admissible set then the arguments for the root argument will never pay off. (b) if the maximally augmented argument graph does correspond to an admissible set then the arguments for the root argument will bring the same payoff with a non-biased strategy.

**Theorem 1.** *For any argument, the admissible set of arguments will be formed or the failure to do so will be proved.*

*Proof.* According to Lemmas 1 and 2, an argument graph will always be formed at every iteration as it brings more payoff to the auditor who augments the argument graph. Hence the maximal augmented argument graph will be formed irrespective of whether it corresponds to an admissible set of arguments or proof of failure to form an admissible set of arguments.

## 3.4    Blockchain Implementation

In this section we will discuss blockchain implementation of the above-mentioned decentralised argumentation protocol. We will use blockchain offline channels to implement the decentralised argumentation protocol for data vetting. Blockchain offline channel allows two peers to securely transfer tokens without updating the blockchain. We will use the protocol developed in [16]. Briefly, it is as follows (illustrated in Fig. 3(a)):

1. Peers $A$ and $B$ will exchange two sets of Hashes of random strings.
2. $A$ sends the Hash $K_A^1$ to $B$ and $B$ sends the Hash $K_B^1$ to $A$.
3. $A$ will send a Hashed-time locked contract $(HTLC_A)$ to $B$ as follows:
   (a) From the multi-signature address $M_{AB}$ 1 token will be given to $A$ and 1 token will be given to another multi-signature address $M'_{AB}$.
   (b) From the multi-signature address $M'_{AB}$ 1 token will be given to $B$ after time $T$ unless $A$ claims these tokens by revealing the key to hash $K_B^1$.
4. $B$ will send a Hashed-time locked contract $(HTLC_B)$ to $A$ as follows:
   (a) From the multi-signature address $M_{AB}$ 1 token will be given to $B$ and 1 token will be given to another multi-signature address $M'_{AB}$.
   (b) From the multi-signature address $M'_{AB}$ 1 token will be given to $A$ after time $T$ unless $B$ claims these tokens by revealing the key to hash $K_A^1$.
5. After exchanging the HTLCs, $A$ and $B$ will fund the multi-signature address $M_{AB}$ by one token. $A$ ($B$) will include $H_A^1, \ldots$ ($H_B^1, \ldots$) in the transaction funding $M_{AB}$.
6. Next, $A$ and $B$ can change the share of the fund in $M_{AB}$ by exchanging new HTLCs. However, they must reveal the key to $K_A^1$ and $K_B^1$.

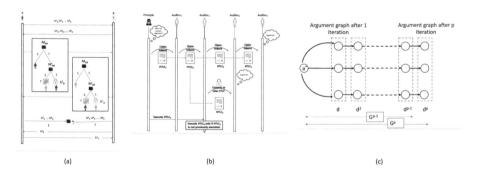

(a)                    (b)                    (c)

**Fig. 3.** (a) Protocol to create offline channel between two peers, (b) explanation of the decentralised argumentation in offline channels, (c) evaluation of strategy of the auditors.

We will use an offline channel network created using this protocol to implement the decentralised argumentation protocol. All actors of the data vetting problem, auditors, principle (who wants to audit) are part of an offline channel network. We will use the following protocol to implement the decentralised argumentation protocol. It is as follows (shown in Fig. 4):

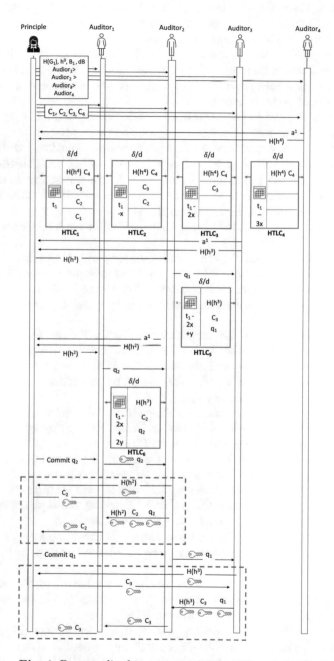

**Fig. 4.** Decentralised argumentation in offline channels

1. The principle will inform all auditors about the graph data stored in the IPFS data sore as Hash of such a graph $(H(G_1))$, the start time of vetting $B_1$, time limit of an auditor $dB$ to construct an argument graph, order among the auditors, and the current argument graph $h^0$.
2. Argument graph $h^0$ can be augmented by finding attacks on the leaf nodes. If $h^0$ has $p$ leaf nodes then all such leaf nodes can be attacked at most $d$ times where $d$ is the average degree of the graph $G_1$. The protocol for augmentation of $h^0$ for a leaf node $a^1$ is as follows:
   (a) The principle will allocate reward $\delta/d$ for every argument $a^1$ which can be attacked at this iteration of argument graph formation.
   (b) Let there are 4 auditors $Auditor_1$, $Auditor_2$, $Auditor_3$, and $Auditor_4$. The principle will send 4 random Hashes $C_1, C_2, C_3, C_4$ to the auditors. Next, it will ask the auditor to augment $h^0$ for the leaf node $a^1$.
   (c) Let $Auditor_4$ first finds an argument to attack $a^1$. It will inform principle by sending Hash of the argument $H(h^4)$.
   (d) Upon receiving the $H(h^4)$ from $Auditor_4$, will form a sequence of HTLCs among all auditors.
   (e) The HTLC between $Auditor_4$ and $Auditor_3$ will state that:
      i. $Auditor_3$ can take $\delta/d$ tokens from the multi-signature address between $Auditor_3$ and $Auditor_4$ after time $t_1 - 3x$ if:
      ii. $Auditor_4$ does not claim these tokens by providing keys to $H(h^4)$ and $C_4$.
   (f) The HTLC between $Auditor_3$ and $Auditor_2$ will state that:
      i. $Auditor_2$ can take $\delta/d$ tokens from the multi-signature address between $Auditor_2$ and $Auditor_3$ after time $t_1 - 2x$ if:
      ii. $Auditor_3$ does not claim these tokens by providing keys to $H(h^4)$ and $C_4$ or by providing key to $c_3$.
   (g) The HTLC between $Auditor_2$ and $Auditor_1$ will state that:
      i. $Auditor_1$ can take $\delta/d$ tokens from the multi-signature address between $Auditor_2$ and $Auditor_1$ after time $t_1 - x$ if:
      ii. $Auditor_2$ does not claim these tokens by providing keys to $H(h^4)$ and $C_4$ or by providing key to $c_3$ or by providing key to $c_2$.
   (h) The HTLC between $Auditor_1$ and $Principle$ will state that:
      i. $Principle$ can take $\delta/d$ tokens from the multi-signature address between $Auditor_1$ and $Principle$ after time $t_1$ if:
      ii. $Auditor_1$ does not claim these tokens by providing keys to $H(h^4)$ and $C_4$ or by providing key to $c_3$ or by providing key to $c_2$ or by providing key to $c_1$.
   (i) Now, it may happen that, other auditors may have found an attack relation with $a^1$. Let $Auditor_3$ finds such a relation. It will inform the Principle about Hash of such argument $H(h^3)$.
   (j) Principle will inform $Auditor_2$ about $H(h^3)$ and a new HTLC between ($HTLC_5$) $Auditor_2$ and $Auditor_3$ will be formed. $Auditor_2$ will send a Hash $q_1$ to $Auditor_3$ as commitment to the new HTLC.
   (k) $HTLC_5$ will state that $Auditor_2$ will take $\delta/d$ tokens from the multi-signature address between $Auditor_2$ and $Auditor_3$ after time $t_1 - 2x + y$ if $Auditor_3$ does not claim it by revealing key $H(h^3)$, $q_1$ and $C_3$.

(l) Similarly $Auditor_2$ may find the argument to attack $a^1$ and corresponding new HTLC between $Auditor_1$ and $Auditor_2$ will be formed.

3. Now execution of these sequence of HTLCs will be guided by the principle as follows:

   (a) In case, all auditors except $Auditor_4$ fails to provide an argument to augment $h^0$ then the principle will send key to $C_4$ to $Auditor_4$ which it will use to claim $\delta/d$ tokens from the multi-signature address between $Auditor_4$ and $Auditor_3$. Same keys will be used by $Auditor_3$, $Auditor_2$, $Auditor_1$ to claim $\delta/d$ tokens as $HTLC_4$, $HTLC_3$, $HTLC_2$, $HTLC_1$ will be sequentially executed.

   (b) In case, other auditors provide arguments to augment $h^0$, the principle will choose auditors according to the announced order among the auditors to reveal their arguments against the argument $a^1$.

   (c) Example: the principle will ask $Auditor_2$ to reveal its argument $H(h^2)$. If $h^2$ is a correct augmentation of $h^0$, i.e., there exists a subgraph isomorphism from $h^0 \cup h^2$ to $G^1$ then it will reveal key to $C_2$ to $Auditor_2$.

   (d) The principle will ask $Auditor_1$ to reveal 'commitment' $q_2$ (key to the Hash $q_2$) to $Auditor_2$. $Auditor_1$ will do so if $HTLC_1$ has not been executed before otherwise, it will lose tokens.

   (e) After $Auditor_1$ reveals key to $q_2$ to $Auditor_2$, $Auditor_2$ will reveal $C_2$ and $h^2$ to $Auditor_1$ and claim $\delta/d$ tokens by the $HTLC_6$. $Auditor_1$ will use these keys to claim $\delta/d$ tokens from $HTLC_1$.

   (f) If $h^2$ is not a valid augmentation of $h^0$ (either $h^2$ has been revealed before by another auditor or there is no subgraph isomorphism from $h^0 \cup h^2$ to $G^1$) then the principle will choose $Auditor_3$ to reveal its argument.

   (g) This process will continue until either a successful HTLC execution occurs or attempts is made to execute all HTLCs.

4. The above procedure to argue against the leaf node $a^1$ will be repeated $d$ times as at most $d$ arguments can be built against $a^1$.

5. After augmenting the argument graph $h^0$ for all leaf nodes, the argument data graph kept in the IPFS blockchain can be updated. In such a case, the principle will inform all auditors about the new argument data graph in the IPFS blockchain. The next iteration of the protocol will use such an updated argument data graph.

An explanation of the above described decentralised argumentation protocol is as follows:

1. The principle will ask all auditors to argue against the argument corresponding to a leaf node of the most recent argument graph. An auditor can argue irrespective of the order among auditors.

2. $Auditor_4$ first finds an argument. The principle guides formation of a sequence of HTLCs to transfer funds from itself to $Auditor_4$.

3. Now, $Auditor_3$ finds an argument and it informs the principle.

4. The principle asks $Auditor_2$ to create a new HTLC with $Auditor_3$ which forks from $HTLC_2$.

5. A malicious principle may allow $Auditor_4$ and $Auditor_3$ to claim tokens as two sequence of HTLCs will be execute: $(HTLC_4 \rightarrow HTLC_3 \rightarrow HTLC_2 \rightarrow HTLC_1)$ and $(HTLC_5 \rightarrow HTLC_2 \rightarrow HTLC_1)$. But it is not possible to execute the sub-sequence of HTLCs $HTLC_2 \rightarrow HTLC_1$ twice. Hence $Auditor_2$ will loose fund as $Auditor_3$ will claim fund twice from $Auditor_2$ and $Auditor_2$ will only claim fund once from $Auditor_1$.

6. To prevent this problem, $Auditor_2$ places a commitment data into $HTLC_5$. It allows execution of $HTLC_5$ only if $Auditor_2$ allows it. Thus $Auditor_2$ will not lose any fund due to double-spending.

**Definition 2.** *The decentralised argumentation protocol is correct if the following holds: (1) An auditor will not lose tokens by participating in the HTLC-based argumentation protocol. (2) Two auditors will not get paid for the same argument. (3) If two auditors have produced the argument then the auditor with the highest priority according to the order among the auditors will get paid.*

**Theorem 2.** *The decentralised argumentation protocol proposed in this paper is correct.*

*Proof.* An auditor will not lose tokens by participating in the HTLC-based argumentation protocol because: (1) The set of HTLCs formed to augment the argument graph $h^0$ for the leaf node $a^1$ will form a tree structure. (2) The principle will only reveal the key to one of the Hashes $C_1, C_2, C_3, C_4$ as it has only assigned $\delta/d$ tokens to be transferred via these sequences of HTLCs. Hence HTLCs will be executed according to only one path among the sequence of HTLCs (shown in Fig. 3(b)).

Two auditors will not get paid for the same argument because: (1) As shown in Fig. 3(b), two auditors may get paid if there is a fork from the first sequence of HTLCs. For example, $HTLC_5$ is forked from $HTLC_2$. (2) In such instance of fork, the common auditor ($Auditor_2$ as shown in this figure.) will prevent execution of both HTLC sequences $(HTLC_4 \rightarrow HTLC_3 \rightarrow HTLC_2 \rightarrow HTLC_1)$ and $(HTLC_5 \rightarrow HTLC_2 \rightarrow HTLC_1)$ by not revealing $q_1$ to $Auditor_3$ if $Auditor_4$ has claimed tokens.

If two auditors have produced the argument then the auditor with the highest priority according to the order among the auditors will get paid, this is because the time constraint in various HTLCs are used according to the order among the auditors. As shown in Fig. 4, time constraints of HTLCs conform to the order among auditors.

## 4    Conclusion

In this paper, we proposed a data vetting method using multi-agent argumentation. We presented a blockchain offline channel-based execution model of such a data vetting method. In the future, we will improve IPFS blockchains for faster retrieval of argument data.

**Acknowledgement.** This publication has emanated from research conducted with the financial support of Science Foundation Ireland (SFI) under Grant Number SFI/12/RC/2289, co-funded by the European Regional Development Fund.

# References

1. Ayaz, Y., Ayaz, N., Erol, I.: Detection of species in meat and meat products using enzyme-linked immunosorbent assay. J. Muscle Foods **17**(2), 214–220 (2006)
2. Bench-Capon, T.J.M.: Value based argumentation frameworks. CoRR cs.AI/0207059 (2002). https://arxiv.org/abs/cs/0207059
3. Benet, J.: IPFS—content addressed, versioned, P2P file system. CoRR abs/1407.3561 (2014). https://arxiv.org/abs/1407.3561
4. Charlebois, S., Sterling, B., Haratifar, S., Naing, S.K.: Comparison of global food traceability regulations and requirements. Compr. Rev. Food Sci. Food Saf. **13**(5), 1104–1123 (2014)
5. Chuah, L.O., He, X.B., Effarizah, M.E., Syahariza, Z.A., Shamila-Syuhada, A.K., Rusul, G.: Mislabelling of beef and poultry products sold in Malaysia. Food Control **62**, 157–164 (2016). https://doi.org/10.1016/j.foodcont.2015.10.030
6. Dung, P.M.: On the acceptability of arguments and its fundamental role in non-monotonic reasoning, logic programming and n-person games. Artif. Intell. **77**(2), 321–357 (1995)
7. Dziuda, W.: Strategic argumentation. J. Econ. Theor. **146**(4), 1362–1397 (2011). https://doi.org/10.1016/j.jet.2011.05.017
8. Grammenos, A., Paramithiotis, S., Drosinos, E.H., Trafialek, J.: Labeling accuracy and detection of DNA sequences originating from GMOs in meat products commercially available in Greece. LWT **137**, 110420 (2021). https://doi.org/10.1016/j.lwt.2020.110420
9. van Hilten, M., Ongena, G., Ravesteijn, P.: Blockchain for organic food traceability: case studies on drivers and challenges. Front. Blockchain **3**, 43 (2020). https://doi.org/10.3389/fbloc.2020.567175
10. Kakas, A., Moraitis, P.: Argumentation based decision making for autonomous agents. In: Proceedings of the Second International Joint Conference on Autonomous Agents and Multiagent Systems, AAMAS '03, pp. 883–890. Association for Computing Machinery, New York, NY, USA (2003). https://doi.org/10.1145/860575.860717
11. Leite, L., Alves, T., Alcântara, J.: Merging argumentation systems. In: 2015 Brazilian Conference on Intelligent Systems (BRACIS), pp. 110–115 (2015). https://doi.org/10.1109/BRACIS.2015.45
12. Litscher, G., Cai, Y., Li, X., Lv, R., Yang, J., Li, J., He, Y., Pan, L.: Quantitative analysis of pork and chicken products by droplet digital PCR. BioMed Res. Int. **2014**, 810209 (2014). https://doi.org/10.1155/2014/810209
13. Munteanu, A.R.: The third party certification system for organic products. Netw. Intell. Stud. (6), 145–151 (2015). https://ideas.repec.org/a/cmj/networ/y2015i5p145-151.html
14. Naaum, A.M., Shehata, H.R., Chen, S., Li, J., Tabujara, N., Awmack, D., Lutze-Wallace, C., Hanner, R.: Complementary molecular methods detect undeclared species in sausage products at retail markets in Canada. Food Control **84**, 339–344 (2018). https://doi.org/10.1016/j.foodcont.2017.07.040

15. Panisson, A.R., Bordini, R.H.: Towards a computational model of argumenta-
    tion schemes in agent-oriented programming languages. In: 2020 IEEE/WIC/ACM
    International Joint Conference on Web Intelligence and Intelligent Agent Technol-
    ogy (WI-IAT), pp. 9–16 (2020). https://doi.org/10.1109/WIIAT50758.2020.00007
16. Poon, J., Dryja, T.: The bitcoin lightning network: scalable off-chain instant pay-
    ments (2016)
17. Rahwan, I., Larson, K.: Pareto optimality in abstract argumentation. In: Fox, D.,
    Gomes, C.P. (eds.) Proceedings of the Twenty-Third AAAI Conference on Artificial
    Intelligence, AAAI 2008, Chicago, Illinois, USA, July 13–17, 2008, pp. 150–155.
    AAAI Press (2008). https://www.aaai.org/Library/AAAI/2008/aaai08-024.php
18. Roth, B., Riveret, R., Rotolo, A., Governatori, G.: Strategic argumentation: a game
    theoretical investigation. In: Proceedings of the 11th International Conference on
    Artificial Intelligence and Law, ICAIL '07, pp. 81–90. Association for Computing
    Machinery, New York, NY, USA (2007). https://doi.org/10.1145/1276318.1276333

# Designing the Chain of Custody Process for Blockchain-Based Digital Evidences

Pablo Santamaría[1], Llanos Tobarra[2(✉)] ⓘ, Rafael Pastor-Vargas[2]ⓘ,
and Antonio Robles-Gómez[2]ⓘ

[1] ETSI Informática, Universidad Nacional de Educación a Distancia (UNED),
Madrid, Spain
psantamar4@alumno.uned.es

[2] Control and Communications System Department, ETSI Informática, Universidad
Nacional de Educación a Distancia (UNED), Madrid, Spain
{llanos,rpastor,arobles}@scc.uned.es

**Abstract.** The design of a chain-of-custody management system must be based on an architecture that guarantees non-manipulation in court, from the very beginning of the evidence is acquired to putting it on judge hands. Knowing, without doubt, who, when, how and why the involved actors were able to access it. This is a necessary condition, although not sufficient, for the court to decide on the admissibility of real the evidence. This way, the Blockchain technology is an optimal solution for issues where integrity must be guaranteed among untrustworthy parties. However, it is necessary to tackle a possible solution in a holistic way, by considering technologies, actors and industry involved during the process. Therefore, three basic architectural designs with different challenges are proposed in this work. In addition to this, we discuss about the need to promote a nationwide smart contract standardization for the chain-of-custody process. It must be open-source and compatible with current top languages used in the Blockchain landscape. These contracts must be deployed in consortium environments, where reliable independent third parties validate the transactions without knowing their specific content. This is possible thanks to the Zero Knowledge Proof protocols.

**Keywords:** Blockchain applications · Chain-of-Custody (CoC) · Platforms for smart contracts · Zero-Knowledge Proof (ZKP)

Authors would like to acknowledge the support of the I4Labs UNED research group, the CiberGID UNED innovation group with the CiberScratch 2.0 project, the SUMA-CITeL research project for the 2022–2023 period, as well as the E-Madrid-CM Network of Excellence (S2018/TCS-4307). The authors also acknowledge the support of SNOLA, officially recognized Thematic Network of Excellence (RED2018-102725-T) by the Spanish Ministry of Science, Innovation and Universities.

J. Prieto et al. (Eds.): BLOCKCHAIN 2022, LNNS 595, pp. 225–236, 2023.
https://doi.org/10.1007/978-3-031-21229-1_21

# 1   Introduction

In today's society, the development of information technology and telecommunications allows people to have available a variety of digital devices. This fact, together with the exponential growth of crimes in which computer and telecommunications technologies are either the protagonists or are present, has brought the concept of digital evidence [18] and its chain of custody to the forefront. However, the volatile nature of digital evidence and improper handling procedures have caused the management of chain of custody is becoming complicated and complex.

According to UNE 71506:2013 [9], the Chain-of-Custody (CoC) is defined as *"as the controlled traceability procedure applied to the evidence, from acquisition to analysis and final presentation"*. This procedure is focused on performing the documentation to the evidence in chronological events [1,10]. So, the chain of custody is the most crucial process of evidence documentation. It must assure that, in event of a crime presented to court of law, the evidence is authentic. In other words, it is the same evidence seized for the crime scene. It should reflect all individuals involved in the process of acquisition, collection, analysis of evidence, time records, as well as contextual information (case identification, unit and laboratory participants).

According to this, the Blockchain technology is been employed in a large range of security applications, in order to store and protect data in a chain structure, as a distributed ledger solution. Encryption also plays a key issue to keep immutable data by digital signatures. This allows a Blockchain architecture to handle digital evidences in a secure and transparent way for forensic investigation.

From our point of view, the use of Blockchain technology can help us to improve the CoC process and the evidence management life-cycle. Blockchain can be seen as a distributed ledger technology with several benefits, such as advancing sustainability, automation and digital transformation. It also offers built-in security control access for information and trace all changes throughout the stored data during its period of life. Blockchain is an emerging technology, which is growing every day, especially, in the supply of chain.

Thus, the main objective of this work is the design and discussion of a general model for the CoC process based on the Blockchain technology that overcomes the challenges derived of handling digital evidences to efficiently transform the law in practice. In particular, a set of architectures for a CoC solution are presented here. As they have in common the employment of safe storage, the paper will also describe the differences in access to the Blockchain network. The paper is illustrated with a proof of concept. This PoC focuses on the system's security properties regarding traceability of actions over the evidences and their integrity within a non-trusted context. The implementation of an associated smart contract with Quorum (based on Ethereum) is included in our CoC solution.

This paper is organized as follows. First, a review of existing proposals is detailed in Sect. 2. Additionally, Sect. 3 describes the different proposed architectures for the CoC process: *distributed*, *central*, and *multi-blockchain*. Section 4 discusses the first results obtained in this work. Finally, Sect. 5 presents our initial conclusions and further works.

## 2    Related Work

From the beginning of Blockchain development, the CoC problem has been observed as a classic example to be solved with its intrinsic characteristics [13, 16]. The most specific studies on the use of Blockchain as a support for the digital custody process date from 2011 [10]. This research line has been growing up to with related concept and approaches.

Bonomi et al. proposed the B-CoC [7] framework aimed at guaranteeing the audit integrity of digital evidence collection, and its traceability by third parties. However, multiple stakeholders have not been considered nor complemented by an adequate evidence management system. As main contribution, this work lays out the main components of a CoC system.

The work of [15] presents a CoC architecture in the context of Blockchain-based forensics. However, it does not offer a complete vision of the transactions nor put special attention on the preservation of evidences or the integrity guarantee mechanisms avoiding any modification. This work presents an evolution of a previous work [14].

In [2], the creation of a Blockchain-based incident response system is combined for easing a subsequent forensic analysis. They propose managing the evidence in a distributed ledger and facilitating the chronological storage of the associated documentation. However, it is not based on the fundamental aspect of guaranteeing all the necessary properties for a CoC process.

The LEChain [12] framework is one of the more complete proposals found in the literature. It presents a Blockchain-based evidence management model that includes mechanisms for the supervision of evidence collection and the protection of the identities of witnesses and jurors (and their specific votes). It even addresses the management of evidence in the cloud, although it does not offer full-support for transparency and audit mechanisms. Additionally, it does not cover the full-life cycle of digital evidence.

On the other hand, there are existing commercial applications, such as Khipus [6]. This app offers a personal evidence custody service over the Alastria [3] network. For instance, with WhatsApp or email. According to their website, they are a service for the issuance of electronic stamps that are not qualified in time. The purpose of this solution differs from the principal objective of this work, since it intends to act as a custody and third party service, but it is an example of the possibility that it can offer society. In 2017, Alastria was established as the Spanish Blockchain consortium, with more than 511 adhered members (public administrations and companies), and supporting the Quorum technology.

At a technological level, an evolution of Hyperledger, which is interoperable with current Ethereum smart contracts, is postulated for the future as the base technology provider. It is also possible that this technology will be offered to consortiums in service mode through large cloud providers. For example, Google or AWS already offer Hyperledger networks in service mode based on Hyperledger Composer. It is only a matter of time before they offer new versions of Hyperledger and other technologies that emerge in the future.

After reviewing previous works, we think there are still several needs that are not covered by them, which have been addressed below.

# 3    Our Proposed Solutions

## 3.1    The General Model

The investigation process of a crime relies heavily on physical and digital evidence. Judicial system has to include digital evidences with a handling mechanism to treat them as the same way as a physical evidence. As the digital forensics field is growing, it requires digital forensic experts with proven skills to capture the crime scene, call data records, search collected records, recover hidden or lost data, and engage with the forensics process.

The most important problems that investigators found related to digital nature of digital information are:

- Digital proofs are easily duplicated or reproduced.
- Guarantee the integrity of the evidence, since the alteration or the modification with new data or removal of important information could be crucial.
- Guarantee a registry of the accessibility to the evidence, so it is documented clearly how and by whom the evidence is treated/managed, and what level of access control has been granted.
- Secure storage of the evidence.
- Depending on the case, evidences can be time-sensitive.
- It is possible that the evidence can be transmitted, since the case can cross countries and different legal jurisdictions.

Thus, it is clear that handling the CoC of digital evidence is more complex that handling physical evidence in general. Figure 1 shows the general process for the management of evidences inside the CoC process. The chain of custody begins in the evidence acquisition stage. Once it is collected, the evidence creation process must be instantiated in the chain of custody. In the analysis and reporting phases, the process of access to the evidence by participants is invoked. Finally, in case destruction is requested, it will be necessary to invoke the evidence destruction method.

Participant means any natural person who contribute in one way or another to the CoC process. It is essential the participant is registered in the chain of custody. This circumstance is regulated. The roles of the participants are basically divided into the creator of the evidence (collects evidence from the scene and registers it into the framework); the owner of the evidence (manage the access to the evidence and guarantee the custody of evidence copies); and the participants with the capacity to examine the evidence (mainly forensic analysts, judicial stuff or state security forces). Additionally, taking into account

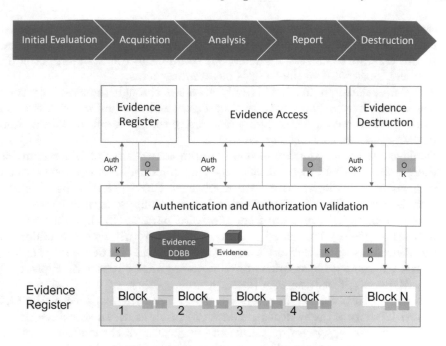

**Fig. 1.** CoC proposed general model

the roles linked with Blockchain management, it is necessary the role of "master" of the CoC, who is in charge of deploying the contract and the maintenance of participants during the whole process.

To sum up, the CoC process implies the definition of the services of evidence registration, its invalidation, its access, transference of its ownership, as well as the log of the activities related to a particular evidence.

## 3.2 Design Principles

Due to its inherent characteristics, Blockchain technology is optimal for problems where integrity among parties that do not trust each other must be guaranteed, such as the process of CoC of digital evidence from a court. This technology will ease for parties to validate the traceability of every circumstance of the trial without the need to know the specific content and by preserving the privacy and confidentiality of the judicial process.

Our solution takes into account the following design principles:

– A smart contract that implements the logic of a secure, simple and standard process of CoC. An example of smart contract can be found in [4]. Programming languages, such as Chaincode (Hyperledger) or Solidity (Ethereum), are the most suitable for this task. But, this is a delicate issue, because it is prone to vulnerabilities as it is reported in [5,17]. Ideally, these smart contracts should be open-source, standardized where they get the endorsement and scrutiny of the entire community, and the acceptance by the competent authorities of being an accredited smart contract to generate a reliable CoC. In this case, we should always adapt our solutions to this standard, even if it does not meet 100% of the entity's own requirements.

– The data related to the CoC is only accessed through the smart contract. The architecture in these networks are layers and interactions with the contract, but it should never maintain or manage the chain information without invoking a smart contract function.

– Deployment of the contract in one or more consortiums of Blockchain. So, there is an external independent entity that could validate the transactions.

– Permissive technologies with Zero-Knowledge Proof (ZKP) consensus support is needed. That is the participation of public or private consortiums that guarantee independence from the third parties. On the other hand, a licensed technology is essential to allow, if necessary, privacy and confidentiality without losing the network's consensus on the validity of the transactions. Currently, compatible solutions for companies must support ZKP protocols over Hyperledger or Ethereum (Corda), among other.

– Custody of the evidence in a separated way from the evidence in the CoC record. The size of the evidence makes it unfeasible to use the Blockchain itself as a secure custody repository. But, the fingerprint of the evidence is stored in the Blockchain registry. This way, any tampering with the evidence would be detected. As a consequence, the resulting architecture is more complex.

– Robust identity management and authentication method. The NIST Guide Digital Identity Guidelines 800-636 [11] is the main reference for this property.

– The use of safe, reliable, durable and low-cost warehousing. The evidence must last as long as it is involved in an open judicial period. Thus, evidence of even terabytes must be maintained and guaranteed to be analyzed even years after its acquisition.

– Electronic evidence format following an open standard. Most evidence management guidelines and standards offer suggestions but no clear standard for storing evidence. Advanced Forensic File Format 4 (AFF4) [8] and the format followed by the EnCase tool tend to be very popular formats. Since there is no consensus, we will focus more on the process of generating the trace than on imposing a standard evidence format.

– Periodic accreditation of external security audits.

## 3.3    Architectures

The final choice of a specific architecture must be determined by each entity, or the service provider, by assessing the most suitable one for their strategy. It works both for an entity that desires to deploy its own CoC solution and for providers offering the service to clients that choose to abstract from the costs or complexities of their own deployment. As observed in Figs. 2, 3, and 4, the three proposed architectures (*distributed, centralized,* and *multi-blockchain*) have in common several elements to be detailed next.

This way, the *Safe Storage* of evidences is protected, and only accessible to authorized CoC participants, as well as user tracking is considered for monitoring. Another element is an *Identity Node* per participant, and the associated *Smart Contract* for the CoC process, within a *Blockchain Consortium* (Alastria, R3 ...) to guarantee the integrity and immutability of each evidence. An *Identity Manager* for authentication and authorization purposes is also included. The *Gateway Safe Storage* and *User Interface* is for additional controls and interactions in the CoC application, being connected to the *Browser* and *Wallet*. There is a *Contract Development and Debugging* environment and, additionally, a *Crypto Service* (Key Management System, KMS) for a secure communication channel among the CoC participants and Identity Nodes.

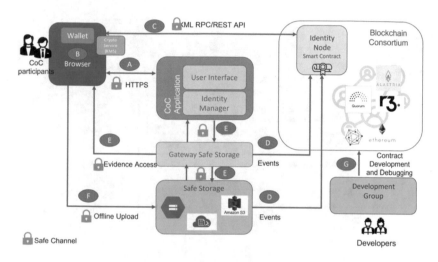

**Fig. 2.** Distributed proposed architecture

The three options contemplated for the implementation of the CoC process are detailed next:

– *Distributed architecture.* This proposed Blockchain-based architecture inherits from public networks (see Fig. 2). Clients access to the network through a wallet app that stores the public cryptographic key-pair. The network stores the public key of the account and verifies the transactions signed from that wallet. Connections to the network depend on the protocol used (XML-RPC in Ethereum networks; and REST API in the case of Hyperledger). This client must interact with the browser in order to integrate network data into a distributed application.
– *Centralized architecture.* This architecture is service-oriented (see Fig. 3). Access to the Blockchain network is shielded with a layer of services or microservices that access the Blockchain network by delegation. In this case, it is necessary to implement a robust authentication and identity management

system for the Blockchain network. Once authenticated, the private key can be retrieved on the server side (from a KMS) to be able to sign the network transactions. However, any third party with a permissioned account could interact with the contract and the network to verify the chain of custody independently.

– *Multi-blockchain centralized architecture*. It is a variation of the centralized solution by abstracting the network technology used (see Fig. 4). The advantage is the redundancy of the chain of custody in different networks. It is an optimal solution, for example, to provide a CoC service to third parties. This redundancy in different networks provides a higher degree of reliability and validation.

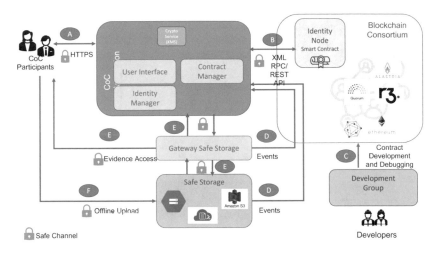

**Fig. 3.** Central proposed architecture

The multi-blockchain architecture derives from the three architectures do not allow self-manipulation, since all the information and traces of the CoC process reside in the Blockchain network. It is not manipulable unless the active agreement of many of the consortium nodes.

The multi-blockchain architecture is the most oriented approach to being a CoC service provider. However, it is applicable to institutions that want redundancy at the Blockchain and Smart Contract network level to avoid problems associated with having only one service provider. The difference with the distributed architecture is the need to separate the connection services layer with the different *Smart Contracts* and *Identity Managers* and the corresponding nodes from the application, as observed in Fig. 4.

The last difference would be, in the case of providing services to several entities, the need for the application and the entire infrastructure to be multi-tenant. That is, the possibility of instantiating different entities, whose information and access is kept isolated between them on the same infrastructure.

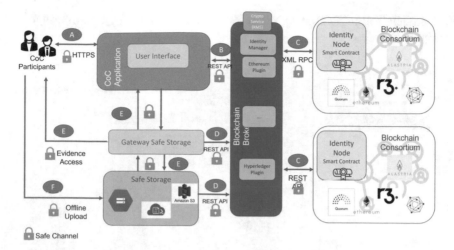

**Fig. 4.** Multi-blockchain proposed architecture

# 4    Results and Discussion

Each of the proposed architectures presented above has been implemented using different available technologies, which have been tested from a prototyping point of view. From our experience, we observe the need to promote smart contracts for a CoC process, which are open, secure, and robust, as our case is. They must also be endorsed by the scientific community and adopted by the competent authorities.

The proposed distributed architecture is optimal for mobile devices and native applications, which will be employed for our proof of concept. Browser extensions are compromise solutions, but it is demanding a prior knowledge of the user, and a browser compatibility. Therefore, In the case of implementing the CoC with a distributed mode, it is advisable to develop a native client for the corresponding operating system, where the smart contract is deployed.

A proof of concept has been carried out under Quorum, as it was the technology chosen by Alastria when developing this work, as well as advancing one of the defined design principles of implementing solutions covered by a consortium. Quorum is a version of Ethereum oriented to business environments with the privacy features not available in the parent version. Quorum includes a completed permissions and privacy system based on ZKP. It is possible to limit who participates in transactions, who can write on the chain, which nodes can know the content of the transactions, and which cannot, among others.

A code example of the implemented Smart Contract is given in Fig. 5, which has been implemented in the Remix environment with the Solidity programming language. It is a high-level programming language that generates compiled code in Bytes code to run on Ethereum virtual machines. The source-code file is https://unedo365-my.sharepoint.com/:u:/g/personal/arobles_scc_uned_es/

EcU_4nQQCVdIueDM4UsVQOIBjH-ZIEdouWInv8S-J3yxXA?e=rCgPb8
*ChainOfCustody.sol.*

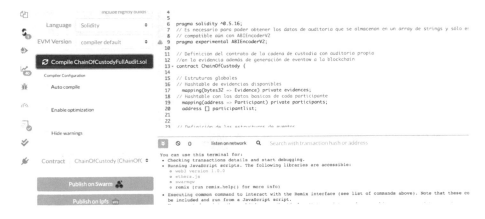

**Fig. 5.** Example of a smart contract implemented in Quorum

In addition to this, the main functionalities of the implemented Smart Contract are the following ones:

- The contract is defined in such a way that it can be used for multiple entities as well as for a single entity. The recommendation is that each entity uses its own smart contract, but it is open to being used by multiple entities that will not be able to consult evidence that is not theirs.
- There is access control both for the evidence, its specified data, and the updating of the properties only for authorized participants.
- There are two main entities on which the methods are focused, the evidence participant type. The first one stores the information about the evidence and its acquisition document, and the second contains the information of the participants in the CoC process.
- All evidences modifications of the entities are audited, and events are issued to the Blockchain network that becomes transactions included in the blocks.
- Access to information (if the account is authorized) leaves no trace on the Blockchain network, but it could be audited by the distributed application.

On the other hand, it is necessary to include the non-functional requirements of reliability, as error and exception management, as well as security issues from the conceptualization stage. The attack surface is large and, in the CoC architecture, it is even more noticeable to reduce this surface at the minimum level.

Finally, it should be emphasized that the management of evidence in secure warehouses should contemplate the process of migration of offline data, and/or even, with specific utilities from these warehouses. This is very advisable, since the gateway (via HTTP protocol) is not the optimal solution for large volumes of information.

# 5    Conclusions and Further Works

The main conclusion of this paper is that the current Blockchain technology can be seen as a viable solution to implement a chain-of-custody process with all the guarantees of integrity and reliability features. Thus, there is a great opportunity to standardize smart contracts for this CoC process, with open-source code, as well as endorsed by both scientific community and competent authorities. This will also satisfy the required security features. It should not be forgotten the need to deploy these smart contracts with a consortium. The CoC process requires a holistic approach, not only at the technological level of process, but also with participants. Even though, Blockchain is a technology that fits perfectly CoC, it would not provide greater assurance if it is not accompanied by a robust process of evidence and participant management.

This way, participation as a node in a network of independent actors that seal selfless and independent transactions brings greater value. Otherwise, the scenario would be not different than using other database technology, where record the chain of custody. The reason is that the use of Blockchain autonomously does not eliminate the risk of intentional self-manipulation of the same. This is only achieved by being part of a consortium.

Another challenge of Blockchain, as its current lack of interoperability, is related to the security of smart contracts and the possibility of interacting directly with external entities. There are advances in the use of gateways, called "Oracles", but there is not yet a final solution. This is one of the reasons where standardization and scrutiny public is very important.

# References

1. Adam, I., Varol, C.: Intelligence in digital forensics process. In: Proceedings of the 2020 8th International Symposium on Digital Forensics and Security (ISDFS), pp. 1–6 (2020). https://doi.org/10.1109/ISDFS49300.2020.9116442
2. Al-Khateeb, H., Epiphaniou, G., Daly, H.: Blockchain for Modern Digital Forensics: The Chain-of-Custody as a Distributed Ledger, pp. 149–168. Springer International Publishing (2019). https://doi.org/10.1007/978-3-030-11289-9_7
3. Alastria, C.: Alastrial (2019). Available at: https://www.alastria.io/. Accessed on 27th April
4. Anne, V.P.K., Ayyadevara, R.C., Potta, D., Ankem, N.: Storing and securing the digital evidence in the process of digital forensics through blockchain technology. In: Proceedings of the International Conference on Data Science, Machine Learning and Artificial Intelligence, pp. 272-276. DSMLAI '21', Association for Computing Machinery, New York, NY, USA (2021). https://doi.org/10.1145/3484824.3484899
5. Arroyo, D., Rezola, A., Hernández, L.: Principales problemas de seguridad en los smartcontractsde ethereum. In: XII Jornadas STIC CCN-CERT. CCN-CERT (2019). https://www.youtube.com/watch?v=r3HruAORpz0
6. ioBUILDERS Blockchain Tech & Ventures: Khipus, deja huella con tu móvil (2018). Available at: https://khipus.io/. Accessed on 27th April
7. Bonomi, S., Casini, M., Ciccotelli, C.: B-CoC: a blockchain-based chain of custody for evidences management in digital forensics. CoRR abs/**1807.10359** (2018). http://arxiv.org/abs/1807.10359

8. Cohen, M., Garfinkel, S., Schatz, B.: Extending the advanced forensic format to accommodate multiple data sources, logical evidence, arbitrary information and forensic workflow. Digit. Invest. **6**, S57–S68 (2009)
9. Española, U.N.: Information Technologies (IT). Methodology for the digital evidences forensic analysis. Standard, UNE. Normalización Española (2013). https://www.une.org/encuentra-tu-norma/busca-tu-norma/norma?c=N0051414
10. Giova, G.: Improving chain of custody in forensic investigation of electronic digital systems. Int. J. Comput. Sci. Netw. Secur. **1** (2011)
11. Grassi, P.A., Garcia, M.E., Fenton, J.L.: NIST special publication 800-63. Digital identity guidelines (2017). Available at: https://pages.nist.gov/800-63-3/. Accessed on 27th April
12. Li, M., Lal, C., Conti, M., Hu, D.: LEChain: a blockchain-based lawful evidence management scheme for digital forensics. Future Gener. Comput. Syst. **115**, 406–420 (2021)
13. Li, X., Jiang, P., Chen, T., Luo, X., Wen, Q.: A survey on the security of blockchain systems. Future Gener. Comput. Syst. **107**, 841–853 (2020)
14. Lone, A.: Forensic-chain: Ethereum blockchain based digital forensics chain of custody. Sci. Pract. Cyber Secur. J. **1** (2017)
15. Lone, A.H., Mir, R.N.: Forensic-chain: blockchain based digital forensics chain of custody with PoC in hyperledger composer. Digit. Invest. **28**, 44–55 (2019)
16. Lusetti, M., Salsi, L., Dallatana, A.: A blockchain based solution for the custody of digital files in forensic medicine. Forensic Sci. Int.: Digit. Invest. **35**, 301017 (2020)
17. Manning, A.: Solidity security: comprehensive list of known attack vectors and common anti-patterns (2018). Available at: https://blog.sigmaprime.io/solidity-security.html. (Accessed 27th April)
18. Stringer-Calvert, D.W.: Digital evidence. Commun. ACM **45**(4), 128 (2002). https://doi-org.bibliotecauned.idm.oclc.org/10.1145/505248.505280

# The Devil Hides in the Model: Reviewing Blockchain and BFT Protocols

Antoine Durand[✉] and Gérard Memmi

LTCI, Télécom Paris, Institut polytechnique de Paris, Paris, France
antoine.durand@telecom-paris.fr, antoine@durand.pw,
gerard.memmi@telecom-paris.fr

**Abstract.** Recent advances in blockchains and Byzantine Fault Tolerant protocols have been numerous and varied in nature. However, making a fair and consistent comparison of existing protocols is a difficult task that must begin right at the execution model. In this work, we undergo a review of several prominent blockchain protocols including their models, *i.e.*, network synchrony, cryptographic assumptions, corruption, as well as latency and communication cost figures. This review illustrates the issues that can arise due to the lack of standardization on blockchain terminology. For example, we show that in two prominent blockchain protocols, seemingly minor technical details in the formulation leads to the execution model being strictly different from the intended one.

**Keywords:** Blockchain · Distributed ledger technology · Benchmark · Complexity

## 1 Introduction

In recent years, blockchains, and distributed ledger technologies in general, experienced a surge of interest, spawning an impressive number of research directions and concurrent works. This proliferation somehow contrasts with the widely recognized difficulty of designing and proving secure distributed cryptographic algorithms. Thorough evaluations of fault-tolerant Byzantine protocols such as blockchains are therefore becoming more complex. As a result, there is currently no recognized evaluation procedure for such protocols, which makes formal comparisons challenging; this also partially explains the lack of standardization and the increased difficulty of designing advanced blockchain applications, particularly when multiple protocols are involved.

In this paper, we propose an evaluation strategy that is centered around the execution model of each blockchain. We selected a collection of prominent and published blockchain protocols. Our contribution is that we interpreted their execution models through a common formulation supporting a fair comparison of their various assumptions, primitives, and, by extension, their performance. This approach allowed us define multiple common variables to carry out a detailed asymptotic evaluation of communication complexity and latency. As a result we

J. Prieto et al. (Eds.): BLOCKCHAIN 2022, LNNS 595, pp. 237–248, 2023.
https://doi.org/10.1007/978-3-031-21229-1_22

are able to offer an assessment summary in Sect. 4 that is clear, rigorous, and comprehensive without being dependent on any one type of protocol.

We chose the following protocols that are representative of different techniques and working principles of blockchains, however, we could only consider sufficiently formalized protocols to enable a fair assessment of their model and properties. Nakamoto agreement [11] is certainly one of the most prominent, and we believe its analysis applies to similar protocols such as Ethereum. We also chose Phantom [17] in order to have a second PoW-based protocol to compare with the Nakamoto agreement. Ouroboros Praos [4] is a non-PoW-based synchronous protocol, and Algorand [12] is somewhat similar to Ouroboros but in a partially synchronous setting. Tendermint [1] operates similarly to classical state machine replication protocols [5], and HoneyBadgerBFT [15] is asynchronous.

Existing benchmarks in published literature focus on qualitative comparisons and experimental performance evaluations [2]. In light of this, the closest work to ours is a taxonomy by Garay and Kiayias [10], for which we make a more enhanced contribution in this work with the inclusion of newer protocols, but most notably, the addition of a generic, multivariate performance analysis.

*Outline.* This paper is organized as follows. Section 2 is a concise introduction to the notations used to capture all of the protocol models. We analyze each protocol in Sect. 3, and discuss the resulting comparison in Sect. 4. Finally, we conclude in Sect. 5.

## 2    Model

To enable a generic discussion of the protocol's models and properties, we adopt the notations from Durand [7]. They are briefly summarized below.

*Background* The distributed system contains a set of nodes $\Pi$, as well as modules which serves as abstract representations of the components present in the system. A module $m$ provides an interface to interact with nodes and also describes its behaviour through this interface; concretely, $m$ is represented by its input and output domains and a validity predicate on sequences of inputs/outputs. Given a list of ideal primitives $M$, expressed as modules, the notation $P \in \mathcal{P}[M]$ means that $P$ is a protocol that may interact with the modules in $M$. Such a protocol, $P$, is represented as a state machine run by each (honest) node and annotated with inputs/outputs. The set of executions of $P$ that are conforming to its model are noted $\mathcal{M}_P^{\mathcal{A}}$, where $\mathcal{A}$ is a parameter that specifies the adversary's capabilities. $\mathcal{A}$ contains three components to indicate (1) whether the adversary is computationally bounded, (2) whether node corruption is dynamic, weakly dynamic, or static, and (3) the corruption structure [13] that must be respected. More specifically, a corruption structure $\mathcal{C}$ contains all sets of nodes that may be corrupted. This generalization of corruption thresholds is well suited for models that leverage Proof-of-X for Sybil resistance.

The notation $\mathcal{M}_P^A \models^\lambda S$ indicates that protocol $P$ satisfies specification $S$ with probability $1 - O(2^{-\lambda})$, where $\lambda$ is the security parameter, and the specification $S$ is also represented by a module.

*Specification* The selected protocols slightly vary in the formulation of their specification. Thus, we chose Atomic Broadcast (ABC) [6] as a common specification based on the state machine replication paradigm. An ABC protocol is a protocol where nodes may input and output values from a message set $M$, such that,

- *ABC-Consistency*: For any honest nodes $p_1$ and $p_2$, for any $k \in \mathbb{N}$ such that $p_1$ and $p_2$ both made at least $k$ outputs, then $p_1$ and $p_2$' $k$-th outputs equal.
- *ABC-Liveness*: All honest players' inputs are eventually output by all honest nodes.

*Primitives* All models include a point-to-point network module, however there are three possible variants depending on synchronicity level, namely:

- "sync_net": Any sent message $m$ takes at most $\delta$ time to be received, and $\Delta \geq \delta$ is given as a parameter to the nodes.
- "weak_net": Any sent message $m$ takes at most $\delta$ time to be received, but $\Delta$ is unknown to the nodes.
- "async_net": There is no bound on the message delays.

The models include a few cryptographic primitives, also modeled as modules. We do not give explicit module definitions here and refer instead to standard cryptographic definitions [14]. These are: random oracle "RO", digital signatures "signatures" (including a public key infrastructure), public parameters "params", and Verifiable Random Functions "VRF".

The performance evaluation is a multivariate asymptotic evaluation of communication cost and latency. Communication cost ($\mathcal{CC}$) is the number of bits sent by honest nodes, and latency ($\mathcal{L}$) is the time elapsed between submission of a message an its delivery by all honest nodes. The common input variables of the asymptotic analysis are:

- $\Delta$ is the known upper bound on network delays, for "sync_net" networks.
- $\delta$ is the unknown upper bound on network delays, for "weak_net" networks.
- $n$ is the number of nodes.
- $b$ is the amortized size of the protocol output payload.
- $\text{poly}_v(\lambda)$ is a function polynomial in the security parameter $\lambda$, and arbitrary in $v$.

It's important to note that the above definitions only require the presence of an ABC protocol and a network. Thus these variables can be evaluated and analyzed for any blockchain protocol.

To simultaneously model corruption for Proof-of-Work (PoW), Proof-of-Stake (PoS), and classical BFT protocols, we use weighted corruption thresholds. Given the nodes' weights $W : \Pi \mapsto [0,1]$ such that $\sum_{p \in \Pi} W(p) = 1$ and a threshold $t \in [0,1]$; then the adversary respects the corruption structure $\mathcal{C}_{W,t}$ iff at all times during the execution, the set of corrupt nodes $C \subseteq \Pi$ is such that $\sum_{p \in C} W(p) < t$.

# 3    Protocol Analysis

## 3.1    Using the Bitcoin Backbone Protocol

For Nakamoto agreement[1] and Ouroboros Praos, we rely on the backbone protocol formalism from Garay et al. [4,11]. They present two analyses, one with a synchronous network and one extended to a partially synchronous network, the latter being similarly defined in Ouroboros Praos [4]. They describe a network offering a global round clock to all nodes. Then, sent messages are delivered after up to $\Delta$ rounds, meaning that such rounds do not have a prescribed duration and act more as "real time step" or "time slot".

However, we believe there is a slight issue in the formulation used. First of all, it's important to note that both protocols are stated to be secure assuming that less than *half* of the stake/hashpower is owned by malicious nodes. This is surprising since consensus cannot be solved in partially synchronous networks with more than a third of the honest nodes. On the other hand, if there is an honest majority, a strongly synchronous network is required to solve consensus.

Moreover, the only difference between partial and strong synchrony is that the $\Delta$ bound is made available to nodes in the strongly synchronous case. Therefore, with careful examination, we observe that both Nakamoto and Ouroboros Praos actually requires the value $\Delta$ to be known to execute correctly, and as a result they could be better described as synchronous protocols. Essentially, for both cases, the protocols can be proven secure assuming they are fed with the correct parameters, but if nodes are able to output transactions with a known probability of error, *i.e.*, if they *assume* the protocol is secure, then they are also able to compute $\Delta$.

For Ouroboros Praos, the protocol takes a parameter $f$ that tunes the probability for a node to be eligible to multicast the current block. To know whether a given node is eligible to multicast a block, a procedure taking $f$ as a parameter is executed *within the protocol*. Then, to prove the security of the protocol, the authors require that $f$ satisfies an inequality [4, Theorem 9 Eq. 12], which encodes the "Majority of Honest Stake" assumption but also depends on $\Delta$. Notably, if $f$ and all the other protocol parameters are known to the nodes, then assuming that the inequality holds, this implies that nodes can solve it to obtain (an upper bound on) $\Delta$.

We adopt a similar reasoning with Nakamoto. Here, $\Delta$ is related to other known parameters through an inequality [11, Honest Majority Assumption (Bounded Delay)], which must hold to prove the security of the protocol. Therefore, nodes can solve it to compute $\Delta$. Formally, these remarks can be expressed with the following theorem.

---

[1] We use the term "Nakamoto agreement" or "Nakamoto" for short to refer to Bitcoin's underlying *ABC* protocol.

**Theorem 1.** (Nakamoto and Praos are strongly synchronous).

$$\exists P \in \mathcal{P}[M_{\text{Praos}}], \ \mathcal{M}_P^{\mathcal{A}_{\text{Praos}}} \models \text{strong\_net}$$
$$\exists P' \in \mathcal{P}[M_{\text{Nakamoto}}], \ \mathcal{M}_{P'}^{\mathcal{A}_{\text{Nakamoto}}} \models \text{strong\_net}$$

where $(M_{\text{Nakamoto}}, \mathcal{A}_{\text{Nakamoto}})$ and $(M_{\text{Praos}}, \mathcal{A}_{\text{Praos}})$ are the models of Nakamoto agreement and Ouroboros Praos, respectively.

*Proof.* To implement the strong\_net module, $P$ and $P'$ work similarly. Initially, nodes compute a bound $D \geq \Delta$ using the procedure described above. Upon request of the $\Delta$ value through the interface of strong\_net, the value $D$ is returned. Any other interaction is proxied to the underlying weak\_net module. Knowing that a weakly synchronous network with a known bound on $\delta$ is exactly a synchronous network; therefore, the conclusion is reached.

### 3.2 Nakamoto Agreement

*Model.* We have already shown that the model from Garay et al. [11] is strongly synchronous, with a computational adversary. Their model is completed with a Random Oracle, which is modified to integrate the Proof-of-Work mining primitive. That is to say, the Random Oracle is given the ability to answer "mining" queries from the nodes, with a limit of $q$ queries per node per network round (or 1 in the "partially" synchronous analysis). Then, an additional interface is added to be able to verify the result of a query without having to make the same query again and be limited by $q$.

Regarding the corruption structure, they state an "Honest Majority Assumption" such that the proportion of hashpower (expressed in RO queries per round) available to the malicious nodes is lower than $\frac{1-d(\lambda,\Delta)}{2-d(\lambda,\Delta)}$ with $d$ a function bounded between 0 and 1. This "Honest Majority Assumption" is easily modeled as a weighted corruption structure with each node being weighted by its hashpower. As expected, $d$ is an increasing function, meaning that the amount of tolerated malicious hashpower gets further from the optimal value $(1/2)$ down to 0 as the security parameter is increased. Node corruptions take effect immediately, therefore the adversary is dynamic.

*Metrics.* In the Bitcoin Backbone analysis, Garay et al. articulate their proofs on the assumption of a *typical execution* that roughly states that parties produce blocks at a rate close enough to their expected value (*i.e.*, hashpower). Then, they show that any execution of $k$ rounds is typical with probability $1 - e^{-\Theta_\epsilon(k)}$, where $\epsilon$ is a variable that quantifies how close the block production rate is to its expected value. In turn, the proofs will rely on $\epsilon$ being appropriately bounded, an assumption that is integrated in their version of "honest majority assumption". In particular, this assumption gives a bound on $\epsilon$ that depends on $\Delta$, hence, an execution which is longer than $\text{poly}_\Delta(\lambda)$ rounds is typical with overwhelming probability.

Assuming a typical execution, Garay et al. prove that it takes $\frac{4k}{1-\epsilon}$ rounds for a transaction to be confirmed by $k$ blocks. Thus, if $k = \text{poly}(\lambda)$ the latency of transaction confirmation is $\mathcal{L}(\text{NAKAMOTO}) = \text{poly}_\Delta(\lambda)$.

The analysis regarding communication complexity is simpler: All $b$ bits from the output are blocks that have been multicast once, hence the communication cost is at least $bn$. Furthermore, this cost is only increased if the adversary forces orphaned blocks. Since all messages on the network are blocks with a valid PoW, it is clear that the overall number of blocks received by honest nodes is a constant factor of the number of blocks that will end up in the blockchain. Hence, $\mathcal{CC}(\text{NAKAMOTO}) = \Theta(bn)$.

### 3.3 Ouroboros Praos

*Model.* We have already shown that the model from Bernardo et al. [4] is strongly synchronous, with Byzantine faults and a computational adversary. Like with the Nakamoto agreement, we now complete the model.

The existence of the Random Oracle and digital signatures is assumed. Node corruptions take immediate effect and the adversary is fully dynamic. This is possible due to the usage of VRFs, on a principle similar to how the Bitcoin miner can be corrupted just after sending a block without issue. Ouroboros nodes use the VRF to learn locally whether they are randomly selected to multicast a block, and once the block is sent they no longer have a privileged role. The verifiability of the VRF output ensures that nodes cannot cheat their eligibility.

Regarding the corruption structure, Bernardo et al. assume that the proportion of stake owned by honest nodes is higher than $\frac{1}{2}d(\lambda, \Delta)$, where $d$ is an (increasing) function lower bounded by 1. This assumption is modeled by a weighted corruption structure using stake as weights.

*Metrics.* In Ouroboros Praos, for every time slot, each node can be independently elected to multicast a new block, and the probability to be elected is only a function of the amount of the owned stake. Generating the randomness to seed the election of block leaders is implemented by having each leader include a VRF output in their block, and periodically concatenating the VRF output to form a random seed for the future VRF evaluation.

Given this method of block production, Ouroboros's properties can be proven using only combinatorial arguments on the distribution of honest and malicious leaders in the block tree. This is done through the analysis of "characteristics strings", which are an encoding of the schedule of honest and malicious leaders. Bernardo et al. show that a string of length of $k$ time slots is "forkable" with probability $\text{negl}(k)$, for an appropriate notion of "forkable" that is later used to prove the protocol security. In particular, the distribution of characteristics strings only depends on the (stakewise) proportion of malicious leaders, and therefore $k$ is only a function of the security parameter, *i.e.* $k = \text{poly}(\lambda)$ and $\mathcal{L}(\text{PRAOS}) = \Theta(\Delta k)$

## 3.4   Tendermint

Our discussion of Tendermint is based on an analysis by Amoussou-Guenou et al. [1].

*Model.* Tendermint follows more classical approaches to State Machine Replication [5] and as such fits very nicely into the ABC formulation. Amoussou-Guenou et al. assume a partially synchronous network with a formalism based on a Global Stabilization Time (GST), which is equivalent to having an unknown network delay $\delta$. All nodes sign their messages with a digital signature algorithm. Tendermint uses a hash function that can be modeled with a Random Oracle. The corruption threshold is $\lfloor n/3 \rfloor$ nodes, which is simply represented as a corruption structure with equal weights for all nodes. The adversarial adaptivity is not explicitly specified, however it is straightforward to see that it can be dynamic; the leader is not expected to be honest, and the schedule of leaders may as be well known when the adversary choose corruptions.

*Metrics.* Tendermint's normal case operation is reminiscent of PBFT [5], except that leaders are always changed after each broadcast, whether they be successful or not. For each block, there is a leader that will execute a reliable broadcast protocol implemented through two all-to-all voting rounds. To optimize the bandwidth, only the hash of the block is included in the voting messages. This requires the $b$ bytes of the block content to be sent to all $n$ nodes, and the additional cost of $n^2$ bytes for the reliable broadcast, taking three communication steps. That is, a single leader attempting to append a block takes $\Theta(bn + n^2)$ bit complexity and $\Theta(1)$ latency.

Malicious leaders can ensure that their attempt does not succeed, and because the leader order is arbitrary, all malicious nodes may be leaders first, thus increasing latency by a factor $\max_{C \in \mathcal{C}} \#C = \Theta(n)$. This does not impact the overall communication cost however, because all honest leaders will also append a block and only a constant fraction of nodes are malicious. Tendermint is optimistic in the sense that, in failure free executions, transactions are immediately appended and the latency becomes $\Theta(\delta)$.

## 3.5   HoneyBadgerBFT

*Model.* In HoneyBadger BFT [15](HBBFT), the network is asynchronous, up to $1/3$ of the nodes may be corrupt, and the adversary is static. The protocol makes use of a hash function and digital signatures. Miller et al. explicitly assume a "Purely asynchronous network", "Static Byzantine faults" and "Trusted setup", and they aim to solve Atomic Broadcast. The setup only serves to implement a PKI for the digital signatures and a common coin subprotocol.

*Metrics.* HoneyBadgerBFT uses the reduction to a reliable broadcast and a binary consensus from Ben-Or et al. [3] to implement a variant of the consensus problem, namely an asynchronous common subset.

The reliable broadcast protocol terminates in three rounds and has a bit complexity of $\Theta(bn+n^2 \log(n) \operatorname{poly}(\lambda))$. The binary consensus is an asynchronous probabilistic protocol with $\Theta(n^2 \operatorname{poly}(\lambda))$ bit complexity. At each round it has $1/2$ probability to terminate. HBBFT must wait for all their consensus instances to terminate, and the time required is $\Theta(\log(n))$ rounds on average.

However, one of the achievements of HBBFT is that at the end of this procedure, it commits data from all node inputs. That is, if all the nodes have an input of size $B$, then the batch committed will be of size $\Theta(nB)$ bits. Hence, for $b$ bits committed, there are $\Theta(bn + n^3 \log(n) \operatorname{poly}(\lambda))$ bits received by honest players.

Miller et al. do provide a complexity analysis, but they state their results in terms of *overhead*, *i.e.*, the total cost divided by $b$. Furthermore, by the specifying minimum input size (batching policy) of $\Omega(n^2 \log(n) \operatorname{poly}(\lambda)) = O(b)$, they obtain an $O(n)$ figure which is a constant per-node overhead, which essentially amounts to looking at the communication complexity *assuming the block contents is the dominating cost*. This emphasis on the low overhead is less visible in our results, although it is translated by the fact that HBBFT complexity on the $b$ factor is $bn$ instead of $bn^2$ for other BFT-style protocols, *i.e.*a reduction factor of $n$ is consistent with the authors' analysis.

### 3.6 Phantom

*Model.* Phantom [17] is a Proof-of-Work protocol whose mining operation is very similar to Nakamoto, except that blocks may have more than one parent. As a result, the model is essentially the same as for Nakamoto, however Sompolinsky and Zohar use their own formulation based on an earlier proposal [16]. Phantom's security statement (Theorem 4) merges the ABC-Consistency and ABC-Liveness properties into a single one, although its formulation is specifically tied to the blockchain structure. Additionally, because the authors explicitly aim for a generalized version of Nakamoto, we are confident in choosing ABC to faithfully capture Phantom's properties.

The network is strongly synchronous; the authors state "if an honest node $v \in \mathcal{N}$ sends a message of size $b$ MB at time $t$, it arrives at all honest nodes by time $t + D$ the latest.", where a bound one $D$ is known to the protocol. Up to $\frac{1}{2}(1 - d(\lambda, \Delta))$ of the hashpower may be corrupt, with $d$ an (increasing) function lower bounded by 0. Although not mentioned explicitly, the adversary is dynamic for the same reason as with Nakamoto agreement.

The Proof-of-Work mining is modelled by a Poisson process. Although common modeling of the mining process results in a Binomial distribution, this is not an issue since this distribution converges towards the Poisson distribution whenever the number of tries goes to infinity. As with Nakamoto, we also require the Random Oracle to answer hash queries unrelated to mining.

*Metrics.* In Phantom, the block structure is a Directed Acyclic Graph (DAG) which allows blocks to be linked to any number of the previous blocks instead of one. A key technique of Phantom is that PoW merely serves as a network-level synchronization primitive, thus decoupling the mining hardness from the protocol security. And indeed the mining rate is independent from $\lambda$ and only required to be high enough so that blocks are not mined faster than they can be transmitted through the network. As such, there is little restriction on the mining process. More precisely, the entirety of the online protocol consists of mining on top of all the blocks with no successor in the block graph.

The online protocol maintains the DAG that grows over time, and another sub-protocol independently determines which values from the DAG will be the ABC output.

Phantom implements a procedure *Risk* that provides a bound on the probability that safety will not hold for this transaction. The authors then show that the bound returned is smaller than a given $\epsilon$ after $O(\log(\frac{1}{\epsilon}))$ honest blocks are created. Hence, with $\epsilon = O(e^{-\lambda})$, and since this procedure requires the upper bound $\Delta$, we have that the number of honest blocks required is $\text{poly}_\Delta(\lambda)$. The time taken for this to happen is obtained by multiplying it by the rate of honest block production, which only depends on $\Delta$. Then, in order for all the nodes to be aware of these blocks, an additional $\Delta$ overhead is added, resulting in $O(\Delta + \text{poly}_\Delta(\lambda))$ latency.

On the other hand, communication complexity is simpler. In fact, the same analysis as the other blockchain style algorithms is applicable. Blocks all require a valid PoW to be sent through the network, and the PoW creation rate is bounded according to the corruption structure, thus reaching an optimal $\mathcal{CC}(\text{PHANTOM}) = \Theta(bn)$.

### 3.7 Algorand

*Model.* Algorand [12] uses a partially synchronous network, the Random Oracle, forward secure signatures, and "VRFs". As expected in a partially synchronous network, a proportion $h > \frac{2}{3}$ of the stake is assumed to be owned by honest nodes. However, since Algorand uses an election mechanism, the protocol failure probability increases as $h$ goes to $\frac{2}{3}$. Like with Ouroboros Praos, their usage of VRFs gives the possibility to tolerate dynamic adversaries.

*Metrics.* Like Ouroboros Praos, Algorand relies on a public randomness computed in previous blocks. It is used to elect a committee (instead of a leader) *at each round* that will have sufficiently many honest nodes, with overwhelming probability. These committees run a consensus protocol, which does not require a private state from the nodes (except from their private key), since committees would not be able to pass it on to the next committee. Except from this property, common techniques from BFT algorithms can be used to reach agreement in a constant number of rounds. In particular, since it has been ensured that committees have a constant fraction of honest nodes, standard quorum-based arguments are still valid. More precisely, at each round, each user has a fixed probability $p$ to be part of the committee. To bound the number of malicious nodes in a committee, the authors leverage the fact that $\forall t' < t$; the probability of having at most $t'$ malicious nodes in a uniformly sampled committee of size $k$ is $1 - \mathrm{negl}(k)$. Thus, the committees expected size is $\mathrm{poly}(\lambda)$.

As a result, each round that would be equivalent to a $n^2$ all-to-all communication in a traditional BFT algorithm is now a "committee-to-all", $\Theta(n\,\mathrm{poly}(\lambda))$ communication. Additionally, Algorand optimizes bandwidth through a block proposing step similar to a leader election. The average committee size for that step is the smallest such that there is at least one proposer with overwhelming probability, *i.e.*, $\mathrm{poly}(\lambda)$ asymptotically.

# 4   Discussion

Table 1 summarizes the above results. Concretely, For each protocol $P$ in Table 1, we have made the following claim.

$$P \in \mathcal{P}[\mathrm{params}, \mathrm{RO}, M], \quad \mathcal{M}_P^{(\mathrm{computational}, \mathcal{C}_{(W,\alpha)}, (E,i,p) \mapsto \top)} \models^\lambda ABC \qquad (1)$$

where: $W(p) := h_p(\sum_{q \in \Pi} h_q)^{-1}$ if the protocol uses the PoW module, or, if the protocol has a *stake* public parameter, $W(p)$ is $p$'s relative stake. $\alpha$ is the contents of the "Adversary" column, $M$ is the contents of the "Modules" column. We also claimed that $\mathcal{L}(P)$ is equal to the contents of the "Latency" column and that $\mathcal{CC}(P)$ is equal to the contents of the "Communication Cost" column.

**Table 1.** Comparison summary.

| Algorithm | Corruption threshold | Modules | Latency | Communication cost |
|---|---|---|---|---|
| Nakamoto [11] | $1/2 - \epsilon(\lambda, \Delta)$ | sync_net, PoW | $\mathrm{poly}_\Delta(\lambda)$ | $\Theta(bn)$ |
| Phantom [17] | $1/2 - \epsilon(\lambda, \Delta)$ | sync_net, PoW | $\Theta(\Delta + \mathrm{poly}_\Delta(\lambda))$ | $\Theta(bn)$ |
| Ouroboros praos [4] | $1/2 - \epsilon(\lambda, \Delta)$ | sync_net, signatures, VRF | $\Theta(\Delta\,\mathrm{poly}(\lambda))$ | $\Theta(bn)$ |
| Algorand [12] | $1/3 - \epsilon(\lambda)$ | weak_net, signatures, VRF | $\Theta(\delta)$ | $\Theta(bn\,\mathrm{poly}(\lambda))$ |
| Tendermint [1] | $1/3$ | weak_net, signatures | $\Theta(\delta n)$, f.f. $\Theta(\delta)$ | $\Theta(bn + n^2)$ |
| HoneyBadgerBFT [15] | $1/3$ | async_net, signatures | $\Theta(\delta \log(n))$ | $\Theta(bn + n^3 \log(n)\,\mathrm{poly}(\lambda))$ |

This table clearly outlines the relation between the protocol's models and their performance metrics. The major drawback of Bitcoin is that its latency has an arbitrary dependence on $\Delta$, which could very well be exponential. The same concern also applies to Phantom, and this is due to the fact that the confidence in a transaction depends on the time it takes for a PoW string to be transmitted. Moreover, it's worth remembering that $\Delta$ must hold for all messages, *of any length*. Therefore, increasing the block size may change $\Delta$ and results in a theoretically unknown impact on the protocol latency. These considerations point towards analyzing how latency depends on $\Delta$ and would be a great step forward in the assessment of PoW-based protocols.

Interestingly, we can see that all three synchronous protocols have their corruption threshold dependent on $\Delta$, and for the same reason for each of them, the "honest majority" assumption is stated as an inequality that depends on

$\Delta$. As expected, the 1/2 corruption threshold is synonymous with strong synchrony, and the PoW oracle is able to replace digital signatures. The presence of "$-\epsilon(...)$" in the corruption threshold is indicative of protocols making use of the election mechanism, which also imply a factor poly($\lambda$) in its costs. Algorand is interesting in this regard, as it uses a large elected committee instead of leaders, thus removing the $\Delta$ factor in the threshold reduction $\epsilon$, but gaining the poly($\lambda$) in communication cost. Reducing this communication cost by a constant factor is possible, *e.g.*, through sharding [8,9]. On the other hand Tendermint and HoneyBadger BFT doesn't use the election mechanism and bear a communication cost at least quadratic in $n$.

## 5    Conclusion

We presented a brief review of a selection of prominent published blockchain protocols. This review aimed at putting the protocol execution models on a common footing and at analyzing and comparing communication cost and latency as a function of multiple common variables. Our work puts an emphasis on having a streamlined terminology that fits the wide range of executions models. We were able to highlight and solve confusions and non-trivial issues that arise due to the lack of standardized formal definitions for common distributed primitives. We believe that this work can serve as a basis for facilitating the comparison of blockchain protocols as well as alleviating both the difficulties of standardization and fine-grained interoperability.

## References

1. Amoussou-Guenou, Y., Pozzo, A.D., Potop-Butucaru, M., Tucci Piergiovanni, S.: Dissecting tendermint. In: NETYS 2019, June 19–21, Marrakech, Morocco (2019)
2. Bano, S., Sonnino, A., Al-Bassam, M., Azouvi, S., McCorry, P., Meiklejohn, S., Danezis, G.: Sok: consensus in the age of blockchains. In: AFT 2019, Zurich, Switzerland (2019)
3. Ben-Or, M., Kelmer, B., Rabin, T.: Asynchronous secure computations with optimal resilience (extended abstract). In: PODC, Los Angeles, California, USA, August 14–17 (1994)
4. Bernardo, D., Gaži, P., Kiayias, A., Russell, A.: Ouroboros praos: an adaptively-secure, semi-synchronous proof-of-stake blockchain. In: EUROCRYPT (2018)
5. Castro, M., Liskov, B.: Practical byzantine fault tolerance and proactive recovery. ACM Trans. Comput. Syst. **20**(4), 398–461 (2002)
6. Cristian, F., Aghili, H., Strong, H.R., Dolev, D.: Atomic broadcast: from simple message diffusion to byzantine agreement. Inf. Comput. **118**(1), 158–179 (1995)
7. Durand, A.: Byzantine consensus and blockchain: models unification and new protocols. Theses, Polytechnic Institute of Paris (2021)
8. Durand, A., Anceaume, E., Ludinard, R.: Stakecube: combining sharding and proof-of-stake to build fork-free secure permissionless distributed ledgers. In: NETYS 2019, Marrakech, Morocco, June 19–21, 2019. LNCS, vol. 11704, pp. 148–165. Springer (2019)

9. Durand, A., Hébert, G., Toumi, K., Memmi, G., Anceaume, E.: The stakecube blockchain : instantiation, evaluation amp; applications. In: 2020 Second International Conference on Blockchain Computing and Applications (BCCA), pp. 9–15 (2020)

10. Garay, J.A., Kiayias, A.: Sok: a consensus taxonomy in the blockchain era. In: CT-RSA 2020, San Francisco, CA, USA, February 24–28. LNCS (2020)

11. Garay, J.A., Kiayias, A., Leonardos, N.: The bitcoin backbone protocol: analysis and applications. In: EUROCRYPT 2015, Sofia, Bulgaria, April 26–30, 2015. Updated version on IACR Cryptol. ePrint Arch. 765 (2014)

12. Gilad, Y., Hemo, R., Micali, S., Vlachos, G., Zeldovich, N.: Algorand: scaling byzantine agreements for cryptocurrencies. In: SOSP 2017, Shanghai, China, October 28–31 (2017)

13. Hirt, M., Maurer, U.M.: Complete characterization of adversaries tolerable in secure multi-party computation. In: PODC '97, Santa Barbara, California, USA, August 21–24. ACM (1997)

14. Katz, J., Lindell, Y.: Introduction to Modern Cryptography. CRC Press (2014)

15. Miller, A., Xia, Y., Croman, K., Shi, E., Song, D.: The honey badger of BFT protocols. In: ACM SIGSAC 2016, Vienna, Austria, October 24–28 (2016)

16. Sompolinsky, Y., Lewenberg, Y., Zohar, A.: SPECTRE: A fast and scalable cryptocurrency protocol. IACR Cryptol. ePrint Arch. 1159 (2016)

17. Sompolinsky, Y., Zohar, A.: PHANTOM and GHOSTDAG: a scalable generalization of Nakamoto consensus. IACR Cryptol. ePrint Arch. 104 (2018)

# Blockchain-Based Business Process Management (BPM) for Finance: The Case of Loan-Application

Galena Pisoni[1]([⊠])(iD), Meriem Kherbouche[2](iD), and Bálint Molnár[2](iD)

[1] Université Cote d'Azur, Polytech Nice Sophia, Campus ShopiaTech, 930 Route des Colles, 06410 Biot, Nice, France
galena.pisoni@univ-cotedazur.fr
[2] Faculty of Informatics, Information Systems Department, Eötvös Loránd University (ELTE), Pázmány Péter 1/C, Budapest 1117, Hungary
meriemkherbouche@inf.elte.hu, molnarba@inf.elte.hu

**Abstract.** Because of the competitive economy, organizations today seek to rationalize, innovate and adapt to changing environments and circumstances as part of business process improvements efforts. The strength of blockchain technology lies in its usage as an apt technology to enhance the efficiency and effectiveness of business processes; furthermore, it prevents the use of erroneous or obsolete data, and allow sharing of confidential data securely. The use of superior technology in the execution and automation of business processes brings opportunities to rethink the specific process itself as well. Business processes modeling and verification are essential to control and assure organizational evolution, therefore, the aim of this paper is three-fold: firstly, to provide business process management patterns in finance, based on blockchain, specifically for the loan-application process, that could be used and customized by companies; secondly, to critically analyze challenges and opportunities from the introduction of such approach for companies; and thirdly to outline how companies can implement the loan business process as a web service.

**Keywords:** Blockchain · Business process · Business process modelling (bpm) · Finance · Loan · Technology management · Service-oriented computing

## 1 Introduction

BPM (Business Process Management and Modeling) as a discipline tries to increase the performance of companies by studying business processes, and possibilities for business process optimizations [26]. In the course of the development of the research branch, it has matured as an academic and professional discipline. The contemporary BPM is a widely adopted and deployed scientific field on both academic and practitioner sides. BPM tries to respond to and address pressing issues in all sectors of Business IT (Information Technology) and to foster value creation for companies.

However, with the rise of the large number of opportunities connected with digitization, the established BPM methods are not fully capitalizing on the latest opportunities that are offered by contemporary technology. For instance, blockchain technology has gained widespread traction, yet its use for business process aims is in its infancy [8].

A blockchain is a ledger of all transactions which are stored in a distributed database. One or single transaction can constitute a block. Each time, a transaction is made, it is linked to a chain, with every block containing a hash of the previous blocks in order of sequence. The fact that the data are stored in a distributed way makes it impossible to change its data. Changing the data would require that one changes the data in each block in each of the distributed databases at the same time. Thus, this solution makes such scenarios impossible within the currently available technologies.

Some examples of current applications of blockchain in the financial industry include Corda platform [4]. This platform is a collaborative effort of 40 banks to redesign the current settlement process by enabling financial agreements to be stored on the blockchain and thereby help improve reconciliation processes to become more optimal. Australian Stock exchange implements a system based on a blockchain ledger [25]. In the UK welfare program, the welfare-payment process is on blockchain technology [6]. In Sweden, land records are put on the blockchain so that banks (for purposes of delivering loans, and mortgages), government agencies, buyers, and sellers can track changes in real-time [5].

The first approach is to use blockchain in finance has proliferated as was said previously, yet it is not yet fully explored the domain of what does it mean to use blockchain for business processes improvements, thereby, this article brings a new view on this issue, and especially focuses on the loan-application process. It aims to provide a multidimensional analysis of the use of blockchain for BPM in Finance, and focuses on scenarios of exploitation and getting this technology acquainted in the domain of BPM, furthermore improvement of existing processes.

The loan business process usually consists of five activities or work steps as follows: submitting a loan application, assessment of the loan application (through loan risk assessment and estimation of property value), and either approval of the loan application or rejection of the loan application.

If the process exploits blockchain technology, the different information can be automatically checked and verified. Firstly, then the whole process would become more transparent for all; secondly, it would increase the speed of response of the bank, moreover, it would minimize the cost of such operations for the bank.

In this scenario, the use of blockchain would provide better, faster, and less erroneous data checks. It yields speedier and more reliable processing of loan claims, and automation of policy enforcement. Especially in such cases, the chain of decision activities is clear enough, so that is no need for a final checkup by a clerk. Therefore, in the loan scenario, blockchain would help resolve some current market challenges, such as:

- all the parties involved in the loan application, revision, and approval will have a single view of the data in a distributed ledger and blockchain platform;
- manual processes can be automated, via checks of the data coded into blocks of blockchain and the use of "smart contracts";
- the whole process is immutable, and everyone can see changes in status in the application, there is clear visibility of ownership, which makes it easy also to be a subject of audits, and the whole process is transparent.

The rest of the paper is organized as follows: Sect. 2 presents background work and related literature on use of blockchain use in BPM literature, Sect. 3 presents the loan business process pattern and introduction of blockchain in it, advantages and challenges for financial companies, as well as how the loan business process can be implemented as a web service; Sect. 4 discussed future challenges and concludes the paper.

## 2   Blockchain Related Literature Overview

The use of blockchain technology is rapidly expanding [17]. Many researchers are looking into it from various angles. It is a technique for storing and transmitting information in the form of a chain of blocks in a decentralized, secure, and transparent manner. Many well-known companies, such as Google, Microsoft, and Black-rock, utilize blockchain.

The blockchain has the ability to bring Business Process Management back to life. Blockchain can make processes involving numerous parties in systems transparent by employing the ability of auditing distributed ledger technology and smart contracts, as well as obeying some specific regulatory regulations. To connect data, circulate documents, and track multiple stages of a transaction. The authors of the paper [16] presented a blockchain-based Business Process Management System (Caterpillar), which uses smart contracts to encode the status of business process instances and conduct workflow routing. Another paper, [9], shows how blockchain technology may help firms reduce transaction costs, improve collaborative business processes, and reinforce trustability and transparency while also assisting with data consistency and security governance and reconciliation.

Wattana et al. [28] looked at an automated BPM system for selecting and composing services in an open business environment in another study. They investigated and proposed blockchain technology (BCT) as a means of transferring and verifying the trustworthiness of enterprises and partners. They created a BPM framework to show how blockchain technology may be used to assist in quick, accurate, and cost-effective evaluation and transfer of service quality in workflow composition and management. Various studies have been conducted on the use of blockchain technology in BPM. The study by [1] discusses the theories, problems, and important success aspects of using Blockchain to handle business processes. For the interoperation of business processes in smart supply chains [27] and business process monitoring [7], blockchain technologies are being used.

The survey paper by Rafael et al. [2] presented a business process view integration of the past, present, and future applications to Blockchain. The work by Pierre [9] also presented the benefits, costs, and barriers of improving BPM with Blockchain. Blockchain solutions are used and combined with M2P solutions, [13,14].

# 3    Loan Business Process

## 3.1    Loan Process Business Process Pattern with Blockchain

First, we present the current 'AS-IS' implementation traditionally seen in financial institutions (Fig. 1), after which, we also present the 'TO-BE' design of process as discussed in the rest of the paper (Fig. 2).

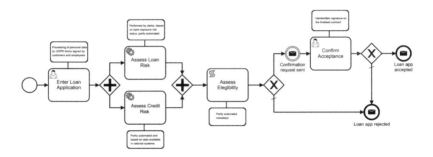

**Fig. 1.** AS-IS loan application business process.

The MIT Business Process Handbook contains a comprehensive collection of Business Process patterns that could be customized for purpose [18]. We did an adaptation of the original patterns to represent the loan application pattern and Table 1 and Fig. 2 show the loan business process where we included blockchain technology into the pattern. The patterns provided in the book are a set of patterns a digitally transformed enterprise can use for their business process management, however, they are to be complemented by the modern security mechanisms to safeguard the integrity, consistency, and trustfulness of the business processes. Therefore, we inserted the components of blockchain technology into the patterns and the patterns can be used during business process realization.

The workflow remains the same in both, the 'AS-IS' and 'TO-BE' loan application processes, yet, the business process is improved on time, quality, costs, and flexibility factors in the 'TO-BE' version. This is due to fact that we replace, for instance, the GDPR-compliant processing of private data from paper-based

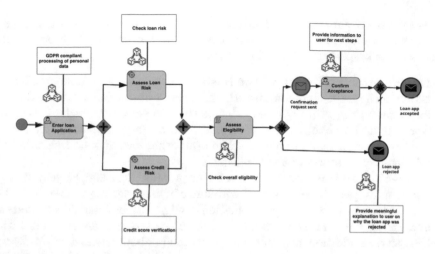

**Fig. 2.** TO-BE loan application process where the different steps of the workflow are performed on the block-chain.

**Table 1.** Loan application pattern.

| | Loan application pattern |
| --- | --- |
| Description | – The process pattern represents the tasks from the beginning of the loan application to the acceptance or rejection; |
| | – The verification starts after receiving the loan request (verification of the applicant's details, residential address, bank statements, income, etc.); |
| | – The request is transferred to the responsible person; |
| | – The step of verification includes authentication, confirmation, and validation of all data that are given by the applicant; |
| | – The result of the verification stage could be acceptance or rejection |
| Goal | To describe a sequence of tasks in a model that depicts the approval or rejection of the loan request |
| Problem | This pattern obliges the verification and approval that ought to be done at the beginning, and during the execution of the process when different personal and confidential data is used proper security systems ought to be applied (e.g., blockchain) subsequently |

to the blockchain-based solution. We use a permissioned enterprise blockchain that supports GDPR in a way that there is only a hash of private data stored on the blockchain, whereas the actual data is stored in a private database, thus improving the process on time and cost factors. In addition, calculations and checks performed partly manually nowadays, are again put on the blockchain, which improves the flexibility and quality of the business process. Our analysis and simulation of the TO-BE process are much superior on all the key factors that are influenced by the restructuring of the business process: time, quality, costs, and flexibility.

We use the framework introduced in [3] to describe how and what it will mean to blockchain-based business process management for loan application business processes for banks and financial institutions.

- **Introduction and story of the business process case**: Loan application is one of the most common operations banks perform. Banks face a demand for credit so the primary role of their staff is to screen applicants and monitor outstanding loans. Loan officers perform an important strategic task in the process of a loan application, like evaluating loan risk and checking the previous credit scores of the applicant.
- **Current situation faced**: Current loan applications and lengthy in time and require specialized personnel with lots of experience to process them. The current loan processes foresee evaluation of eligibility by loan officers, and this analysis may often not be visible to other managers and departments within the bank. In addition, customers often complain about the lack of clarity and transparency in the processes of the bank, and the lack of clear understanding of the steps and checks that the bank performs regarding their requests. It may not be understandable to the end customers why their loan was rejected. If this is the case, what the reason was (was it due to low credit score risk? or does the risk seem too big for the bank to undertake?). Lastly, data must be treated in accordance with GDPR [12]
- **Action taken**: Underlying technological solution like the one that we propose in this paper, based on blockchain technology, can improve the execution of the process and make the process more transparent
- **Lessons learned**: Based on our previous works [20, 21], we understood that financial companies are the slow adaptors of new technologies. Technical teams, working on loan technology solutions, would need to perform a mapping between their current solutions and solution as described in our article, and analyze critically the feasibility of the transformation for their company from different points of view, technical and managerial.

## 3.2   Advantages and Challenges for Financial Companies

In this section, we discuss open challenges and recommendations for the development of blockchain loan solutions for companies.

Some of the advantages for companies as reported by previous literature [24]:

- **Agility** it gives companies the ability to assemble teams around such developments freely, which can include "smart contracts", and allow the company to automatize business processes;
- **Actions** that were created in relation to past and current loan applications will be visible to all, e.g. new employees of companies can also see past actions on blockchain and act in line with previous decisions made, it helps passing knowledge in the company;
- **Institutional memory** that is there are indelible extracts of all transactions and past actions of the company.

Financial companies' business models will be disrupted as more and more blockchains are put in place.

The primary challenge in this scenario is the absence of regulation and operating frameworks that haven't been put in place, specifically regarding the use of blockchain technology in this domain. Although blockchain is often used in the financial world for cryptocurrencies, applications of blockchain for optimizing and supporting financial operations are still on the rise, therefore, such regulation is still missing.

The second challenge is in terms of the maturity of the technology. According to previous literature, there are a number of open challenges related to scalability, interoperability, standardization, and energy consumption, that may still need to be taken into account before implementing such a solution [11].

Another challenge comes from the cybersecurity standpoint, although in the beginning it was claimed that blockchains provide no point of failure, lately, blockchain attacks are on the rise, like in the example of 51% attack [31]. Checks and prevention techniques are already ideated [23] for such attacks, yet in solutions such as loan applications, such defense mechanisms must be seriously analyzed, and put in place obligatory.

### 3.3   Loan BPM Implementation as a Web Service

The authors in Ref. [30] describe design patterns that can be applied in software and service architecture to incorporate blockchain technologies. In Ref. [15], there is a framework that can be used for the analysis of whether the targeted domain is appropriate for the application of blockchain technologies. In the banking environment, the emphasis was/is on payment transactions. However, other essential business processes are already automated entirely or partly in finance. Furthermore, these business processes use algorithms that are linked to a discipline that is called AI (Artificial Intelligence) succinctly. The complex activities are realized through Web services in the style of Service Oriented Architecture, naturally, the services that occur could be microservices too. There are use cases and templates for business processes that are related to finance, and especially to the bank industry [18, 19, 22]. These models and templates of business processes make it possible that beyond the patterns of workflows, the patterns for application of blockchain technologies can be included in the generic models and the invocation of the function of blockchain technologies can be implemented through Web services [10].

- **Conformance** examination aims to test whether the behavior of the instance of a process model is in line with the prescribed requirements for the single process model;
- **Compliance** investigation contains techniques to audit whether the regulation, rules, prescriptions, and constraints are satisfied in the instance of the business process;
- **XAI, explainable AI** The AI (ML, Machine Learning, D.Sc., Data Science) algorithms give the foundation for decision-making. In a business process, some humans play active roles and there are who play passive roles or

only monitor the results. The AI algorithm should provide information that is interpretable to humans. This interpretable information might have been created through specific activities that might consist of Web services invoking blockchain patterns, services that are a façade of AI algorithms, and conveys the basic information that can be translated into an interpretable format.

The blockchain patterns implemented as Web services should care for data quality, the volume of data and data removability to keep in hand the conformance with non-functional requirements and GDPR [29]. Software patterns are defined that can be used in a service-oriented architecture and business process models [30].

- **Oracle and Reverse Oracle pattern** provides the communication channel between the closed blockchain environment and the execution of business process models;
- **Off-chain data storage** Since the representation of the instance of a business process and the documents that are involved in transactions may generate a volume of data that are hard to store in blocks and that are stored in external storage;
- **Contract patterns** The smart contracts yield the tool and interface to describe the business rules that are realized in web services that are devoted to representing *business entities* and *activities of business processes*. The relevant smart contracts can interpret the business roles codified in web services.

## 4   Conclusions

In this paper, we showed how blockchains improve the quality of the loan business process, as blockchain-based tracking of information can be beneficial for financial companies from different points of view. The pattern and business process descriptions provided in this paper can be valuable first input for practitioners and can be used as a comprehensive starting point to be adjusted for the specific needs of the financial company, and to be able to support the company properly. In conclusion, the benefits of the use of blockchain technology in the financial industry become more apparent, which is considered of big importance for companies to improve business processes. The identified research challenges, serve as a baseline for future implementation of such systems. Our next steps foresee work on the workflow rules construction for blockchain and business processes patterns, implementation of the sequence of workflow patterns, development of proof of concept, and conducting of software experiment.

**Acknowledgements.** The paper was supported by the "Application Domain-Specific Highly Reliable IT Solutions" project, implemented with the support provided by the National Research, Development and Innovation Fund of Hungary, financed under the Thematic Excellence Programme TKP2020-NKA-06, TKP2021-NVA-29 (National Challenges Subprogramme) funding scheme, and by the COST Action CA19130 - "Fintech and Artificial Intelligence in Finance Towards a transparent financial industry" (FinAI).

# References

1. Al-Rakhami, M.S., Al-Mashari, M.: Blockchain and internet of things for business process management: theory, challenges, and key success factors. Int. J. Adv. Comput. Sci. Appl. **11**(10), 552–562 (2020)
2. Belchior, R., Guerreiro, S., Vasconcelos, A., Correia, M.: A survey on business process view integration: past, present and future applications to blockchain. Bus. Process Manag. J. (2022)
3. Brocke, J.V., Mendling, J.: Frameworks for business process management: a taxonomy for business process management cases. In: Business Process Management Cases, pp. 1–17. Springer (2018)
4. Brown, R.G., Carlyle, J., Grigg, I., Hearn, M.: Corda: an introduction. R3 CEV, August **1**(15), 14 (2016)
5. Chavez-Dreyfuss, G.: Sweden tests blockchain technology for land registry. Reuters, June 16 (2016)
6. Cho, S., Lee, K., Cheong, A., No, W.G., Vasarhelyi, M.A.: Chain of values: examining the economic impacts of blockchain on the value-added tax system. J. Manag. Inf. Syst. **38**(2), 288–313 (2021)
7. Di Ciccio, C., Meroni, G., Plebani, P.: On the adoption of blockchain for business process monitoring. Soft. Syst. Model. 1–23 (2022)
8. Du, W.D., Pan, S.L., Leidner, D.E., Ying, W.: Affordances, experimentation and actualization of fintech: a blockchain implementation study. J. Strateg. Inf. Syst. **28**(1), 50–65 (2019)
9. Edrud, P.: Improving BPM with blockchain technology: benefits, costs, criteria and barriers (2021)
10. Erl, T.: Service-Oriented Architecture?: Analysis and Design for Services and Microservices. Prentice Hall, Service Tech Press, Boston (2017)
11. Fraga-Lamas, P., Fernández-Caramés, T.M.: Leveraging blockchain for sustainability and open innovation: a cyber-resilient approach toward EU green deal and UN sustainable development goals. In: Computer Security Threats. IntechOpen (2020)
12. Herian, R.: Regulating disruption: blockchain, Gdpr, and questions of data sovereignty. J. Internet Law **22**(2), 1–16 (2018)
13. Kherbouche, M., Molnár, B.: Formal model checking and transformations of models represented in UML with alloy. In: International Workshop on Modelling to Program, pp. 127–136. Springer (2020)
14. Kherbouche, M., Molnár, B.: Modelling to program in the case of workflow systems theoretical background and literature review. In: Proceedings of the 13th Joint Conference on Mathematics and Informatics, Budapest, Hungary, pp. 1–3 (2020)
15. Lo, S.K., Xu, X., Chiam, Y.K., Lu, Q.: Evaluating suitability of applying blockchain. In: 2017 22nd International Conference on Engineering of Complex Computer Systems (ICECCS). IEEE (2017). https://doi.org/10.1109/iceccs.2017. 26
16. López-Pintado, O., García-Bañuelos, L., Dumas, M., Weber, I.: Caterpillar: a blockchain-based business process management system. BPM (Demos) **172** (2017)
17. Lu, Y.: Blockchain and the related issues: a review of current research topics. J. Manag. Anal. **5**(4), 231–255 (2018)
18. Malone, T.W., Crowston, K., Herman, G.A.: Organizing Business Knowledge: The MIT Process Handbook. The MIT Press, Illustrated edn. (2003). https://process. mit.edu/Default.asp

19. Business Innovation Through Blockchain. Springer, Cham (2017). https://doi.org/10.1007/978-3-319-48478-5
20. Pisoni, G.: Going digital: case study of an Italian insurance company. J. Bus. Strategy (2020)
21. Pisoni, G., Molnár, B., Tarcsi, Á.: Data science for finance: best-suited methods and enterprise architectures. Appl. Syst. Innov. 4(3), 69 (2021)
22. von Rosing, M., von Scheel, H., Scheer, A.W.: The Complete Business Process Handbook: Body of Knowledge from Process Modeling to Bpm, vol. 1. MORGAN KAUFMANN PUBL INC (2014)
23. Sayeed, S., Marco-Gisbert, H.: Assessing blockchain consensus and security mechanisms against the 51% attack. Appl. Sci. 9(9), 1788 (2019)
24. Thomas, D.: Blockchain as a backbone to asset and wealth creation. In: The WealthTech Book: The FinTech Handbook for Investors, Entrepreneurs and Finance Visionaries, pp. 170–172 (2018)
25. Thuvarakan, M.: Regulatory changes for redesigned securities markets with distributed ledger technology. Knowl. Eng. Rev. 35 (2020)
26. Van Der Aalst, W.M., La Rosa, M., Santoro, F.M.: Business process management. Bus. Inf. Syst. Eng. 58(1), 1–6 (2016)
27. Viriyasitavat, W., Bi, Z., Hoonsopon, D.: Blockchain technologies for interoperation of business processes in smart supply chains. J. Ind. Inf. Integr. 100326 (2022)
28. Viriyasitavat, W., Da Xu, L., Bi, Z., Sapsomboon, A.: Blockchain-based business process management (bpm) framework for service composition in industry 4.0. J. Intell. Manuf. 31(7), 1737–1748 (2020)
29. Voigt, P., von dem Bussche, A.: The EU General Data Protection Regulation (GDPR). Springer International Publishing (2017)
30. Xu, X., Weber, I., Staples, M.: Blockchain patterns. In: Architecture for Blockchain Applications. Springer International Publishing (2019)
31. Ye, C., Li, G., Cai, H., Gu, Y., Fukuda, A.: Analysis of security in blockchain: case study in 51%-attack detecting. In: 2018 5th International Conference on Dependable Systems and Their Applications (DSA), pp. 15–24. IEEE (2018)

# Objective-Aware Reputation-Enabled Blockchain-Based Federated Learning

Samaneh Miri Rostami$^{(\boxtimes)}$, Saeed Samet, and Ziad Kobti

University of Windsor, Windsor, Canada
{miriros,saeed.samet,kobti}@uwindsor.ca

**Abstract.** Federated Learning (FL) is a distributed-collaborative machine learning framework to overcome the challenges of data silos and data privacy. The conventional FL enhances data owners' privacy by maintaining their data on devices. However, assuring the quality of data and model verification is challenging since no one contributor can see the others' data. The problem becomes even more challenging if data is not independent and identically distributed (Non-IID). This study proposes a multi-channel objective-aware validation method to identify valuable contributors to the FL model on a blockchain network with Non-IID data. We implement the proposed system on Hyperledger Fabric by utilizing a multi-layer deep learning model. The evaluation results demonstrate the effectiveness of the proposed approach.

**Keywords:** Federated learning · Blockchain · Validation · Non-IID

## 1 Introduction

The long-debated privacy issues in conventional Machine Learning (ML) have motivated researchers to study and design privacy-preserving ML techniques. In 2017 Google introduced Federated Learning (FL) [1] as a distributed collaborative ML framework. FL has privacy by design, which means data owners' raw data never leaves their devices. However, studies showed that the conventional FL could not provide sufficient privacy guarantees. Since the introduction of FL, several studies have been undertaken towards improving the framework in both security and efficiency [2]. Researchers have employed different techniques like Secure Multiparty Computation (MPC) [3], Homomorphic Encryption (HE) [4], and Differential Privacy (DP) [5] to tackle the problem related to the possibility of learning from the intermediate local updates. Other efforts have also been made to improve FL's efficiency, including, but not limited to, handling the size of communicated data [6], reducing communication rounds [7], training procedure and convergence [8,9], client sampling [10], and learning from Non-IID data [11].

© The Author(s): under exclusive license to Springer Nature Switzerland AG 2023
J. Prieto et al. (Eds.): BLOCKCHAIN 2022, LNNS 595, pp. 259–268, 2023.
https://doi.org/10.1007/978-3-031-21229-1_24

In FL, a centralized server is an essential component of the framework that has responsibilities, including broadcasting models and parameters, client sampling, intermediate local parameter collecting, and aggregation [8]. However, having single coordination has drawbacks such as single-point-of-failure, bias, and security problems [12]. Consequently, recent studies have followed decentralized schemes like blockchain [13,14]. Blockchain technology [15] is compatible with the nature of FL systems. Its decentralized infrastructure and the consensus protocols can help remove the centralized coordination and synchronize updates among peers. As a complementary, blockchain has shown the potential for improving FL in four aspects: decentralization [16], management [17], security [18–20], and motivation [21,22].

Although blockchain-based FL (BFL) brings some advantages, having a robust and reliable FL models is still challenging. For example, without seeing contributors' data, it is not straightforward to identify whether or not contributors to the FL model are trusted to participate in the training process. Additionally, rating the reliability of the contributors is even more challenging with Non-IID data. This study aims to create networks of similar contributors to address the evaluation problem in FL with Non-IID data. For this purpose, we leverage Hyperledger Fabric channels. Hyperledger Fabric is a permissioned blockchain that helps secure the interactions among entities that have a common goal but do not fully trust each other [23]. A Fabric network supports channels that accept different peers as members. A channel in Fabric forms a logical blockchain with a specific configuration. To the best of our knowledge, this is the first work using fabric channels to connect similar datasets to solve the model validation problem in FL with Non-IID.

The contribution of this paper is threefold: 1- Improving the efficiency of the model validation process in blockchain-based FL with Non-IID data, 2- Mitigating poisoning attacks, and 3- Increasing the performance of the Blockchain-based FL framework by filtering out high-quality updates.

The structure of this paper is as follows: Sect. 2 reviews relevant literature on Blockchain-based FL. Section 3 introduces the proposed framework. Section 4 presents the experimental settings and evaluation results. Finally, Sect. 5 concludes the paper and describes future work.

## 2    Related Work

Research on fully decentralized FL has been started with the potential of integrating blockchain technology into the FL framework. With its decentralized infrastructure, blockchain can be a substitute for the central server in the FL framework [24]. BlockFL [25] was one of the first studies that leveraged blockchain to overcome bias and single-point-of-failure problems. The authors studied the feasibility of their approach by extending FL tasks to untrustworthy devices in a blockchain network. Because consensus tasks are computationally intensive, a more flexible topology was proposed by [26] to separate FL from the blockchain network.

To improve FL model performance, BMFL [19] tried to avoid the influence of false updates using a self-reliability filter. The authors suggested recording filtering results on the blockchain ledger for audit purposes. Another study [9] investigated how contributors' reliability can enhance the global model's performance. The authors utilized the reputation metric to select trusted participants for the subsequent training rounds.

FL follows the data minimization principle [27]; therefore, the model validation process is a challenging task since no one contributor can see the others' data. DAM-SE [22] is a double-layer BFL platform that verifies the FL model using a cluster-based approach and voting system. In another study [28], a validation scheme was proposed to track the errors of intermediate updates. The authors assumed the availability of a validation dataset where each contributor is given a portion of it. Authors in [18] suggested monitoring and evaluating contributors' activities (e.g., their intermediate updates) throughout the training procedure. The contributors' updates can be used as a metric to promote honest behaviour and penalize dishonest behaviour for mitigating attacks [9]. Other studies [20,29] also used a committee to evaluate the contributors' updates.

The existing approaches assume a standard validation dataset and share it between contributors, which contradicts the nature of FL. Additionally, the characteristics of the validation data have not been considered. The differences in feature space or label space make the evaluation process challenging. For example, a dataset that contains face images cannot be used as a validation dataset for a model trained with hand gesture images.

# 3   The Proposed Framework

The proposed approach aims to connect contributors with similar objectives to handle the model validation problem. The contributors then become endorsers to each other to verify who can contribute to the global model.

## 3.1   Components of the Proposed Framework

Our proposed system has three main components: Admin, FL, and Blockchain Network.

**Admin** The admin is responsible for channel creation and connecting peers with similar objectives. We assume no collusion, meaning that the admin and the peers do not collude to break others' privacy. Admin also cannot learn information from others. Figure 1a shows the channel creation process for training a multiclass supervised FL model. Peers A, B, and C have their own private labelled data samples. Before starting the channel creation process, they agree on a hash algorithm they will use.

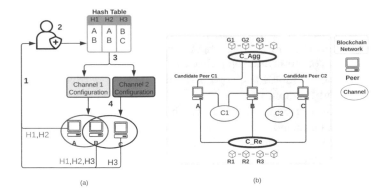

(a)                                                              (b)

**Fig. 1.** System architecture. (a) Diagram represents the channel creation process, (b) blockchain network.

The channel creation process works as follows:

1. A, B, and C send the hash of their labels to the admin.
2. Admin creates a dictionary of key-values (key: hash of label, value: address of peers who reported the hash).
3. Admin uses the hash map to initialize channels and endorsement policies.
4. Peers are informed about the channel(s) and their endorsement policies.

**Federated Learning** We assume there are $K$ peers who want to collaboratively train a standard multi-class supervised learning objective function $f_i(x) = l(x_i, y_i, w)$ that is the loss function for the prediction of a sample $(x_i, y_i)$ made with model parameters $w$. Each client stores its data $\mathcal{P}_k$ locally and do not share it with others.

The goal is to minimize the loss function $f(w)$ [30]:

$$f(w) \stackrel{def}{=} \sum_{k=1}^{K} \frac{n_k}{n} F_k(w)$$

$$F_k(w) = \frac{1}{n_k} \sum_{i=1}^{n_k} f_i(w)$$

(1)

where $n_k = |\mathcal{P}_k|$ is the number of data samples that client $k$ holds, and $n$ is the total number of all samples.

**Blockchain Network** Hyperledger Fabric [23] is used in the proposed framework shown in Fig. 2b. Fabric separates a transaction flow into three phases: endorsing, ordering, and validation. Its flexible endorsement policy specifies which peers need to confirm the correctness of a given smart contract. A smart contract (SC) operates as a trusted distributed application. We use four SCs for learning, evaluation, reputation, and aggregation.

The initialized model at iteration #0 is placed in block number zero (genesis block) of the global ledger (G) and is accessible to all peers. However, only the peers in the aggregation channel (C_Agg) have the write access to the ledger. The peers train the model with their local data and send their model parameters to their endorser in the same channel (C1 and C2). At iteration #0 the endorsers are selected randomly. They have two main tasks. First, they check to verify if the update is from an authorized peer in the channel. Second, the endorsers evaluate the model updates with their local data. The model's accuracy is the indicator to score the updates and find whether the updates are qualified. After receiving required endorsements, the updates are allowed to be added to the ledger of that channel. Additionally, the peers use the reputation SC for adding scores to the reputation ledger. Upon receiving enough updates, the aggregation SC will be triggered for aggregating the updates of the current iteration.

The reputation ledger is accessible to all peers. After each iteration, the average score a peer received is calculated as follows:

$$ScoreP_i = \frac{1}{N} \sum_{j=1}^{N} S_{P_j \to P_i} \tag{2}$$

where $S_{P_j \to P_i}$ is the score party $j$ gave to $P_i$, and $j = \{1, .., N\}$ indicates that the performance of $P_i$ is evaluated by $N$ endorsers. For example, in Fig. 2, B is evaluated by two peers A and C and receives two scores from different channels. The higher the score a peer has, the higher its reliability will be. The scoring system's novelty comes from how the scores are given. Compared to other studies, the scores are more reliable as peers responsible for the evaluation process have appropriate validation data to evaluate intermediate updates of other peers in the same channel.

## 3.2  Threat Model

In this study, a permissioned blockchain network is used, the contributors are known, and access is managed by access control lists. Therefore, a 51% attack and network partitioning attacks are not as significant of a threat. However,

| Channel #1 | | | | Channel #3 | | | |
|---|---|---|---|---|---|---|---|
| Label | P6 | P8 | P9 | Label | P2 | P3 | P4 | P10 |
| 1 | 500 | 100 | 50 | 0 | 300 | 100 | 600 | 5 |
| 2 | 10 | 900 | 200 | 3 | 50 | 500 | 8 | 100 |
| Channel #2 | | | | 4 | 50 | 90 | 20 | 20 |
| Label | P1 | P5 | P7 | 5 | 200 | 190 | 10 | 40 |
| 7 | 300 | 100 | 600 | 6 | 100 | 200 | 100 | 200 |
| 8 | 50 | 500 | 8 | 9 | 100 | 50 | 200 | 300 |

Fig. 2. Summary of data distribution between channels and peers.

as contributors do not share their data with others, they may intentionally or unintentionally act dishonestly and send malicious updates to poison the shared model. For this purpose, we assume the adversary may control multiple peers, but it does not exceed more than 30% of all contributors. When adversaries perform a poisoning attack, we assume that their goal is to control the behaviour of model training to reach some goals defined by attackers (here, degrade the performance of the final global model). We aim to filter out malicious updates that are sufficiently different from honest behaviour. In this study, we limit the adversaries to the label flipping poisoning attack [31]. It is a train-time attack where labels are manipulated.

## 4  Implementation and Evaluation

### 4.1  Experimental Setup

We use Hyperledger Fabric (version 2.4) as a private blockchain network to implement the proposed framework. Peers participate in both training and auditing the model. The FL environment is implemented with TensorFlow (2.4.0) and Python (version 3.8.8). We use the MNIST[1] dataset, which contains samples of handwritten digits. The class labels values are 0–9. We have four experimental scenarios: conventional FL [1], blockchain-based FL (BFL) [25], committee-based BFL [29], and proposed framework (ORBFL). In all scenarios, we follow the Non-IID data distribution (label distribution skew and imbalanced datasets). A summary of data distribution between channels is shown in Fig. 2. As seen, there are three channels. Each channel has different peers, labels, and data samples. For example, in channel 1, peer 6 has 10 data samples for label 2. It is worth noting that clients can be members of different channels based on the labels they have.

We add Gaussian noise to some samples to decrease their quality. It's worth noting that FL and BFL do not support the intermediate updates evaluation. For the committee-based BFL, one training channel is considered and 40% of parties are selected to establish a committee for the evaluation process. We use a multi-layer neural network as the training model with the following structure: an input layer, two hidden layers, and an output layer. The learning rate is set to 0.01, and the batch size is 32.

### 4.2  Results and Discussion

We evaluate the performance of the proposed framework through three different aspects, including accuracy, evaluation efficiency, and attack success rate.

---

[1] http://yann.lecun.com/exdb/mnist/.

**Evaluation of accuracy** We measure the classification accuracy on the test dataset with the final model at each iteration. Figure 3 shows the classification accuracy of the four scenarios. The X-axis represents the number of iterations. The result shows that ORBFL reaches a higher accuracy of 0.96, while the maximum accuracy for FL, BFL, and the committee-based BFL is 0.86, 0.84, and 0.59, respectively. Fluctuation in the model's accuracy for the committee-based BFL is because the data is Non-IID, and committee members may discard many helpful updates during evaluation process. You can also understand that ORBFL has higher accuracy in the beginning. The reason is that ORBFL only includes the intermediate updates if they pass the evaluation step.

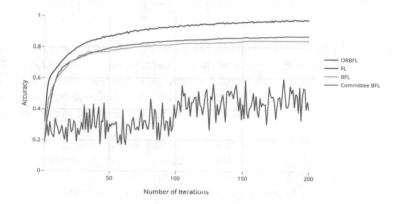

**Fig. 3.** Accuracy versus communication rounds on Non-IID MNIST data.

**Evaluation of efficiency** For evaluating the performance of the proposed evaluation approach, we compare our work with the committee-based BFL where 40% of training peers are involved in the evaluation process. The number of evaluations needed at each iteration for the committee-based BFL and ORBFL with 10 peers is 21 and 7, respectively. In fact, in the former, the number of evaluations is $\frac{N}{4}(N - \frac{N}{4})$, while for ORBFL it depends on the number of channels $N - \#channels$, where N is the number of all peers. The number of channels does not exceed the number of class labels.

**Evaluation of attack success rate** To evaluate the effectiveness of the proposed method in detecting poisoning attacks, we conduct the label flipping attack where malicious parties have poisoned datasets of digit 7s labelled as 1s. The adversary can control multiple peers. If a malicious node is selected as leader of a group, it will give random high score to malicious updates. We assume that malicious parties join the network after some iterations. Therefore, they will not be selected to be a leader of the channel until they prove their honesty. Table 1 presents the summary of the attack scenarios. The number of parties is 10 in all scenarios. The attack success rate means the proportion of 7s predicted to be 1s by the model in the test set.

**Table 1.** Summary of the attack scenarios.

| #Malicious parties | Conventional FL | | | ORBFL | | |
|---|---|---|---|---|---|---|
| | 1 | 2 | 3 | 1 | 2 | 3 |
| F1-score (digit 7) | 0.79 | 0 | 0 | 0.95 | 0.95 | 0.95 |
| F1-score (other digits) | 0.82 | 0.75 | 0.73 | 0.98 | 0.98 | 0.98 |
| Attack success rate | 22.66% | 95.81% | 96% | 0 | 0 | 0 |

The conventional FL is susceptible to the poisoning attacks, and malicious parties can start deteriorating the global model performance. By controlling 20% of training parties, the F1-score for digit 7 reaches to 0. On the other hand, ORBFL can detect the malicious parties at the first step (in the related channel) and put their IDs on the blocklist. Therefore, the performance of the global model will not be affected.

Considering the malicious parties joining the network from the beginning, the adversary needs at least two malicious nodes to start the attack. The adversary cannot harm more than one channel. Here malicious parties have poisoned datasets of digit 7s labelled as 1s. Therefore, they will join channel 1. The probability of choosing a malicious node as the leader of a channel is $\frac{M}{N_P}$; $M$ is the number of malicious nodes and $N_P$ is the number of all peers in the channel.

For this evaluation, we excluded the committee-based BFL because the evaluators discard many updates because their data distributions are different not because the updates are malicious or have lower quality.

## 5   Conclusion and Future Work

In this work, we introduced ORBFL to address the problem of model evaluation in FL with Non-IID data. We measured the effectiveness of the proposed framework through three perspectives: model accuracy, evaluation process efficiency, and attack success rate. The results showed the feasibility of the proposed evaluation process with Non-IID data. By connecting contributors with similar objectives and filtering out the low-quality updates, ORBFL showed better accuracy than others. The proposed evaluation process also outperforms the committee-based BFL approach. Additionally, the proposed method can mitigate the poisoning attacks. For future work, we will further evaluate the framework's performance by exploring different datasets and the scalability by considering a larger number of contributors.

# References

1. McMahan, B., Ramage, D.: Federated Learning: collaborative machine learning without centralized training data. Google Research Blog 3 (2017)
2. Bonawitz, K., Ivanov, V., Kreuter, B., et al.: Practical secure aggregation for privacy-preserving machine learning. In: Proceedings of the 2017 ACM SIGSAC Conference on Computer and Communications Security, pp. 1175–1191 (2017)
3. Mohassel, P., Zhang, Y.: Secureml: a system for scalable privacy-preserving machine learning. In: 2017 IEEE Symposium on Security and Privacy (SP), pp. 19–38. IEEE (2017)
4. Chai, D., Wang, L., Chen, K., Yang, Q.: Secure federated matrix factorization. IEEE Intell. Syst. **36**(5), 11–20 (2020)
5. Bittau, A., Erlingsson, Ú., Maniatis, P., et al.: Prochlo: strong privacy for analytics in the crowd. In: Proceedings of the 26th Symposium on Operating Systems Principles, pp. 441–459 (2017)
6. Konečný, J., McMahan, H.B., Yu, F.X., et al.: Federated learning: strategies for improving communication efficiency. arXiv Preprint (2016). arXiv:1610.05492
7. Yao, X., Huang, T., Wu, C., et al.: Towards faster and better federated learning: a feature fusion approach. In: 2019 IEEE International Conference on Image Processing (ICIP), pp. 175–179. IEEE (2019)
8. Kairouz, P., McMahan, H.B., Avent, B., et al.: Advances and open problems in federated learning. Found. Trends® Mach. Learn. **14**(1–2), 1–210 (2021)
9. Kang, J., Xiong, Z., Niyato, D., et al.: Incentive mechanism for reliable federated learning: a joint optimization approach to combining reputation and contract theory. IEEE Internet Things J. **6**(6), 10700–10714 (2019)
10. Lai, F., Zhu, X., Madhyastha, H.V., Chowdhury, M.: Oort: efficient federated learning via guided participant selection. In: 15th USENIX Symposium on Operating Systems Design and Implementation (OSDI 21), pp. 19–35 (2021)
11. Ghosh, A., Chung, J., Yin, D., Ramchandran, K.: An efficient framework for clustered federated learning. Adv. Neural Inf. Process. Syst. **33**, 19586–19597 (2020)
12. Ma, C., Li, J., Ding, M., et al.: On safeguarding privacy and security in the framework of federated learning. IEEE Network **34**(4), 242–248 (2020)
13. Lu, Y., Huang, X., Dai, Y., Maharjan, S., Zhang, Y.: Blockchain and federated learning for privacy-preserved data sharing in industrial IoT. IEEE Trans. Ind. Inform. **16**(6), 4177–4186 (2019)
14. Qu, Y., Gao, L., Luan, T.H., et al.: Decentralized privacy using blockchain-enabled federated learning in fog computing. IEEE Internet Things J. **7**(6), 5171–5183 (2020)
15. Yaga, D., Mell, P., Roby, N., Scarfone, K.: Blockchain technology overview. arXiv Preprint (2019). arXiv:1906.11078
16. Beloglazov, A., Abawajy, J., Buyya, R.: Energy-aware resource allocation heuristics for efficient management of data centers for cloud computing. Future Gener. Comput. Syst. **28**(5), 755–768 (2012)
17. Jiang, C., Xu, C., Zhang, Y.: PFLM: privacy-preserving federated learning with membership proof. Inf. Sci. **576**, 288–311 (2021)
18. Desai, H.B., Ozdayi, M.S., Kantarcioglu, M.: Blockfla: accountable federated learning via hybrid blockchain architecture. In: Proceedings of the Eleventh ACM Conference on Data and Application Security and Privacy, pp. 101–112 (2021)
19. Wang, R., Li, H., Liu, E.: Blockchain-based federated learning in mobile edge networks with application in internet of vehicles. arXiv Preprint (2021). arXiv:2103.01116

20. Toyoda, K., Zhang, A.N.: Mechanism design for an incentive-aware blockchain-enabled federated learning platform. In: 2019 IEEE International Conference on Big Data (Big Data), pp. 395–403. IEEE (2019)

21. Peng, Z., Xu, J., Chu, X., et al.: VFChain: enabling verifiable and auditable federated learning via blockchain systems. IEEE Trans. Netw. Sci. Eng. **9**(1), 173–186 (2021)

22. Xuan, S., Jin, M., Li, X., et al.: DAM-SE: a blockchain-based optimized solution for the counterattacks in the internet of federated learning systems. Secur. Commun. Netw. (2021)

23. Androulaki, E., Barger, A., Bortnikov, V., et al.: Hyperledger fabric: a distributed operating system for permissioned blockchains. In: Proceedings of the Thirteenth EuroSys Conference, pp. 1–15 (2018)

24. Shayan, M., Fung, C., Yoon, C.J., Beschastnikh, I.: Biscotti: a blockchain system for private and secure federated learning. IEEE Trans. Parallel Distrib. Syst. **32**(7), 1513–1525 (2020)

25. Kim, H., Park, J., Bennis, M., Kim, S.L.: Blockchained on-device federated learning. IEEE Commun. Lett. **24**(6), 1279–1283 (2019)

26. Kasyap, H., Tripathy, S.: Privacy-preserving decentralized learning framework for healthcare system. ACM Trans. Multimedia Comput. Commun. Appl. (TOMM) **17**(2s), 1–24 (2021)

27. Goldsteen, A., Ezov, G., Shmelkin, R., et al.: Data minimization for GDPR compliance in machine learning models. AI Ethics 1–15 (2021)

28. Martinez, I., Francis, S., Hafid, A.S.: Record and reward federated learning contributions with blockchain. In: 2019 International Conference on Cyber-Enabled Distributed Computing and Knowledge Discovery (CyberC), pp. 50–57. IEEE (2019)

29. Li, Y., Chen, C., Liu, N., et al.: A blockchain-based decentralized federated learning framework with committee consensus. IEEE Network **35**(1), 234–241 (2020)

30. McMahan, B., Moore, E., Ramage, D., et al.: Communication-efficient learning of deep networks from decentralized data. In: Artificial Intelligence and Statistics, pp. 1273–1282. PMLR (2017)

31. Paudice, A., Muñoz-González, L., Lupu, E.C.: Label sanitization against label flipping poisoning attacks. In: Joint European Conference on Machine Learning and Knowledge Discovery in Databases, pp. 5–15. Springer, Cham (2018)

# Enhancing Smart Contract Quality by Introducing a Continuous Integration Pipeline for Solidity Based Smart Contracts

Hauke Precht[(⊠)][iD], Florian Schwarm, and Jorge Marx Gómez

Carl von Ossietzky University of Oldenburg, Ammerländer HeerstraSSe 114-118,
Oldenburg 26129, Germany
{hauke.precht,florian.schwarm,jorge.marx.gomez}@uol.de

**Abstract.** Vulnerability concerns and quality issues lead to devastating and high-risk exploits in Smart Contracts. Several tools and approaches emerged to identify known errors and vulnerabilities in an early stage of Smart Contract development. However, those tools and approaches are currently considered stand-alone approaches and serve specific tasks that must be bonded manually. From traditional software engineering, the concept of Continuous Integration (CI) is well known for combining several tools and functions to ensure high code quality throughout every development stage. In this paper, we analyze the possibility of leveraging CI for the Smart Contract development. We identify requirements and design a general pipeline for Ethereum-based Smart Contracts. After two iterations, we developed a containerized approach that can be used independently on the used CI platform without the need to configure tools or test environments. We are able to show that we can link multiple tools and thus provide an automatic check and verification of the Smart Contract through different development stages. This approach can aid practitioners and researchers alike in creating safe and secure Smart Contracts of high code quality, thus contributing to the research area of testing in the context of blockchain oriented software engineering.

**Keywords:** Blockchain · Smart contract vulnerabilities · Continuous integration · Blockchain oriented software engineering

## 1 Introduction

With the rising of Smart Contracts (SCs), initially coined by *Nick Szabo* [40] but revived in the context of the Ethereum blockchain, a whole new set of possibilities for developing Distributed Application (dApps) are given. Currently, developers tend to make mistakes and introduce bugs to the SC [10,17], which can lead to potentially high and expensive risks. In recent history, several notable SC design flaws and bugs caused trouble in blockchain systems, e.g. [13,37]. *Destefanis et al.* analyzed the *Parity Wallet hack* and which resulted in the call for

© The Author(s): under exclusive license to Springer Nature Switzerland AG 2023
J. Prieto et al. (Eds.): BLOCKCHAIN 2022, LNNS 595, pp. 269–278, 2023.
https://doi.org/10.1007/978-3-031-21229-1_25

Blockchain Oriented Software Engineering (BOSE) as flaws and vulnerabilities in SCs are issues that could have been detected and avoided. In traditional Software Engineering (SE), Continuous Integration (CI) pipelines are a well-known and established concept to automatically execute unit and integration tests to ensure code quality and correctness, automatically binding several tools and functionalities together. In this paper, we aim to show that, based on existing approaches and tools, a general CI pipeline can be created that supports the development of SCs for Ethereum, detecting and preventing common mistakes in SC. By developing such CI concept for SC we will contribute to the research area of BOSE [38], focusing on the specific area of testing and tools. We further aim to provide a CI pipeline that can be used without requiring extensive manual configuration and dependency management. Such an approach can be utilized by practitioners and researchers alike to secure and enhance the quality of their SC.

## 2   Related Work

In 2017, *Liao et al.* identified a lack of sufficient toolchains for general concepts such as CI as the currently common practice for Blockchain-oriented Software (BOS) is to bond the toolchain manually [34]. In his blog post from 2019, Camilletti described a first approach of a CI pipeline for SCs based on the Microsoft Azure DevOps Pipelines [9]. Nevertheless, his approach focuses only on the Azure DevOps Pipelines and lacks generalizability and discussion. Another recent work from 2021, analyzing the applicability of DevOps in blockchain development, is presented by Wohrer and Zdun. They analyzed existing gray literature and GitHub projects to systematically analyze practices and approaches for DevOps approaches in the context of blockchain development [46]. In this regard, they discussed usages of CI as part of the DevOps approach in existing projects and showed that CI tools are already used to some degree in the development phase [46]. This shows that the idea to apply CI concepts to the SC development is worth to be further analysis, thus complementing the research of [46]. In the following subsections, formal verification approaches and tools for supporting SC development are briefly presented as they will be incorporated into the CI pipeline.

### 2.1   Formal Verification

In 2016, [5] outlined a framework for translating Solidity, a SC language for the Ethereum blockchain, to $F^*$ which is a functional programming language used for formal verification. *Bai et al.* proposed to model the SC in Process Meta Language (PROMELA) and using Simple PROMELA Interpreter (SPIN) for verification [2]. *ZEUS* is another tool for formal verification of SCs [31] which translates the SC together with a (user-generated) policy specification into a low-level intermediate representation (e.g., LLVM). Similar to *ZEUS*, [41] present a static analysis tool for Solidity called *SmartCheck* which translates Solidity SCs into an XML representation and verifies it against XPath patterns [41]. Further tools that follow similar approaches are *VERISOL* [45], *EASYFLOW* [18], *Securify* [43] or *MadMax* [23]. In 2019, *Feist et al.* presented a static analysis framework called *Slither*, which converts Solidity SCs to an intermediate representation that uses Static Single Assignment (SSA) to reduce the initial set of instructions for a more straightforward analysis while maintaining the original semantic [15]. While most of the approaches focus on the evaluation of Solidity SCs, several works emerged focusing the evaluation of the Ethereum Virtual Machine (EVM) itself e.g., [1,36]. We would like to refer the keen reader to the work presented in [26] for an additional overview.

## 2.2 Tools for Supporting Smart Contract Development

A survey conducted by [7] shows that developers consider the overall ecosystem for blockchain technology as immature. They see a lack of tool support for debugging and testing. But especially in the Ethereum ecosystem, a range of tools exists, for example, *Remix, Vim Solidity, Truffle, HardHat* or *Waffle*. Truffle is a well-known framework (over five million downloads [42]. and around 10300 stars on GitHub [29]) for unit testing of SC. In the same realm, Waffle is establishing itself for writing and testing SCs [44]. Hardhat provides a development environment for Ethereum, enabling developers to execute tests, code analysis and, SC interaction [25]. *Hegedűs* evaluates the possibilities to gather software code metrics (based on [11]) for Solidity SCs to aid developers and Quality Assurance (QA) processes in improving the quality of SCs [27]. Within the Ethereum space, Solidity-coverage is a popular example [3] of a free-to-use coverage solution [20] which presents, after executing, a coverage report of how well the SC code is covered by tests. The tool *SmartInspect*, which enables developers to analyze the current state of an already deployed contract on the Ethereum blockchain without being forced to redeploy the contract, is proposed in [8] to tackle the problem of debugging runtime states. Another emerging field that deals with the generation of SC code to enable a larger number of people to create SCs. One approach was already presented in 2016, evaluating the usage of a Domain-specific Language (DSL) [16]. A similar approach is proposed by *Choudhury et al.* by leveraging the Web Ontology Language (OWL) [12]. *Garamvolgyi et al.* also deal with the idea of generating SCs based on Unified Modeling Language (UML) statecharts [19]. *Mavridou and Laszka* propose a more technical approach as they present *FSolidM*, a framework that allows developers to create a finite state machine as a contract [35]. Note that also elements, discussed in 2.1 are developed as tools, e.g. Slither. For a more in-depth discussion of available approaches, the authors refer to [14,32,39].

## 3   Methodology

To ground our assumption of benefits when applying CI to the SC development, we follow the design science research approach [28] as we aim to develop a unique and novel CI pipeline for supporting SC development and enhancing security. Within this approach, we choose to follow the guidelines for design science introduced by [24]. First, we identified requirements for a CI pipeline based on unstructured interviews with two experts from the area of research and innovation and SE from a German software and infrastructure provider. In addition to these interviews, we analyzed an existing document containing software development guidelines based on the ISO 27001. With these requirements in mind, we designed and implemented a general concept of a CI pipeline for SCs. The given artifacts, the design and the respective implementation were created through iterations. Our research contribution is a general CI pipeline that supports the development of solidity-based SCs while being usable with multiple CI tools due to our containerized approach developed in our second iteration. Our results are presented in this paper, supporting the general findings on the usage of CI in blockchain development presented in [46].

## 4    Requirements for a CI Pipeline Considering Smart Contracts

In order to gather general requirements for a CI pipeline, we conducted two unstructured interviews with experts from the area of research and innovation and SE from a German software and infrastructure provider. They pointed out that they see a CI pipeline as a four-step process: (1) Build; The code will be compiled. (2) Check; Unit and integration tests are executed. (3) Package; The software is packaged and ready for deployment, e.g., into a JAR or WAR. (4) Deploy; The software is deployed to the respective stage, e.g., development or even production stage. Based on a document for general development guidelines based on ISO 27001, further requirements have been identified. Those requirements were grouped into functional (FR) and non-functional (N-FR) requirements. Within this paper, we focus on the SC development specific requirements for a CI system, discarding general requirements, e.g., notification of developers upon build failures. Based on our interviews and analysis, only one SC specific requirement for a CI pipeline has been identified: *If a developer pushes code changes to the code repository or creates a merge request, the code must go through automated vulnerability analysis.* This is due to the criticality of vulnerabilities in SCs as pointed out in the introduction section. The other requirements for a CI pipeline extracted from general SE considering the execution of unit/integration tests or running code coverage apply to SC development too. It seems that the CI concept from traditional SE can be transferred to the BOSE.

## 5    Developing a General CI Pipeline For Smart Contracts

Based on the before-identified requirements, we developed a general CI pipeline for supporting SC development. We developed our approach in two iterations described in the following while presenting the final CI pipeline in Fig. 1.

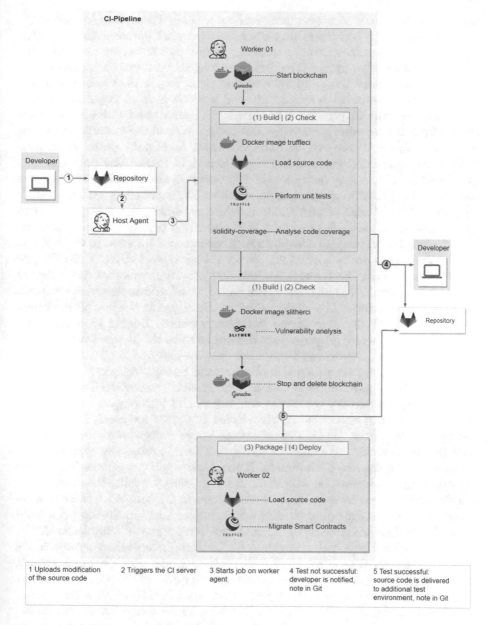

**Fig. 1.** A CI pipeline for solidity-based SC based on Docker. Source: own illustration.

As a code repository, we choose GITLAB. GitLab also provides an CI tool [21], but we used JENKINS as a dedicate CI tool as it is widely used and well known. Jenkins itself is open-source as well and supports all major operating systems, and can be used in the form of a Docker image. The proposed architecture for Jenkins is a host/worker architecture [30]. The host agent serves the User Interface (UI) and distributes the respective work packages to the worker agents that perform the actual execution. In our prototype, we used one host agent and two worker agents. The first worker agent performs the actual execution of the necessary tests and vulnerability detection, while the second agent performs the deployment once all tests and analysis have been performed successfully. Once the CI pipeline is triggered, the source code with all changes is pulled from the repository, compiled and tested. Within the testing phase, unit and integration tests are executed and validated. As SCs are prone to vulnerabilities, an automated vulnerability check is part of this testing phase as well. As shown in section 2, several tools and frameworks are available that perform vulnerability checks. In our prototype, we chose TRUFFLE along with GANACHE [42], SOLIDITY-COVERAGE and SLITHER, which are among the most popular and used tools [3]. Ganache-cli provides a local Ethereum blockchain to run tests on it, providing the fastest and cheapest test environment compared to the mainnet or testnets. As shown in section 2.1 Slither[15] leverages formal verification for vulnerability checks One limitation of our first prototype was that dependencies and requirements had to be manually installed for all said tools and frameworks on each used machine. This is described by *Boettiger* as the so-called 'dependency hell' [6]. Within our second iteration, we chose Docker to containerize the components as also [6] states that it helps to solve the 'dependency hell'. As shown in Fig. 1 we created two Docker images. The first one, for Truffle, is based on our Docker image truffleci which also contains Solidity-coverage. The second one is for Slither, based on the Docker image slitherci. The created Docker images can be used to create any number of containers, and thus our test environment is able to scale test execution. As the Docker container are created based on a defined Docker image, this is the only place where changes must be applied in case the test environment should be changed. This minimizes the required effort when adjusting the test environment. If the Jenkins job was successful, the job for deploying is triggered on worker 02. If the Jenkins job fails, the developer is notified of the result by email and a note is made in GitLab. With these tools in place, the design of the initial pipeline can be implemented in the form of a build script for Jenkins. As described, our containerized approach is implemented for Jenkins, but we also aim to provide a CI platform-independent approach. As GitLab, which we already used as a code repository, also provides as CI feature, we introduced our pipeline in this environment to evaluate the portability of our design and implementation. We only had to connect GitLab to a Docker runner, which creates and runs the containers [22]. This shows that with our containerized approach, dependencies and manual configuration is kept at a minimum. With this approach, we can provide a portable CI solution for SC development, independent of used or preferred CI platforms, while providing great flexibility to add or adjust tools in the CI process.

# 6    Discussion and Future Work

By performing two iterations, we optimized the first draft of a CI pipeline to minimize configuration effort by introducing containerization via docker. This enables us to define the test environment independent of the system environment or CI platform. We minimized the manual effort to set up and configure the respective tools and frameworks. This way, our CI pipeline implementation can be easily adapted and ported, as we have shown by porting our pipeline from Jenkins to GitLab CI. We showed that the CI concept from 'classical' SE could be transferred to the emerging area of BOSE without issue, supporting findings from [46] We conducted a preliminary workshop with practitioners in SE, receiving positive feedback and interest. However, to gain further insights and to validate the feasibility of the developed approach, interviews with SC engineers must be conducted as well. Possible further SC specific requirements, e.g., gas cost estimations, are not yet included. Within our current research, we did not encounter requirements that would require a dedicated CI concept for SC development, but instead, we can leverage already well-known and used CI approaches with SC specific tooling, supporting the finding of Wohrer and Zdun.

For future work, it is also necessary to test SCs in conjunction with the network itself, which is not considered yet in this approach. Such test environments are essential to test the BOS thoughtfully, which can be seen in the scope of integration testing [33]. Our current implementation is limited as it was conducted under laboratory conditions using only three tools in the CI pipeline. But as shown by [3,14,32,39] a range of different tools exists which need to be evaluated for usefulness in context of a CI pipeline. For example, *Bhardwaj et al.* provide a new penetration testing framework for SC showing newly detectable vulnerabilities [4]. To gain further insights in usability and suitability, the approach must also be introduced into a real SC development process to gain insights from SC engineering by conducting interviews. However, with this paper, we aim to make the first step towards quality enhancements of SCs via the introduction of a general CI pipeline for SCs.

# References

1. Albert, E., Correas, J., Gordillo, P., Román-Díez, G., Rubio, A.: Safevm: a safety verifier for ethereum smart contracts. In: Zhang, D., Møller, A. (eds.) Proceedings of the 28th ACM SIGSOFT International Symposium on Software Testing and Analysis-ISSTA 2019, pp. 386–389. ACM Press, New York, USA (2019). 10.1145/3293882.3338999
2. Bai, X., Cheng, Z., Duan, Z., Hu, K.: Formal modeling and verification of smart contracts. In: Unknown (ed.) Proceedings of the 2018 7th International Conference on Software and Computer Applications-ICSCA 2018, pp. 322–326. ACM Press, New York, USA (2018). 10.1145/3185089.3185138
3. Beer, N.: Whitepaper why and how to test smart contract. https://q-leap.eu/wp-content/uploads/2018/10/Why-and-How-to-test-Smart-Contracts-Whitepaper-q-leap-1.pdf (2018). Accessed 29 Apr 2022

4. Bhardwaj, A., Shah, S.B.H., Shankar, A., Alazab, M., Kumar, M., Gadekallu, T.R.: Penetration testing framework for smart contract blockchain. Peer-to-Peer Netw. Appl. **5**(2), 303 (2020). https://doi.org/10.1007/s12083-020-00991-6

5. Bhargavan, K., Swamy, N., Zanella-Béguelin, S., Delignat-Lavaud, A., Fournet, C., Gollamudi, A., Gonthier, G., Kobeissi, N., Kulatova, N., Rastogi, A., Sibut-Pinote, T.: Formal verification of smart contracts. In: Murray, T., Stefan, D. (eds.) Proceedings of the 2016 ACM Workshop on Programming Languages and Analysis for Security-PLAS'16, pp. 91–96, ACM Press, New York, USA (2016). 10.1145/2993600.2993611

6. Boettiger, C.: An introduction to docker for reproducible research. ACM SIGOPS Oper. Syst. Rev. **49**(1), 71–79 (2015). https://doi.org/10.1145/2723872.2723882

7. Bosu, A., Iqbal, A., Shahriyar, R., Chakroborty, P.: Understanding the motivations, challenges and needs of blockchain software developers: a survey. Empirical Software Engineering (2019). http://arxiv.org/pdf/1811.04169v2

8. Bragagnolo, S., Rocha, H., Denker, M., Ducasse, S.: Smartinspect: solidity smart contract inspector. In: 2018 International Workshop on Blockchain Oriented Software Engineering (IWBOSE), pp. 9–18, IEEE (20032018–20032018). 10.1109/IWBOSE.2018.8327566

9. Camilletti, N.B.: Continuous integration for smart contracts — by nicolás bello camilletti—southworks—medium. https://medium.com/southworks/continuous-integration-for-smart-contracts-4a8b78d387c (2019). Accessed 18 July 2021

10. Chepurnoy, A., Rathee, M.: Checking laws of the blockchain with property-based testing. In: 2018 International Workshop on Blockchain Oriented Software Engineering (IWBOSE), pp. 40–47. IEEE (20032018–20032018). 10.1109/IWBOSE.2018.8327570

11. Chidamber, S.R., Kemerer, C.F.: A metrics suite for object oriented design. IEEE Trans. Softw. Eng. **20**(6), 476–493 (1994). https://doi.org/10.1109/32.295895

12. Choudhury, O., Rudolph, N., Sylla, I., Fairoza, N., Das, A.: Auto-generation of smart contracts from domain-specific ontologies and semantic rules. In: 2018 IEEE International Conference on Internet of Things (iThings) and IEEE Green Computing and Communications (GreenCom) and IEEE Cyber, Physical and Social Computing (CPSCom) and IEEE Smart Data (SmartData), pp. 963–970. IEEE (30072018– 03082018). 10.1109/Cybermatics_2018.2018.00183

13. Cryptopedia Staff: What was the dao (2017). https://www.gemini.com/cryptopedia/the-dao-hack-makerdao

14. Di Angelo, M., Salzer, G.: A survey of tools for analyzing ethereum smart contracts. In: 2019 IEEE International Conference on Decentralized Applications and Infrastructures (DAPPCON), pp. 69–78. IEEE (04042019–09042019). 10.1109/DAPPCON.2019.00018

15. Feist, J., Grieco, G., Groce, A.: Slither: a static analysis framework for smart contracts. pp. 8–15 (2019). 10.1109/WETSEB.2019.00008. http://arxiv.org/pdf/1908.09878v1

16. Frantz, C.K., Nowostawski, M.: From institutions to code: towards automated generation of smart contracts. In: 2016 IEEE 1st International Workshops on Foundations and Applications of Self* Systems (FAS*W), pp. 210–215. IEEE (12092016–16092016). 10.1109/FAS-W.2016.53

17. Gao, J.: Guided, automated testing of blockchain-based decentralized applications. In: 2019 IEEE/ACM 41st International Conference on Software Engineering: Companion Proceedings (ICSE-Companion), pp. 138–140. IEEE (25052019–31052019). 10.1109/ICSE-Companion.2019.00059

18. Gao, J., Liu, H., Liu, C., Li, Q., Guan, Z., Chen, Z.: Easyflow: keep ethereum away from overflow. In: 2019 IEEE/ACM 41st International Conference on Software Engineering: Companion Proceedings (ICSE-Companion), pp. 23–26. IEEE (25052019–31052019). 10.1109/ICSE-Companion.2019.00029

19. Garamvolgyi, P., Kocsis, I., Gehl, B., Klenik, A.: Towards model-driven engineering of smart contracts for cyber-physical systems. In: 2018 48th Annual IEEE/IFIP International Conference on Dependable Systems and Networks Workshops (DSN-W), pp. 134–139. IEEE (25062018–28062018). 10.1109/DSN-W.2018.00052

20. GitHub: sc-forks/solidity-coverage. https://github.com/sc-forks/solidity-coverage (13072021). Accessed 13 July 2021

21. GitLab: Simplify your workflow with gitlab. https://about.gitlab.com/stages-devops-lifecycle/. Accessed 07 July 2021

22. GitLab: Run your ci/cd jobs in docker containers. https://docs.gitlab.com/ee/ci/docker/using_docker_images.html (2021). Accessed 30 Aug 2021

23. Grech, N., Kong, M., Jurisevic, A., Brent, L., Scholz, B., Smaragdakis, Y.: Madmax: surviving out-of-gas conditions in ethereum smart contracts. Proc. ACM Program. Lang. 2(OOPSLA), 1–27 (2018). 10.1145/3276486

24. Gregor, S., Hevner, A.R.: Positioning and presenting design science research for maximum impact. MIS Q. 37(2), 337–355 (2013). 10.25300/MISQ/2013/37.2.01

25. Hardhat: overview (2022). https://hardhat.org/getting-started

26. Harz, D., Knottenbelt, W.: Towards safer smart contracts: a survey of languages and verification methods. http://arxiv.org/pdf/1809.09805v4

27. Hegedűs, P.: Towards analyzing the complexity landscape of solidity based ethereum smart contracts. In: Tonelli, R., Destefanis, G., Counsell, S., Marchesi, M. (eds.) Proceedings of the 1st International Workshop on Emerging Trends in Software Engineering for Blockchain-WETSEB '18, pp. 35–39. ACM Press, New York, USA (2018). 10.1145/3194113.3194119

28. Hevner, A.R., March, S.T., Park, J., Ram, S.: Design science in information systems research. MIS Q. 28(1), 75–105 (2004)

29. Inc., C.S.: Truffle suite. https://github.com/trufflesuite (13072021). Accessed 13 July 2021

30. Jenkins: Distributed builds-jenkins-jenkins wiki. https://wiki.jenkins.io/display/jenkins/distributed+builds. Accessed 07 July 2021

31. Kalra, S., Goel, S., Dhawan, M., Sharma, S.: Zeus: Analyzing safety of smart contracts. In: Traynor, P., Oprea, A. (eds.) Proceedings 2018 Network and Distributed System Security Symposium. Internet Society, Reston, VA (February 18-21, 2018). 10.14722/ndss.2018.23082

32. Kirillov, D., Iakushkin, O., Korkhov, V., Petrunin, V.: Evaluation of tools for analyzing smart contracts in distributed ledger technologies. In: Misra, S., Gervasi, O., Murgante, B., Stankova, E., Korkhov, V., Torre, C., Rocha, A.M.A., Taniar, D., Apduhan, B.O., Tarantino, E. (eds.) Computational Science and Its Applications – ICCSA 2019, Lecture Notes in Computer Science, vol. 11620, pp. 522–536. Springer International Publishing, Cham (2019). 10.1007/978-3-030-24296-1_41

33. Koul, R.: Blockchain oriented software testing-challenges and approaches. In: 2018 3rd International Conference for Convergence in Technology (I2CT), pp. 1–6. IEEE, Piscataway, NJ (2018). 10.1109/I2CT.2018.8529728

34. Liao, C.F., Cheng, C.J., Chen, K., Lai, C.H., Chiu, T., Wu-Lee, C.: Toward a service platform for developing smart contracts on blockchain in bdd and tdd styles. In: 2017 IEEE 10th Conference on Service-Oriented Computing and Applications (SOCA), pp. 133–140. IEEE (22112017–25112017). 10.1109/SOCA.2017.26

35. Mavridou, A., Laszka, A.: Tool demonstration: Fsolidm for designing secure ethereum smart contracts. In: Bauer, L., Küsters, R. (eds.) Principles of Security and Trust. Lecture Notes in Computer Science, vol. 10804, pp. 270–277. Springer International Publishing, Cham (2018). 10.1007/978-3-319-89722-6_11

36. Park, D., Zhang, Y., Saxena, M., Daian, P., Roşu, G.: A formal verification tool for ethereum vm bytecode. In: Leavens, G.T., Garcia, A., Păsăreanu, C.S. (eds.) Proceedings of the 2018 26th ACM Joint Meeting on European Software Engineering Conference and Symposium on the Foundations of Software Engineering-ESEC/FSE 2018, pp. 912–915. ACM Press, New York, USA (2018). 10.1145/3236024.3264591

37. Peterson, B.: The amount of ether frozen in digital wallets is worth $162 million—which is less than initially feared (2017). https://www.businessinsider.com/ethereum-price-parity-hack-bug-fork-2017-11

38. Porru, S., Pinna, A., Marchesi, M., Tonelli, R.: Blockchain-oriented software engineering: Challenges and new directions. In: 2017 IEEE/ACM 39th International Conference on Software Engineering companion, pp. 169–171. IEEE, Piscataway, NJ (2017). 10.1109/ICSE-C.2017.142

39. Sayeed, S., Marco-Gisbert, H., Caira, T.: Smart contract: attacks and protections. IEEE Access 8, 24416–24427 (2020). https://doi.org/10.1109/ACCESS.2020.2970495

40. Szabo, N.: Smart contracts: Building blocks for digital markets (1996). http://www.fon.hum.uva.nl/rob/Courses/InformationInSpeech/CDROM/Literature/LOTwinterschool2006/szabo.best.vwh.net/smart_contracts_2.html

41. Tikhomirov, S., Voskresenskaya, E., Ivanitskiy, I., Takhaviev, R., Marchenko, E., Alexandrov, Y.: Smartcheck: static analysis of ethereum smart contracts. In: Tonelli, R., Destefanis, G., Counsell, S., Marchesi, M. (eds.) Proceedings of the 1st International Workshop on Emerging Trends in Software Engineering for Blockchain-WETSEB '18, pp. 9–16. ACM Press, New York, USA (2018). 10.1145/3194113.3194115

42. Truffle: Truffle dashboard. https://www.trufflesuite.com/dashboard. Accessed 13 July 2021

43. Tsankov, P., Dan, A., Cohen, D.D., Gervais, A., Buenzli, F., Vechev, M.: Securify: practical security analysis of smart contracts. http://arxiv.org/pdf/1806.01143v2

44. Waffle: Waffle documentation (2022). https://ethereum-waffle.readthedocs.io/en/latest/

45. Wang, Y., Lahiri, S.K., Chen, S., Pan, R., Dillig, I., Born, C., Naseer, I.: Formal specification and verification of smart contracts for azure blockchain. http://arxiv.org/pdf/1812.08829v2

46. Wohrer, M., Zdun, U.: Devops for ethereum blockchain smart contracts. In: 2021 IEEE International Conference on Blockchain (Blockchain), pp. 244–251. IEEE (2021). 10.1109/Blockchain53845.2021.00040

# "Are You What You Claim to Be?" Attribute Validation with IOTA for Multi Authority CP-ABE

Aintzane Mosteiro-Sanchez[1,2(✉)] ⓘ, Marc Barcelo[1] ⓘ, Jasone Astorga[2] ⓘ, and Aitor Urbieta[1] ⓘ

[1] Ikerlan Technology Research Centre, Arrasate/Mondragón, Spain
{amosteiro,mbarcelo,aurbieta}@ikerlan.es
[2] University of the Basque Country (UPV/EHU), 48013 Bilbao, Spain
jasone.astorga@ehu.eus

**Abstract.** Ciphertext-Policy Attribute-Based-Encryption (CP-ABE) is a one-to-many encryption scheme that generates secret keys according to attributes held by users. It is assumed that attribute authorities (AAs) know this matching between users' identities and their attributes and always generate valid secret keys. However, this approach is not scalable in systems with numerous users or several AAs. This lack of scalability makes attribute management difficult, leading to attackers abusing this situation to escalate privileges by spoofing attributes. We consider attribute spoofing a security risk and propose an attribute spoofing prevention scheme using a DAG-type DLT. We analyze the proposal by studying its effectiveness against the identified attack vectors. We also apply the proposal to a value-chain use case and provide a qualitative evaluation of the system by analyzing its ability to prevent attribute spoofing and other attack vectors identified in the literature review.

**Keywords:** DAG · IOTA · CP-ABE · Multi authority CP-ABE · Value chain

## 1 Introduction

Traditional encryption schemes guarantee data confidentiality between two endpoints. However, their performance is reduced when the same information is intended for multiple users. To address this limitation, in 2005, Sahai and Waters proposed Attribute-Based Encryption (ABE) [15]. ABE schemes correlate encryption and decryption with access policies and attributes. Thus, the same ciphertext can be decrypted by multiple users if the attributes and the policy match. However, this approach is limited: In CP-ABE, secret key generation is centralized, which creates a potential single-point failure.

To decentralize secret key generation, Rouselakis and Waters proposed a decentralized CP-ABE scheme [14]. Instead of relying on a single Key Generation Center (known in ABE schemes as Attribute Authorities, AAs), decentralized

J. Prieto et al. (Eds.): BLOCKCHAIN 2022, LNNS 595, pp. 279–288, 2023.
https://doi.org/10.1007/978-3-031-21229-1_26

CP-ABE generates each secret key combining multiple AAs. This makes it a suitable scheme for distributed environments, e.g., value chains. Value Chains are composed of participants who do not trust each other and exchange sensitive information. Therefore, implementing several AAs makes the system robust by avoiding single-point failure.

However, managing multiple AAs creates new risks and vulnerabilities that must be patched before deploying them in production environments. Traditional ABE schemes assume that AAs know which attributes belong to which users. However, this assumption is not scalable in distributed environments like value chains, with several AAs and many users. Overloading AAs with the management and maintenance of that much information is unfeasible. In this context, we identify a new risk: attribute spoofing. Spoofing is caused by users who take advantage of the low scalability of the system to demand keys from AAs that reflect privileges they do not possess.

Literature review has shown that solutions like DLTs can provide a secure and auditable attribute management system capable of avoiding attribute spoofing. They also guarantee integrity, immutability, and auditability, building trust among all chain members. In addition, they relieve the AAs from storing and managing this information internally.

**Problem Statement.** Our previous work [9] established a secure E2E data exchange in value chains by applying CP-ABE. However, it did not consider the following; (i) Centralized secret key generation causes a bottleneck and creates a one-point-failure. (ii) The risk of attribute spoofing. There is no verification of users' attributes or a system that securely manages user-attributes pairings and generates confidence for value chain members.

**Our Contributions.** This paper goes a step further and extends our previous work with the following contributions.

- We alleviate the bottleneck created by centralization by using the Rouselakis scheme to distribute key generation among multiple authorities.
- We avoid attribute spoofing by using a DAG-type DLT [16] as a tamper-proof record reflecting user attributes. We authenticate users with Federated Identity Management (FIM) to prevent impersonation.

The proposed solution provides E2E security and ensures industrial data integrity and confidentiality. The rest of the paper is organized as follows: Section 2.1 summarises related work, while Sect. 2.2 summarizes the functions of the Rouselakis scheme. Section 3.1 defines the identified attack vectors capable of exploiting attribute spoofing, and Sect. 3.2 introduces the industrial use case. The proposed solution is generic, but the use case allows us to explore it thoroughly. Section 4 discusses the proposed system and Sect. 5 evaluates how the proposal protects against the attack vectors identified in Sect. 3.1.

# 2    Background and Related Work

## 2.1    Related Work

As mentioned in the introduction, cryptographic schemes that allow one-to-many encryption are beneficial for distributed systems. This can be especially relevant in settings where the amount of information to be exchanged is high [8]. Therefore, solutions have been proposed that allow one-to-many encryption [5] but still need to know the identity of each recipient of the information.

Literature presents various combinations of CP-ABE [13] and Blockchain technologies [11] to achieve the secure distribution of information. It has even been studied which DLTs are more suitable in industrial environments [2]. However, no case considers the risk of attribute spoofing when using CP-ABE.

Equation (3) shows that AAs require two user-related parameters: $UID$ and $\mathbb{D}$. Their use binds the key to the DU, mathematically preventing DUs from colluding their keys to forge privileges. However, since the $\mathbb{D}-to-UID$ matching is not verified, malicious users can force this scenario to get privileges they do not have. We call this situation attribute spoofing.

Some solutions [11] consider that AAs should manage and distribute attributes. However, this solution involves AAs managing and storing the $\mathbb{D}-to-UID$ matching for every user in the system. This reduces scalability and leads to trust issues between companies that are collaborators and competitors.

The authors of [3] retrieve attributes from LDAP services, enterprise databases, or SAML attribute authorities. However, they do not detail how this integration occurs or whether a consensus is established about the attributes. In another paper [4], authors still consider that attributes can be retrieved from third parties. However, the risk of attribute spoofing still exists.

Therefore, we can conclude that it is common to assume that the $\mathbb{D}-to-UID$ matching is already known, correct, and accurate. In fact, it is an assumption that recent works like [10] or [12] continue to hold. However, attribute spoofing is a security risk that must be considered to design a holistic security system that ensures E2E data confidentiality between partners in real-world scenarios.

## 2.2    Multi Authority CP-ABE

This section summarizes the functions of the Rouselakis scheme [14], depicted in Fig. 1.

**Global Setup:** Executed by the SM, it takes the security parameter $k$ as input and outputs the system's global parameters $GP$.

$$GlobalSetup(K^k) \rightarrow GP \tag{1}$$

**Attribute Authorities Setup:** Every AA runs this algorithm. Each AA manages an attribute subset $\mathbb{A}_i = \{att_1, att_2, \cdots, att_n\} \in AA$ where $1 \leq i \leq m$; such that $n \geq 1$ and $m$ is the total number of attributes in the system. Thus, the AAs take their identity $AID$ and the $GP$ as inputs. Next, they use these

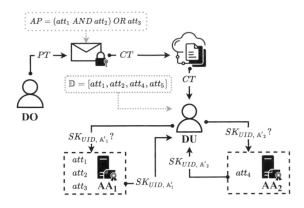

**Fig. 1.** Multi authority CP-ABE scheme.

parameters to generate the public key $PK_{AID}$ and the secret key $SK_{AID}$ related to $\mathbb{A}_i$. Authorities can be set at any time after $GP$ generation.

$$AASetup(GP, AID) \rightarrow PK_{AID}, \ SK_{AID} \tag{2}$$

**User Secret Key Generation:** Data Users, DUs are defined according to their attribute subset, which we define as $\mathbb{D} = \mathbb{A}'_1 \cup \mathbb{A}'_2 \cup \cdots \cup \mathbb{A}'_p$ in which $1 \leq p \leq m$. DUs provide their unique identifier $UID$ to the AAs that fulfill $\mathbb{A}'_p \subset \mathbb{A}_i$. In return, the AAs return $SK_{UID, \ \mathbb{A}'_p}$. By combining those, users obtain the secret key $SK_{UID, \ \mathbb{D}}$.

$$KeyGen(UID, \ AID, \ \mathbb{D}, \ SK_{AID}, \ GP) \rightarrow SK_{UID,\mathbb{D}} \tag{3}$$

**Plaintext Encryption:** Data Owners, DOs, take a plaintext $PT$, an access policy $AP$, the $PK_{AID}$ of the required AAs, and the global parameters $GP$. They output the ciphertext $CT$. An access policy can be defined as $AP = (att_1 \ AND \ att_2) \ OR \ att_3$.

$$Enc(PT, \ AP, \ PK_{AID}, \ GP) \rightarrow CT \tag{4}$$

**Ciphertext Decryption:** Data Users, DUs, take a ciphertext $CT$, their $SK_{UID,\mathbb{D}}$, and the system's global parameters $GP$. $PT$ is obtained if the attributes comply with the policy; if not, it returns $\perp$.

$$Dec(CT, \ SK_{UID,\mathbb{D}}, \ GP) \rightarrow PT \tag{5}$$

## 3    Methodology and Use Case

### 3.1    Attribute Spoofing

We define attribute spoofing as the misuse of attributes by an attacker to force an AA to provide a fake SK. The identified attack vectors are based on attackers providing AAs with fake $\mathbb{D}$ or $UID$ to force them to provide illegitimate SKs. The identified vectors are listed below:

- **V1.** An attacker provides a fake $UID$ to AAs.
- **V2.** An attacker provides a fake $\mathbb{D}$ to AAs.
- **V3.** An attacker provides a fake $\mathbb{D} - to - UID$ mapping, even though $\mathbb{D}$ and $UID$ exist independently in the system.

To prevent **V1**, the system requires user authentication. The scenario considered in this paper requires DUs to authenticate with each AA from which they require attributes. Similarly, AAs must identify DUs coming from different environments. This reduces the system's scalability by adding operations prior to key generation. Therefore, it is necessary to have an efficient and scalable authentication system. In addition, the No Trust Assumption [1] is deemed desirable: it should be unnecessary for different companies in the value chain to establish trust relationships between them.

**V2** and **V3** are related to the number of users and AAs. Therefore, decentralizing knowledge while preserving its confidentiality and integrity can alleviate AAs. Given this goal, DLT technology offers a straightforward solution.

The DLT type deployed in a system is closely related to the use case. Based on our previous work [9], and as mentioned in the introduction, the chosen use case are value chains. Literature [2] shows that DAGs are the best choice for industrial ecosystems, especially IOTA.

Data decentralization through IOTA requires that nodes participating in its network (called Tangle) validate transactions. In the use case, each company is responsible for uploading its data. However, attackers can take control of the company's nodes and try to upload fake data. This leads to the following vector:

- **V4.** An attacker controls the node that stores attribute and user information.

In a distributed environment where companies do not trust each other, solutions that assign a reputation to nodes can be helpful. These anti-spam mechanisms give companies confidence in the chain.

### 3.2   Industrial Use Case

Value chains are the use case chosen to analyze our attribute-spoofing prevention system. Information exchange between various hops in a value chain is currently limited [1], but collaboration improves chain performance [6]. Therefore, the value chain benefits from Rouselakis' deployment for multi-hop data sharing; and from the proposed attribute spoofing prevention system. This use case allows us to explore the nuances of attack vectors in a distributed environment where companies do not trust each other.

Let us consider the following example. $Company_A$ creates $Product_A$ to be consumed by $Company_B$. Sensitive information related to $Product_A$ may only be read by engineers of $Company_A$. However, $Company_B$ uses $Product_A$ to create $Product_B$ for the final customer. The final customer also gets a maintenance service through $Company_C$. Thus, information about $Product_A$ can be protected according to "$AP = (Company_A \; AND \; Engineer) \; OR \; Company_B$

*OR Company$_C$*" and data about *Product$_B$* according to "*AP = (Company$_B$ OR Company$_C$ OR Customer)*."

The example shows that attackers can access sensitive information by spoofing the attribute *Company$_C$*. A holistic security scheme has to address this.

## 4  Proposed System

Our proposal outlines a system that prevents attribute spoofing by guarding against the attack vectors identified in Sect. 3.1.

### 4.1  Attribute Storage and Distribution

According to the literature, and as mentioned in Sect. 3.1, the best DLT for our system is IOTA. However, the system has to protect against the attack vectors identified in Sect. 3.1. For this, we choose the DAG proposed by Stefanescu *et al.* [16]. Their proposal improves IOTA's efficiency with the following measures:

– Data is stored in Interplanetary File System (IPFS). Storing login credentials directly in IOTA requires a high-capacity network. This operation generates several transactions [10], which can be reduced by storing the IPFS datafile hash in IOTA. This system protects against attack vectors **V2** and **V3**.
– Our use case considers the possibility of a node going rogue (identified vector **V4**) and thus benefits from the Proof-of-Reputation (PoR) layer proposed in [16]. This PoR ties users' reputation to the number of validated transactions, detecting rogue nodes and reducing the computational burden on legitimate nodes.

We use [16] to store the $\mathbb{D} - to - UID$ mappings. However, the original proposal does not consider the data retrieval workflow from the IPFS. We extend their work by defining the data recovery workflow for our attribute spoofing prevention system (Fig. 2). To this end, we identify two mechanisms in the secret key generation process: user registration and key recovery.

For user registration, each company stores the $\mathbb{D} - to - UID$ mappings in IPFS. The hash of the IPFS file is stored in IOTA, ensuring data integrity. We use the PoR layer proposed by [16] to prevent spam and compromised nodes from storing fake $\mathbb{D} - to - UID$ matchings. These measures increase the business confidence in the stored data. The step-by-step process is shown below:

1. The company (DO) sends the $UIDs$ to the company members (potential DUs). It also sends the $\mathbb{D} - to - UID$ mapping such that $UID_{data} = (UID\|\mathbb{D})$ to the node as required by [16].
2. The node sends the $UID_{data}$ file to IPFS, which returns $H(UID_{data})$.
3. The node registers $H(UID_{data})$ in IOTA, obtaining its transaction $t_x$.
4. Finally, the node distributes $(t_x\|UID)$ through the same channel it distributed the GPs required by the Rouselakis scheme.

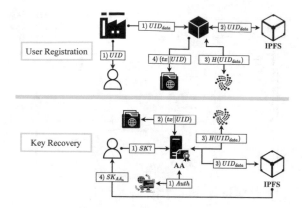

**Fig. 2.** Flow diagrams of the proposed user registration and key recovery mechanisms.

The registration process securely stores the $UID_{data}$ and guarantees its integrity. The AAs retrieve $UID_{data}$ from IPFS and verify the file's integrity through the DAG to avoid attribute spoofing. Thus, any DU who needs a $SK$ can interact with the corresponding AAs to obtain it. The detailed process for key recovery is explained below.

1. DUs query AAs for their $SK_{UID, \mathbb{D}}$. AAs authenticate DUs and obtain $UID$.
2. The AAs use the $UID$ to retrieve the $t_x$ in which the information associated with the DU's attributes is stored.
3. The AAs retrieve $H(UID_{data})$ from IOTA, which they use to obtain $UID_{data}$ from IPFS.
4. The AAs compute $SK_{UID, \mathbb{A}'_p}$ per Eq. (3), and send them to the DU. The DU combines them to form $SK_{UID, \mathbb{D}}$, and uses it for decryption according to Eq. (5).

### 4.2 User Authentication

The attack vector **V1** identified in Sect. 3.1 requires AAs to authenticate DUs and prevent impersonation. Identity verification is crucial in any security system, especially in multi-authority CP-ABE schemes such as Rouselakis'. DU' $SKs$ are generated for specific $UID$ to prevent them from colliding the keys and escalating privileges.. Therefore, before retrieving the $\mathbb{D} - to - UID$ mapping, the system must ensure that there is no identity theft-derived attribute spoofing.

The authentication system chosen for our attribute spoofing prevention system is Federated Identity Management (FIM) [7], whose behavior is shown in Fig. 3. FIM allows users from different companies to use their company credentials to authenticate to different AAs. This also implies that AAs do not have to manage the credentials of multiple users from different companies. Instead, they verify the token issued through an identity provider (IdP). This IdP establishes a trust relationship with the different companies and acts as an intermediary between the AAs and the companies. This way, AAs only have to manage the trust relationship with the IdP.

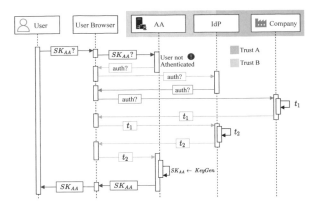

**Fig. 3.** Message exchange of the proposed user authentication process.

# 5    System Evaluation

## 5.1    Protection Against Attribute Spoofing

This paper introduces the security risk posed by attribute spoofing in CP-ABE schemes. And more specifically, it considers it for the Rouselakis multi-authority scheme. This section evaluates the effectiveness of the proposed solution against the attack vectors identified in Sect. 3.1.

- **V1.** AAs authenticate users through FIM, preventing impersonation. Which, in return, prevents attribute spoofing and collusion attacks.
- **V2.** The system guarantees that the AAs obtain $\mathbb{D}$ from IPFS, where they have been stored by the company to which the user belongs. IOTA certifies the immutability of the $\mathbb{D}$ stored in IPFS.
- **V3.** The use of IPFS and IOTA prevents attackers from escalating privileges by combining pre-existing $\mathbb{D}$ and $UIDs$ in the system.
- **V4.** The PoR layer proposed by [16] guarantees the detection of rogue nodes.

## 5.2    Additional Security Properties

The proposed technology combination protects the system against the identified attack vectors. However, Sect. 3.1 mentioned additional security properties that would benefit the system, among them the No-Trust Assumption. Therefore, we can consider that our system achieves the following properties:

- **No-Trust Assumption** [1]: Applying FIM guarantees that companies do not have to establish a trust relationship between them. Instead, the AAs establish a trust relationship with the IdP, which establishes the trust relationship with different companies.
- **Data protection:** IPFS provides distributed data storage, preventing the loss of $\mathbb{D} - to - UID$ mappings. Similarly, storing the hash of the IPFS file in a DAG allows AAs to detect tampered IPFS files.

- **No One-Point-Failures:** The Rouselakis scheme allows various AAs to manage the same attribute. Thus, if an attacker compromises one of the AAs, others may take their place.
- **Collusion attacks:** The Rouselakis scheme avoids collusion attacks by relating users' $SK$ to their $UID$. Thus, our system reinforces this protection by preventing impersonation via FIM.

## 6 Conclusions

This paper identifies the issue of attribute spoofing in CP-ABE. It outlines the basis of the attack and proposes an attribute spoofing prevention system that relies on a DAG.

The combination of IOTA and IPFS ensures secure attribute storage, and using FIM for user authentication prevents impersonation. This way, the system distributes responsibility and trust among each value chain member and protects the system against single-point failures, identity theft, and attribute spoofing. This enhances E2E security and ensures the integrity and confidentiality of industrial data. Therefore the combination of IOTA and IPFS has proven its suitability for attribute validation.

To demonstrate the proposal's suitability, we conducted a qualitative assessment of the system's effectiveness against the identified attack vectors and its additional security properties. We can conclude that the tools used in this paper improve the features of the Rouselakis cryptographic algorithm. FIM improves the algorithm's ability to resist collusion attacks; likewise, IOTA and IPFS protect against attribute spoofing.

**Acknowledgment.** The European Commission financially supported this work through the Horizon Europe program under the COMP4DRONES project (grant agreement N° 826610). It was also partially supported by the *Ayudas Cervera para Centros Tecnológicos* grant of the Spanish Centre for the Development of Industrial Technology (CDTI) under the project EGIDA (CER-20191012) and by the Basque Country Government under the ELKARTEK program, project REMEDY - REal tiME control and embeddeD securitY (KK-2021/00091).

## References

1. Bader, L., Pennekamp, J., Matzutt, R., Hedderich, D., Kowalski, M., Lücken, V., Wehrle, K.: Blockchain-based privacy preservation for supply chains supporting lightweight multi-hop information accountability. Inf. Process. Manag. **58**(3), 102529 (2021). https://doi.org/10.1016/j.ipm.2021.102529
2. Cui, L., Yang, S., Chen, Z., Pan, Y., Xu, M., Xu, K.: An efficient and compacted dag-based blockchain protocol for industrial internet of things. IEEE Trans. Ind. Inf. **16**(6), 4134–4145 (2020). https://doi.org/10.1109/TII.2019.2931157
3. Di Francesco Maesa, D., Lunardelli, A., Mori, P., Ricci, L.: Exploiting blockchain technology for attribute management in access control systems. In: Djemame, K., Altmann, J., Bañares, J.Á., Agmon Ben-Yehuda, O., Naldi, M. (eds.) GECON

2019. LNCS, vol. 11819, pp. 3–14. Springer, Cham (2019). https://doi.org/10.1007/978-3-030-36027-6_1

4. Di Francesco Maesa, D., Mori, P., Ricci, L.: A blockchain based approach for the definition of auditable access control systems. Comput. Secur. **84**, 93–119 (2019). https://doi.org/10.1016/j.cose.2019.03.016

5. Epiphaniou, G., Pillai, P., Bottarelli, M., Al-Khateeb, H., Hammoudesh, M., Maple, C.: Electronic regulation of data sharing and processing using smart ledger technologies for supply-chain security. IEEE Trans. Eng. Manage. **67**(4), 1059–1073 (2020). https://doi.org/10.1109/TEM.2020.2965991

6. Flynn, B.B., Huo, B., Zhao, X.: The impact of supply chain integration on performance: A contingency and configuration approach. J. Oper. Manage. **28**(1), 58–71 (2010). https://doi.org/10.1016/j.jom.2009.06.001

7. Hardt, D.: The OAuth 2.0 Authorization framework. RFC 6749 (Oct 2012). https://doi.org/10.17487/RFC6749

8. Liu, R., Kumar, A.: Leveraging information sharing to configure supply chains. Inf. Syst. Front. **13**(1), 139–151 (2011). https://doi.org/10.1007/s10796-009-9222-8

9. Mosteiro-Sanchez, A., Barcelo, M., Astorga, J., Urbieta, A.: End to end secure data exchange in value chains with dynamic policy update. Preprint arXiv:2201.06335 (2022)

10. Nakanishi, R., Zhang, Y., Sasabe, M., Kasahara, S.: Combining IOTA and attribute-based encryption for access control in the internet of things. Sensors **21**(5053), 1–15 (2021). https://doi.org/10.3390/s21155053

11. Pennekamp, J., Bader, L., Matzutt, R., Niemietz, P., Trauth, D., Henze, M., Bergs, T., Wehrle, K.: Private multi-hop accountability for supply chains. In: 2020 IEEE International Conference on Communications Workshops (ICC Workshops), pp. 1–7 (2020). https://doi.org/10.1109/ICCWorkshops49005.2020.9145100

12. Preuveneers, D., Joosen, W., Bernal Bernabe, J., Skarmeta, A.F.: Distributed security framework for reliable threat intelligence sharing. Secur. Commun. Netw. **2020**, 8833765 (2020). https://doi.org/10.1155/2020/8833765

13. Qi, S., Zheng, Y., Li, M., Liu, Y., Qiu, J.: Scalable industry data access control in RFID-enabled supply chain. IEEE/ACM Trans. Netw. **24**(6), 3551–3564 (2016). https://doi.org/10.1109/TNET.2016.2536626

14. Rouselakis, Y., Waters, B.: Efficient statically-secure large-universe multi-authority attribute-based encryption. In: Böhme, R., Okamoto, T. (eds.) FC 2015. LNCS, vol. 8975, pp. 315–332. Springer, Heidelberg (2015). https://doi.org/10.1007/978-3-662-47854-7_19

15. Sahai, A., Waters, B.: Fuzzy identity-based encryption. In: Cramer, R. (ed.) EURO-CRYPT 2005. LNCS, vol. 3494, pp. 457–473. Springer, Heidelberg (2005). https://doi.org/10.1007/11426639_27

16. Stefanescu, D., Galán-García, P., Montalvillo, L., Unzilla, J., Urbieta, A.: Towards a holistic DLT architecture for IIoT: improved DAG for production lines. In: Prieto, J., Partida, A., Leitão, P., Pinto, A. (eds.) BLOCKCHAIN 2021. LNNS, vol. 320, pp. 179–188. Springer, Cham (2022). https://doi.org/10.1007/978-3-030-86162-9_18

# Privacy-Preserving Energy Trade Using Double Auction in Blockchain Offline Channels

Subhasis Thakur$^{(\boxtimes)}$ ⓘ, John Breslin ⓘ, and Sweta Malik

National University of Ireland, Galway, Ireland
{subhasis.thakur,john.breslin,M.SWETA1}@nuigalway.ie

**Abstract.** Blockchain is a promising tool to implement peer-to-peer energy trade algorithms because it lowers the cost of electricity by eliminating 3rd parties such as the utility companies from energy trade and by creating a secure trade platform. However, the state of the art blockchain-based peer to peer energy trade solutions have privacy and scalability problems. In this paper, we proposed a novel method to execute double auction-based peer to peer energy trade in blockchain offline channels to enhance security, privacy and scalability of peer to peer energy trade. We prove that the proposed decentralised double auction is secure, privacy-preserving and more efficient as McAfee's double auction.

**Keywords:** Peer to peer energy trade · Blockchains · Blockchain offline channels · Bitcoin lightning network · Double auction

## 1 Introduction

Blockchains are recently used to implement peer to peer energy trade algorithms. Blockchain can reduce the cost of electricity by removing utility companies from the local electricity trade. Also, it provides a secure trade platform where parties who do not trust each other can trade electricity. There are many game-theoretic formulations of electricity trade, such as trade models based on cooperative games [10], non-cooperative games, and auctions [12]. Double auction is a popular trade model that allows the sellers and the buyers to choose their desired cost and price for electricity trade. McAfee's double auction [6] is widely used in the energy trade. In this trade model, the seller sets its asking price (minimum price at which it will sell), and the buyer sets its reservation price (maximum price that it will pay). The auction mechanism finds a price that satisfies both the seller and the buyer. The auction mechanism also needs to satisfy few economic properties such as individual rationality, budget-balancing, truth-fullness, and economic efficiency. McAfee's double auction is known to be individually rational and truthful. But it is not budget-balanced and economically efficient.

© The Author(s): under exclusive license to Springer Nature Switzerland AG 2023
J. Prieto et al. (Eds.): BLOCKCHAIN 2022, LNNS 595, pp. 289–302, 2023.
https://doi.org/10.1007/978-3-031-21229-1_27

Double auction-based electricity trade algorithms imple-ment with blockchains [12] have privacy and scalability problems. In such an auction procedure, the prosumers (who may produce and consume electricity) have to inform the auctioneer about its asking price and reservation price. It may reveal private information about the prosumers. For example, if a prosumer wants to sell electricity from 6 PM to 8 PM then, it may indicate that the prosumer may not be in his/her house during this time. Besides the physical security problem from such information, an adversary may use such information to alter electricity prices. For example, if the adversary is a utility company and it finds that its several customers who want to use electricity from peer to peer energy trade for a specific time during a day, i.e., they will not buy electricity from the grid at the price offered by the utility company at these times, then the adversary may alter price of electricity from the grid to make peer to peer energy trade economically insignificant. Hence, in this privacy-preservation problem, an adversary is the entity who wants to reveal such trade patterns of the prosumers. An adversary may control (via cyberattack) a centralised auctioneer to execute such a privacy disruption attack.

Additionally, the current blockchain implementation of double auction-based energy trade may have a scalability problem. Public blockchains have scalability problems. There are double auction-based energy trade solutions that use pub-lic blockchains. Scalability problems may prevent real-time execution of energy trade operations. For example, in the Bitcoin network it may take 10 min to find a new block, this means it may take up to 10 min to record a transaction in the blockchain. However, electricity trade windows may be as short as 5 min. Thus short trade operations may not be executed via public blockchains.

Further, the short window electricity trade has a very low monetary value. If blockchain transactions are created for such trade decisions, transaction fees may be too low to attract miners. Increasing the fees may increase the price for electricity in peer to peer energy trade and may defeat the purpose of the local electricity trade.

Also, blockchains (public blockchains) have a high environmental impact. It is estimated that Bitcoin has annual electricity consumption adds up to 45.8 TWh and produces 36.95 megatons of $CO_2$ annually. Thus the creation of blockchain transactions may increase the carbon footprint of the blockchain-based electricity trade. Hence environment-friendly energy trade should minimise the number of transactions to be created to execute the trade operation.

In this paper, we mitigate these problems with public blockchain-based elec-tricity trade using double auction. We proposed to use blockchain-offline channels to execute the double auction procedure. Offline channels allow secure transac-tions without immediately updating the blockchains. It significantly reduces the number of transactions needed to execute energy trade operations. Our main contributions are as follows: (1) We present a double auction-based energy trade algorithm using offline channels. (2) We prove that the proposed auction mech-anism is privacy-preserving. (3) We prove that the proposed double auction is secure as it prevents double spending of units of electricity to be sold. (3) We show that the proposed double auction is more efficient than McAfee's double auction [6].

The paper is organised as follows: in Sect. 2 we discuss related literature, in Sect. 3 we define the auction problem, in Sect. 4 we discuss the offline channel-based double auction for energy trade, in Sect. 5 we discuss economic properties of the auction, in Sect. 6 we present experimental evaluation on efficiency of the proposed energy trade, and we conclude the paper in Sect. 7.

## 2    Related Literature

Blockchains are recently used to implement trade algorithms for peer to peer energy trade. In [10] authors have used coalitional game theory in peer to peer energy trade which also includes electric vehicles. In [11,14] the authors used coalitional game theory to model blockchain-based energy trade. Double auction is a popular trade mechanism for peer to peer energy trade. In [12] the authors used double auction for peer to peer energy trade using blockchains. In [2] the authors used continuous double auction for peer to peer energy trade. In this paper, we investigated double auction-based energy trade. The Bitcoin lightning network was proposed in [8] which allows peers to create and transfer funds among them without frequently updating the blockchain. Similar networks are proposed for Ethereum [1] and credit networks [4]. Blockchain is a suitable platform for peer to peer energy trade. In [3] authors have analysed the suitability of blockchain network in terms of network size, communication delay, etc on recording transactions for the energy trade. In [13], the authors used blockchain offline channels to implement a cooperative game-based peer to peer energy trade.

In this paper we used proof of work-based blockchains. Proof of work-based blockchains was proposed in [7]. There are several variations of blockchains in terms of consensus protocols. Offline channels for Bitcoin, i.e., Bitcoin Lightning network was proposed in [8] which allows peers to create and transfer funds among them without frequently updating the blockchain. Similar networks were proposed for Ethereum [1] and credit networks [5]. [5,9] proposed a landmark-based routing protocol for fund transfer in a credit network. We advance the state of the art in double auction-based energy trade as follows: (1) We proposed a privacy-preserving double auction which prevents an adversary from identifying the trading parties. (2) We proposed to use blockchain offline channels which allows us to build a high scale double auction protocol Table 3.

## 3    Energy Trade Problem

A double auction for peer to peer energy trade can be described as follows:

**Definition 1.** The double auction for the trade window $t_i$ to $t_{i+1}$ is as follows:

- $A$ be a distinguished entity acting as the auctioneer. All prosumers know $A$ and have a secure communication method with $A$.

**Table 1.** Notations used to model the energy trade problem.

| | |
|---|---|
| $\{p_i\}$ | A set of $n$ prosumers(who may buy and (or) sell electricity from each other.) A prosumer may be a house with renewable energy generator such as asolar panel |
| $\{t_i\}$ | Discrete time instances $dt$ time apart |
| $D_i^j$ and $S_i^j$ | Energy demand and energy supply (through its own energy generators, i.e., solar panels) of the prosumer $p_i$ at time $t_j$ until time $t_{j+1}$ or for time duration $dt$. The energy requirement at $p_i$ at time $t_j$ is $E_i^j = D_i^j - S_i^j$. A positive value of $E_i^j$ means prosumer $p_i$ has surplus energy (i.e., it is generating more than its own consumption) and a negative value of $E_i^j$ will mean the prosumer has an energy deficiency for next $dt$ time duration |
| $d^{i,j}$ | The distance between prosumer $p_i$ and $p_j$ w.r.t the distribution lines |

- Asking prices: $\{P_{i-\epsilon}^x\}$ be the set of asking prices of the sellers received by the auctioneer $A$ at most $\epsilon$ time before time instance $t_i$ and not after $t_i$. $\{p_{i-\epsilon}^x\}$ be amount of electricity a prosumer wants to sell for the time duration $t_i$ to $t_{i+1}$.
- Bids: $\{Q_{i-\epsilon}^x\}$ be the set of bids of the sellers received by the auctioneer $A$ at most $\epsilon$ time before time instance $t_i$ and not after $t_i$. $\{q_{i-\epsilon}^x\}$ be amount of electricity a prosumer wants to buy for the time duration $t_i$ to $t_{i+1}$.
- Outcomes: A set of messages to the prosumers $(W_i^x, C_i^x)$ such that $W_i^x$ is the amount of electricity $x$ will buy or sell (i.e., they will be paid or receive fund for only this amount of electricity) at the price or cost $C_i^x$. $W_i^x$ is positive means $p_x$ will sell $W_i^x$ electricity (in kWh) and it will receive fund $C_i^x$ per unit of electricity (kWh). $W_i^x$ is negative means $p_x$ will buy $W_i^x$ electricity (in kWh) and it will lose fund $C_i^x$ per unit of electricity (kWh).

The economic characterisation of a double auction are as follows:

- Individual rationality: No prosumer should pay more than its bid, no prosumer should get less than its asking price, no prosumer will trade without participating in the auction.
- Balanced Budget: The auctioneer $A$ will collect fund from prosumers who wants to buy and it will pay the prosumers who wants to sell. Using the outcome $(W_i^x, C_i^x)$, we can calculate the fund at the auctioneer as $B^i = \sum_{x \in n: W_i^x > 0} W_i^x \times C_i^x + \sum_{x \in n: W_i^x < 0} W_i^x \times C_i^x$. We will say the double auction is strong budget-balanced if $B^i = 0$ and we will say the double auction weak budget-balanced if $B^i > 0$.

- Truthfulness: We will say a double auction is Nash equilibrium incentive compatible if it is a Nash equilibrium for the prosumers to report true bid or asking price.
- Economic efficiency: We will define economic efficiency in terms of amount of electricity can be traded using the auction. It can be defined as:

$$EE^i = \left( \sum_{x \in N} p_i^x - \sum_{x \in N: W_i^x > 0} W_i^x \right) + \left( \sum_{x \in N} q_i^x + \sum_{x \in N: W_i^x < 0} W_i^x \right) \quad (1)$$

$EE^i$ is the amount summation of the electricity which could not be sold by the auction and the amount of electricity that can not be bought from the double auction.

Further, we can characterise the double auction with privacy-preservation properties. An adversary in a double auction wants to know the asking price, bids, and outcome of a double auction. We assume that the adversary can control a fraction of nodes of the blockchain network which is computing the double auction procedure.

## 4   Energy Trade with Decentralised Double Auction

Briefly, the auction procedure is as follows: (1) There are two types of nodes in the offline channel network, one is the set prosumers, and another is the set of nodes controlled by the DSOs (any of these nodes can be the auctioneer). (2) Each prosumer randomly chooses an auctioneer node to buy or sell electricity on its behalf. (3) The chosen auctioneer can either find a matching prosumer who also wants to buy or sell via it. If there is no such prosumer, then another auctioneer may buy or sell electricity from it. (4) We designed a protocol that allows each auctioneer to buy electricity from other prosumers with the assurance that if it cannot sell the electricity then it can sell it to either another auctioneer or the first prosumer.

### 4.1   Unidirectional Offline Channel

Blockchain offline channels [8] uses multi-signature addresses to open an offline channel among peers of the blockchain. This offline channel [8] is bidirectional and potentially infinite, i.e., it can execute the infinite number of transfers between two peers provided they do not close the channel and each of them has sufficient funds. We construct an offline channel for proof of work-based public blockchain with the following properties: (1) We construct a uni-directional channel between two peers, i.e., only one peer can send funds to another peer of this channel. (2) We construct a uni-directional channel which can be used for a finite number of transfers from a designated peer to another peer.

The procedure for creating the uni-directional channel from $A$ to $B$ ($A$ transfers token to $B$)is as follows: Let $A$ and $B$ are two peers of the channel network $H$. $M_{A,B}$ is a multi-signature address between $A$ and $B$. This is a unidirectional channel from $A$ to $B$.

1. $A$ creates a set of $k$ ($k$ is a positive even integer) random strings $S_A^1, \ldots, S_A^k$. Using these random strings $A$ creates a set of Hashes $H_H^1 = H(S_B^1), H_B^2 = H(S_B^k) \ldots, H_B^k = H(S_B^k)$ where $H$ is Hash function (using SHA256). $A$ creates a Merkle tree order $\lambda$ using these Hashes. Thus there are $k$ leaf nodes and $k - 1$ non-leaf nodes of this Merkle tree. We denote the non-leaf nodes as $H_A'^1, \ldots, H'(k-1)_A$.
2. $B$ creates a set of $k1$ random strings $S^1, \ldots, S^k$ and corresponding Hashes $H_B^1, \ldots, H_B^k$.
3. $A$ sends the Merkle tree to $B$ and $B$ sends the set of Hashes $H_B^1, \ldots, H_B^k$ to $A$.
4. $A$ sends a Hashed time-locked contract $HTLC_A^1$ to $B$ as follows:
   (a) From the multi-signature address $M_{A,B}$, 1 token will be given to $A$ after time $T$ if $B$ does not claim these tokens before time $T$ by producing the key to $H_A'^1$ and 0 token will be given to $A$ if it can produce the key to $H_B^1$.
   (b) $A$ sends $HTLC_A^1$ to $B$.
5. Now, $A$ sends 1 token to $M_{A,B}$. $A$ includes the Merkle tree and $H_B^1, \ldots, H_B^k$ in this transaction. This records the Merkle tree and $H_B^1, \ldots, H_B^k$ in the blockchain and any other peer can verify the existence of these Hashes by checking transactions of the public blockchain. Also, at this stage, $A$'s funds are safe as it can get the tokens from $M_{A,B}$ after time $T$ as $B$ does not know $H_A'^1$.
6. Next to send another $(1/k)$ tokens to $B$, $A$ sends $S_A^1$ to $B$ and $B$ sends $H_B^1$ to $A$. Then $A$ forms the following HTLC:
   (a) From the multi-signature address $M_{A,B}$, $1 - 1/k$ token will be given to $A$ after time $T$ if $B$ does not claim these tokens before time $T$ by producing the key to $H_A'^2$ and $1/k$ token will be given to $A$ if it can produce the key to $H_B^2$.
   (b) $A$ sends $HTLC_A^2$ to $B$.
7. This process continues until all keys of the Hashes of non-leaf nodes are revealed by $A$.

In this model of the unidirectional channel, $A$ is sequentially releasing the keys of the Merkel tree of the HTLCs. Its fund in this channel is decreasing with time. It can not prevent $B$ from obtaining the tokens as only $B$ can publish the HTLCs. $B$ will publish the HTLC where it gets the maximum value. A path-based fund transfer (PBT) is possible in this uni-directional channel network along paths in the channel network. For example, three nodes $p_x, p_y, p_z$ can facilitate token transfer from $p_x$ to $p_z$ as follows:

1. $p_z$ will create a lock $H_z$ and inform $p_x$ about $H_z$.
2. A sequence of two HTLCs will be created. The first HTLC will transfer fund of 1 token from $p_x$ to $p_y$ if $p_y$ can present the key to $H_z$ before time 10 sec. The second HTLC will transfer fund of 1 token from $p_y$ to $p_z$ if $p_z$ can present the key to $H_z$ before time 8 sec.
3. $p_z$ will initiate the execution of these HTLCs by revealing key to $H_z$ to $p_y$. And, $p_y$ will use the same key to take 1 token from $p_x$.

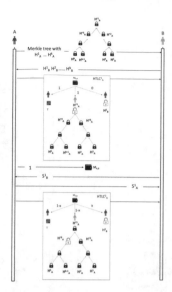

**Fig. 1.** Procedure of creating unidirectional offline channels.

## 4.2   Double Auction Using Offline Channels

We will a blockchain network with $m > n$ peers consisting of prosumers, DSOs, and miners. We assume that the blockchain network uses Bitcoin-like proof of work-based blockchains and there is an offline channel network using a unidirectional network as described in the previous section. The blockchain network will consist of a set of distinguished and recognised (possibly the miners of the blockchain network) as the auctioneers. We denote these peers as the set $\{d_i\}$. A prosumer may establish a uni-directional channel with a subset of auctioneers. Auctioneers may establish a uni-directional channel among themselves. We will denote the channel network as a directed graph $G = (V, E)$ where $V$ is the peers of the channel network and $E$ is the channels. $W(E)$ will denote channel balances, i.e., $W(p_i, p_j)$ is the amount of fund $p_i$ can send to $p_j$ using the channel $p_i \rightarrow p_j$. The blockchain network will also consist of a set of nodes $\{D_i\}$ representing the DSOs of the electricity management networks. They are regulators and their objective is to ensure the integrity of the energy trade and security of the electricity grid. They need to ensure that one unit of electricity to be sold to only one prosumer asking for one unit of electricity.

**Asking Price and Bids:** The channel from the prosumer $p_i$ to the auctioneer node $d_j$ will be used for payment for electricity to be used by $p_i$. The channel to the prosumer $p_i$ from the auctioneer node $d_j$ will be used to pay $p_i$ for its surplus electricity to be sold to other prosumers. The bid and asking price announcement process is shown in Fig. 2(a) and it is as follows:

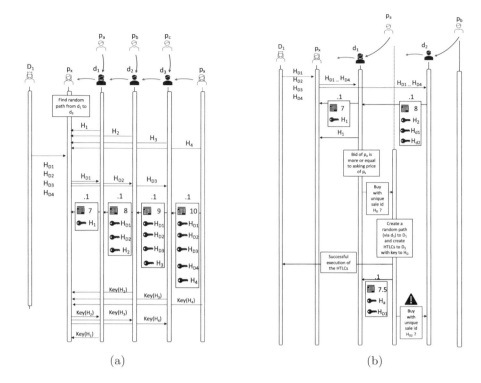

**Fig. 2.** (a) Double auction procedure: Sequence of key distribution and HTLC formation, (b) Double auction procedure: Procedure to ensure uniqueness of a trade.

1. A prosumer can submit its bid to any auctioneer node (with whom it has a offline channel) by creating a HTLC as follows:
   (a) Let $p_a$ wants to submit a bid to $d_1$. The HTLC will state that $p_a$ will pay $d_1$ if $d_1$ can prove that the offered electricity is unique as it will not be sold to another prosumer.
2. Let $p_x$ wants to sell electricity it can submit its asking price as follows:
   (a) $p_x$ contacts a DSO node $D_1$ for a token to be used as a unique identity to its offer to sell electricity for the next trade window (if $p_x$ asks it at time $t_i$ then its trade window is from $t_i$ to $t_i + dt$.). $D_1$ will inform $p_x$ about a set of the locks $H_{D1}, H_{D2}, H_{D3}, H_{D4}$ (number of such locks is equal to the total number of auctioneer + 1) to be used to uniquely identify $p_x$'s asking price for the trade window from $t_i$ to $t_i + dt$.
   (b) $D_1$ will check with all DSO nodes if $p_x$ has applied for another such identifier at the same or overlapping trade window. In such a case $D_1$ will not issue the unique offer identifier to $p_x$.
   (c) We assume that all prosumers are permission-ed, as their identity, location, smart-meter identification numbers are verified by the DSO nodes. This means prosumers can not create a false identity to participate in this trade with multiple identities.

(d) Next, $p_x$ will create a random path among all auctioneer nodes and inform all of them about $H_{D1}$.

(e) All auctioneer node will inform $p_x$ about the Hash in the Merkle tree for the unidirectional channel between pairs of auctioneer nodes according to the path among the auctioneer node created by $p_x$ which can transfer fund equal to the asking price of $p_x$. As shown in Fig. 2(a) such Hashes are $H_1, H_2, H_3$, and $H_4$.

(f) Now, $p_x$ will facilitate the creation of sequence $HTLCs$ from itself to the auctioneer nodes and finally to itself using the Hashes of the Merkle tree of the channel among the auctioneers. As shown in Fig. 2(a), first HTLC states that $d_1$ will give $p_x$ .1 token if $p_x$ can produce key to $H_1$. The second HTLC states that $d_2$ will given $d_1$ .1 token if $d_1$ can reveal $H_2$ and $H_{D1}, H_{D2}$ before time 8. And so on until $p_x$ gives $d_3$ .1 tokens before time 10 for keys to all hashes $H_4$, and $H_{D1}, H_{D1}, H_{D2}, H_{D3}, H_{D4}$. The time mentioned in these HTLCs are just examples, in practice, these times will be few seconds but constantly increasing.

(g) After creating these HTLCs, $d_2$ will reveal key to $H_2$ to $p_x$, and $d_3$ will reveal key to $H_3$ to $p_x$.

(h) Next, $p_x$ will reveal key to $H_2$ to $d_1$, key to $H_3$ to $d_2$, and key to $H_4$ to $d_3$.

(i) Now, the auctioneer $d_1$ will buy the electricity from $p_x$ as follows: $d_1$ reveals the key of $H_1$ to $p_x$ as it purchases the surplus electricity from $p_x$. $d_1$ can either sell the electricity to any other prosumer who has submitted a bid to $d_1$ or sell to $d_2$ if there is no such prosumer.

(j) Similarly, $d_2$ can purchase the electricity it from $d_1$ and may sell it to any prosumer who had submitted a bid to $d_2$ or sell it to $d_3$ otherwise. This process can continue to $p_x$. This means an auctioneer node can always purchase electricity from another auctioneer or a prosumer as it can always resale it.

(k) Thus, using the above protocol $p_x$ can submit asking prices to the auctioneer. Its asking price will be evaluated by the auctioneer in a random sequence chosen by $p_x$. Any rational auctioneer, say $d_1$ may be able to sell this electricity from $p_x$ to another prosumer such as $p_a$ if bid of $p_a$ is more or equal to the asking price of $p_x$. Otherwise it will resale the electricity to another auctioneer $d_2$.

(l) It is possible that the asking price $p_x$ is more than any other bid submitted by other prosumers. In this case, no rational auctioneer will be able to sell electricity from $p_x$ and initial fund given to $p_x$ will be taken from $p_x$ by $d_3$. Thus if it is not possible to sell electricity from $p_x$ then $p_x$ does not get paid.

**Trade uniqueness:** However, in the given protocol it may be possible to double-spend the electricity as follows:

1. It is possible that $d_1$ finds a prosumer $p_a$ whose bid is more than the asking price of $p_x$.

2. $d_1$ will sell electricity from $p_x$ to $p_a$ and also, resale the electricity to the next auctioneer $d_2$. Thus $d_1$ will be able to sell the electricity at least twice.
3. All such auctioneer may do the same and resale the electricity multiple times.

Thus it is necessary to maintain uniqueness of the electricity trade. We allow the prosumer to trade a uniform amount of electricity per its asking price and bid. We solve the uniqueness problem of electricity trade as follows:

1. As shown in Fig. 2(b), before submitting the bid, $p_x$ needs to collect a unique offer identifier from a DSO node $D_1$. Let $D_1$ informs $p_x$ about the unique offer identifiers $H_{D1}, H_{D2}, H_{D3}, H_{D4}$.
2. $p_x$ will inform all auctioneer about this unique offer identifier.
3. After forming the HTLCs for asking prices, an auctioneer $d_1$ may find a prosumer $p_a$ whose bid is more than the asking price of $d_1$.
4. $p_a$ will pay $d_1$ if $d_1$ can prove uniqueness of the trade offer.
5. $d_1$ will inform $p_a$ that unique offer identifier is $H_{D1}$ and it is issued by the DSO node $D_1$.
6. $p_a$ can create and execute a PBT from $p_a$ to $D_1$ via $d_1$ with lock $H_{D1}$. In such a transfer $D_1$ will execute the PBT by revealing the key to $H_D$. Key to $H_{D1}$ will eventually reach $p_a$ and $d_1$ after successful execution of the PBT.
7. If such PBT in unsuccessful then it will prove that the proposed trade is not unique and $p_a$ will not $d_1$.
8. Hence if $d_1$ will not be able to sell electricity from $p_x$ multiple times.

## 5    Analysis

**Theorem 1.** *If a prosumer $p_x$ trades electricity with another prosumer $p_a$ then an adversary may not know the trade between $p_x$ and $p_a$ unless the adversary controls all parties in the path from $p_x$ to $p_a$. For example, such a path will include $p_x \rightarrow d_1 \rightarrow p_a$.*

*Proof.* Note that, submission of bids and asking prices are executed via exchange of HTLCs. For example, as shown in Fig. 2(a), $p_x$ will receive an updated HTLC from $d_1$ as it submits its asking price to $d_1$. Similarly, $d_1$ will receive an updated HTLC from $p_a$ as $p_a$ submits a bid to $d_1$. The updated HTLCs are not visible by other parties and hence the adversary needs to control all entities in a path between two parties who are trading electricity.

**Theorem 2.** *It is not possible to double-spend the electricity in the proposed double auction protocol.*

*Proof.* As discussed in the previous section, a prosumer has to collect a unique offer identifier from a DSO node before it can submit its asking price. DSO nodes want to secure the electricity grid, and hence they will not allow multiple offer identifiers to the same prosumer for overlapping trade window. This is because multiple prosumers may consume electricity simultaneously if there is a double-spending of electricity offered by a prosumer. This will imbalance the electricity grid. Further, as mentioned before, prosumers are permission-ed nodes, i.e., DSO will check their location and identity to ensure that each prosumer has only one node in the blockchain. Thus a prosumer can't use multiple identities to sell the same unit of electricity. Further, before buying electricity a prosumer will seek proof of uniqueness from the DSO using the offer identifier provided by the auctioneer. The DSO node will reveal the key to such an offer identifier. For example, (Fig. 2(b)) $p_a$ may verify offer identifier $H_{D1}$ by creating a PBT from $p_a$ to the DSO node $D_1$ with the lock $H_{D1}$. $D_1$ will reveal the key to this Hash to $p_a$ and $p_a$ will use it to execute a PBT to $D_1$. If $D_1$ has already revealed the key to $H_{D1}$ then it will not participate in such a PBT, and failure of this PBT will cause $p_a$ not to buy electricity from $d_1$. Further, if $d_1$ tries to resale electricity to another auctioneer $d_2$ then $p_b$ (who had submitted the bid to $d_2$) will check if the trade offer is unique by creating a PBT to the DSO node $D_1$. Again $D_1$ can ensure offer uniqueness. And, $p_b$ will not buy if the offer is not unique. Hence $D_2$ will not buy the electricity from $D_1$ as it can not resale the electricity.

**Theorem 3.** *The proposed auction is individually rational, weakly budget balanced, and have the same economic efficiency compared with McAfee's double auction [6].*

*Proof.* The proposed auction is individually rational because

1. a prosumer does not pay more than its bid,
2. a prosumer does not receive less than its asking price,
3. and, if the surplus electricity of a prosumer can not be sold by the auction then the prosumer does not get paid.

(1) and (2) hold because a rational auctioneer $d_1$ will only buy electricity from $p_x$ if it can sell it to another prosumer $p_a$ whose bid is higher than the asking price of $p_x$. Otherwise, the auctioneer will lose funds. (3) holds because using the set of HTLCs as shown in Fig. 2(a), if electricity from $p_x$ can not be sold then although $p_x$ initially gets paid by $d_1$ but $p_x$ pays back the fund to $d_3$.

The proposed double auction is weakly budget-balanced as the auctioneer will only pay the prosumer such as $p_x$ if it can sell electricity from $p_x$ at the price at least equal to the asking price of $p_x$.

The proposed double auction has at least the same economic efficiency as McAfee's double auction because the asking price of $p_x$ is sequentially compared will all bids and it is matched (i..e., corresponding electricity is sold) as soon as there is a bid more than asking price of $p_x$. Any asking price which is matched with a bid (i.e., the bid is more than the asking price) by McAfee's algorithm will also be matched in the proposed double auction method.

The proposed auction can significantly reduce the number of transactions needed to be recorded in the blockchain. A unidirectional channel can be used a finite number of times without updating the blockchain. If such a number of channel updates is $k$ then it can reduce $k-1$ transactions needed to be recorded in the blockchain (one transaction is needed to open then offline channel). Auctioneers will have a non-negative revenue from the proposed auction. This will attract investment in building the blockchain network to execute the energy trade.

## 6    Experimental Evaluation

We used prosumer energy demand and PV generation data from [12] to evaluate proposed decentralised double auction. The data contains energy demand and PV generation data of 100 prosumers for 24 hours (data recorded in every 5 minutes). We used the blockchain simulator developed in [3] to simulate a proof-of-work-based blockchain network and offline channels. First, we used agent-based modelling to implement a centralised double auction and then we implemented the decentralised double auction in the blockchain simulator. In each set of experiments, we executed simulated peer to peer energy trade among these prosumers. We execute four sets of simulations. In each set, first we execute the energy trade simulation for centralised auction, and then, we execute the the energy trade simulation for the decentralised auction with identical asking price and bid data(in the range $[0, 1]$). In these experiments we measured the amount of electricity traded as an indicator of energy trade efficiency. Fig. 3 show that the decentralised auction is more efficient than centralised double auction as more electricity is traded with decentralised double auction.

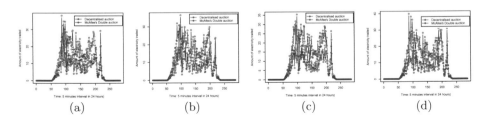

   (a)            (b)            (c)            (d)

**Fig. 3.** (a) It shows the performance of the proposed decentralised double auction and centralised McAfee's double auction, (b) It shows the performance of the proposed decentralised double auction and centralised McAfee's double auction, (c) It shows the performance of the proposed decentralised double auction and centralised McAfee's double auction, (d) It shows the performance of the proposed decentralised double auction and centralised McAfee's double auction.

## 7    Conclusion

In this paper, we proposed a secure and privacy-preserving decentralised double auction using blockchain offline channels. The proposed method will be useful if energy trade is executed in public blockchains. Public blockchains have scalability problems, and there is a significant carbon footprint for creating a transaction in public blockchains such as Bitcoin or Ethereum. However, these public blockchains can be valuable platform to implement decentralised energy trade due their easy access and high token valuation. Our solution implements decentralised energy trade with a minimum number of transactions. Hence it is not only a highly scalable solution but also reduces the carbon footprint of using public blockchains.

**Acknowledgement.** This publication has emanated from research conducted with the financial support of Science Foundation Ireland (SFI) under Grant Number SFI/12/RC/2289, co-funded by the European Regional Development Fund.

# References

1. Raiden network. https://raiden.network/. Accessed 2018
2. Chen, K., Lin, J., Song, Y.: Trading strategy optimization for a prosumer in continuous double auction-based peer-to-peer market: a prediction-integration model. Appl. Energy **242**, 1121–1133 (2019)
3. Hayes, B., Thakur, S., Breslin, J.: Co-simulation of electricity distribution networks and peer to peer energy trading platforms. Int. J. Electric. Power Energy Syst. **115**, 105419 (2020)
4. Malavolta, G., Moreno-Sanchez, P., Kate, A., Maffei, M.: Silentwhispers: enforcing security and privacy in decentralized credit networks. IACR Cryptology ePrint Arch. **2016**, 1054 (2016)
5. Malavolta, G., Moreno-Sanchez, P., Kate, A., Maffei, M.: Silentwhispers: enforcing security and privacy in decentralized credit networks. IACR Cryptology ePrint Arch. **2016**, 1054 (2016)
6. McAfee, R.: A dominant strategy double auction. J. Econ. Theory **56**(2), 434–450 (1992). https://doi.org/10.1016/0022-0531(92)90091
7. Nakamoto, S.: Bitcoin: A peer-to-peer electronic cash system. https://www.bitcoin.org (2008)
8. Poon, J., Dryja, T.: The bitcoin lightning network: scalable off-chain instant payments. https://lightning.network/lightning-network-paper.pdf
9. Roos, S., Moreno-Sanchez, P., Kate, A., Goldberg, I.: Settling payments fast and private: efficient decentralized routing for path-based transactions. CoRR abs/1709.05748 (2017). https://Darxiv.org/abs/1709.05748
10. Thakur, S.: A unified model of peer to peer energy trade and electric vehicle charging using blockchains. In: IET Conference Proceedings, pp. 77(6 pp.)(1).https://digital-library.theiet.org/content/conferences/10.1049/cp.2018.1909
11. Thakur, S., Breslin, J.G.: Peer to peer energy trade among microgrids using blockchain based distributed coalition formation method. Technol. Econ. Smart Grids Sustain. Energy **3**(1), 1–17 (2018). https://doi.org/10.1007/s40866-018-0044-y
12. Thakur, S., Hayes, B.P., Breslin, J.G.: Distributed double auction for peer to peer energy trade using blockchains. In: 2018 5th International Symposium on Environment-Friendly Energies and Applications (EFEA), pp. 1–8 (Sep 2018). doi: https://doi.org/10.1109/EFEA.2018.8617061

13. Thakur, S., Breslin, J.G.: Real-time peer to peer energy trade with blockchain offline channels. In: 2020 IEEE International Conference on Power Systems Technology (POWERCON), pp. 1–6 (2020). doi: https://doi.org/10.1109/POWERCON48463.2020.9230545
14. Tushar, W., Saha, T.K., Yuen, C., Azim, M.I., Morstyn, T., Poor, H.V., Niyato, D., Bean, R.: A coalition formation game framework for peer-to-peer energy trading. Appl. Energy **261**, 114436 (2020). https://doi.org/10.1016/j.apenergy.2019.114436

# Interoperable Industry 4.0 Plant Blockchain and Data Homogenization via Decentralized Oracles

Denis Stefanescu[1,2]([envelope])[iD], Leticia Montalvillo[1][iD], Patxi Galán-García[1][iD], Juanjo Unzilla[2][iD], and Aitor Urbieta[1][iD]

[1] Ikerlan Technology Research Centre, Arrasate-Mondragon 20500, Spain
distefanescu@ikerlan.es, lmontalvillo@ikerlan.es, pgalan@ikerlan.es
aurbieta@ikerlan.es
[2] University of the Basque Country (UPV/EHU), Bilbao 48013, Spain
juanjo.unzilla@ehu.eus

**Abstract.** The use of Distributed Ledger Technologies (DLTs) in the field of Industry 4.0 has been constantly increasing over the last years. In a complete Industry 4.0 ecosystem there are multiple smart factory clusters that belong to one or more companies and openly interact with their business partners, clients and suppliers. In such a complex ecosystem, multiple interoperable DLTs are typically required in order to ensure the integrity of the data throughout the whole process from when the data is generated up until it is processed at higher levels. Furthermore, industrial data is commonly heterogeneous and needs to be homogenized, without neglecting security. Therefore, in this work we address the aforementioned issues by (1) providing a production line data homogenization approach via secure blockchain oracles and (2) designing an interoperable plant blockchain for trustworthy homogenized data management across several industrial plants. Finally, we also present a prototype implementation of our proposal and discuss our results.

**Keywords:** Blockchain · Industry 4.0 · Industrial Internet of Things · IIoT · Oracles · Data homogenization · Directed Acyclic Graphs

## 1 Introduction

Industry 4.0 conceptualizes the application of a wide range of modern approaches and technologies of the 21st century in the field of manufacturing industry. The most relevant technologies that are related to Industry 4.0 are Big Data, Artificial Intelligence (AI), advanced robotics, edge computing, 5G, the Industrial Internet of Things (IIoT) and overall digitalisation [1]. Over the last few years, Distributed Ledger Technologies (DLTs) such as blockchain or Directed Acyclic Graphs (DAGs) have started to become increasingly relevant in the field of Industry 4.0 due to their ability to process and store data in a decentralized and trustworthy manner [2].

© The Author(s): under exclusive license to Springer Nature Switzerland AG 2023
J. Prieto et al. (Eds.): BLOCKCHAIN 2022, LNNS 595, pp. 303–313, 2023.
https://doi.org/10.1007/978-3-031-21229-1_28

A typical DLT-based Industry 4.0 scenario is presented in [3], where there is a broad ecosystem of inter-connected smart plants clusters. Figure 1 depicts an Industry 4.0 scenario with two cluster plants, where each of the cluster plant has two production lines (e.g. Plant A has production lines "PL1-A1" and "PL2-A2"). Within each production line, several IIoT devices operate and generate raw production line data that measures, among other data, machines' performance, process productivity and quality, and machines' End-Of-Life. This production data generated by IIoT devices is securely stored by means of DAG type DLTs, which are the appropriate DLTs for IIoT devices [4]. As IIoT devices generate high amount of data every second (or minute), in order to reduce the storage burden of the DAGs it is a good practice to leverage appropriate storage, such as an InterPlanetary File System (IPFS)[1] storage system, where the actual IIoT raw data is stored, whilst the DAGs store the data hashes in an immutable and trustworthy manner. This provides data integrity keeping the DAG lightweight.

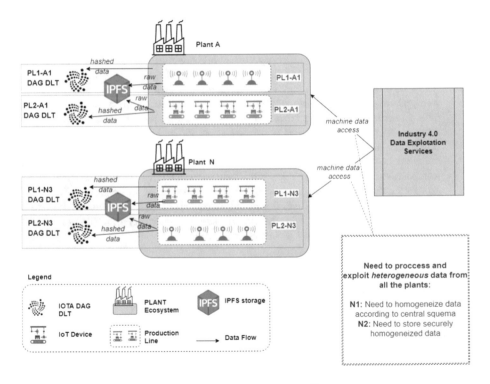

**Fig. 1.** Industry 4.0 motivating Scenario

So far, the data generated by IIoT devices is securely gathered and stored by the means DAG type DLTs and IPFS storage. However, Industry 4.0 does not stop at machine data level, and this data that is being gathered at "the lowest

---

[1] https://ipfs.io/.

level" needs to be exploited and processed by "higher" levels to derive and build actual information, such as, machine and IIoT fleet status, machines predictive maintenance (by AI algorithms), compute overall process productivity, etc. Hence, these "upper" processes **need to access and process heterogeneous data from all cluster plants**. However, accessing and processing all the raw machine data from all production level DAG type DLTs is not straightforward, mainly due to:

- Lack of homogeneous machine data. Industrial data is typically heterogeneous. For instance, data could be expressed in different units of measure depending on the machine provider and machine version; they could have a different number of decimal places; and they can include certain errors or variations. This problem stems from the fact that, and according to Jirko et al. [5], *"machines within a complex system are produced by different manufacturers with different data models and interfaces"*. As a consequence, this problem hinders industrial interoperability and integration, and has a detrimental impact on the ability to process data using disruptive technologies such as Big Data or AI.
- Lack of efficiency and security to access machine data [6]. It is not efficient nor secure to leave the responsibility to "data exploitation services" to access and transform the raw machine data into "readable" plant level data. Accessing machine data means that each "data exploitation service" needs to be a client of every production line DLT that wants to access data from. Additionally, these services would need to process all the data from all machines and homogenize it accordingly. This is not efficient as the data exploitation services will spend high amount of time to access and homogenize data before exploiting it. This is not entirely secure either, since it breaks data custody chain and mixes responsibilities, as in each "data exploitation service" it becomes obscure the actual data format being used to be exploited, and the traceability and integrity is lost.

Hence, to solve aforementioned problems there is a need to provide means so that:

- The raw machine data that resides in DAG type DLTs is securely and consistently homogenized by a secured and traceable process. This will ensure that the data conforms to a common data model, i.e. interoperability, so that processes at higher levels can exploit the data consistently.
- The homogenized data is securely stored and accessed, ensuring its integrity and availability. This will ensure trust on the data throughout the whole process, i.e. from where the data is generated at a production line level to where it is processed at a "higher levels". Processing raw IIoT data through a DAG DLT is not sufficient if at a higher level we have a centralized and non-persistent data structure where the data can be easily tampered with [7].

This work presents the following contributions for the described problem:

1. A "data homogenization" process for solving data interoperability issues that relies on the use of Smart Oracles as the source of trust for the target data model scheme the data needs to conform to.
2. A blockchain-based solution to securely store the homogenized "plant level" data.
3. A prototype that implements a secure "data homogenization" process, that (i) accesses machine raw data stored in IOTA DLTs, (ii) gets the target data model schema from the ChainLink Smart Oracles, (iii) performs the data homogenization from the source data scheme to the target data schema, and (iv) stores the homogenized data into a Hyperledger Fabric blockchain network so that it can be consistently accessed by other services.

The remainder of the paper is organized as follows. In Sect. 2 we analyze the existing related work in this field and outline our contributions. In Sect. 3 we describe our proposed solution for solving the Industry 4.0 data interoperability and security challenges. In Sect. 4 we present our prototype based on IOTA, ChainLink and Hyperledger Fabric technologies. Finally, Sect. 5 includes the conclusion of the paper and future work insights.

## 2    Related Work

In this section we analyze the most relevant works that are related to multiple DLT usage in IoT and industrial environments. We also perform a comparison study between this work and the related works.

P. Bellavista *et al.* [8] propose a relay scheme based on Trusted Execution Environment (TEE) to provide secure interoperability between blockchains in Industry 4.0. The proposed solution employs an off-chain secure computation element that is invoked by smart contracts. However, secure interoperability between blockchains is achieved at high performance costs and reduced scalability. Furthermore, the proposed solution relays on specific hardware requirements. Finally, other relevant types of DLTs such as DAGs, which were specifically designed for industry, were not taken into account.

E. Scheid *et al.* [9] propose Bifröst, a highly modular blockchain interoperability API that is based on a notary scheme. However, the proposed solution incurs high latency and several critical security issues.

Y. Jiang *et al.* [10] try to integrate a DAG type DLT with a consortium blockchain via sidechains. The consortium blockchain is used as a control station that manages several sidechains of IoT data. There are notaries that act like gateways between the DLTs. However, the proposed mechanism adds a high grade of complexity, energy consumption and latency to the network, since it employs the PoW mechanism in order to avoid centralization.

Z. Gao *et al.* [11] propose a data migration architecture using an oracle that acts like a trusted notary that connects two blockchains. However, this approach creates centralization and a single point of failure, thus making the use of blockchain pointless.

C. Wiraatmaja *et al.* [12] propose a custom centralized oracle service made in JavaScript to transfer data between decentralized storage solutions such as IPFS or IOTA and the Ethereum blockchain. However, this architecture uses a centralized oracle service that can be easily compromised.

Unlike the aforementioned works, we design an efficient and trustworthy industrial scheme where we aim to achieve data integrity throughout the whole process from where data is generated up until it is standardized and managed at a higher level. We start from a scenario where raw IIoT data is processed by efficient DAG DLTs and design a decentralised oracle scheme for data homogenization and an interoperable blockchain for trustworthy homogenized data management at a plant level. Table 1 shows a comparison between the related works and our own.

**Table 1.** Related work comparison.

| | Approach | Oracles | Decentralized | Efficient | Data integrity | Support |
|---|---|---|---|---|---|---|
| [8] | TEE | No | No | No | Yes | Blockchain |
| [9] | Notary | No | No | Yes | No | Many DLTs |
| [10] | Sidechains | No | Yes | No | Yes | Blockchain & DAG |
| [11] | Notary | Yes | No | Yes | No | Blockchain |
| [12] | Notary | Yes | No | Yes | No | Blockchain & DAG |
| **Our work** | Notary | Yes | Yes | Yes | Yes | Blockchain & DAG |

## 3   Interoperable Plant Blockchain for Homogenized Data via Smart Oracles

In this section we described the proposed solution for machine data interoperability and trustworthy storage of plant level data. Figure 2 depicts the proposed solution where the different components of the solution are depicted.

As presented in a previous work [3], the actual IIoT data is stored inside an IPFS storage system, while the data-source DAGs would only store the hashes in order to reduce the storage burden of the DLTs. After receiving and storing the raw IIoT data hashes from IPFS, a data homogenization service (that is executed periodically) would make a call to an external decentralized oracle service in order to retrieve the data model used for the data homogenization process. A blockchain oracle is an external third-party that provides trustworthy data from outside in a decentralized and trustworthy manner [13]. Hereafter, the data homogenization process is executed. The aforementioned process consists of transforming raw IIoT data to a standardized data scheme. Finally, the data homogenization service would then send the homogenized data to an

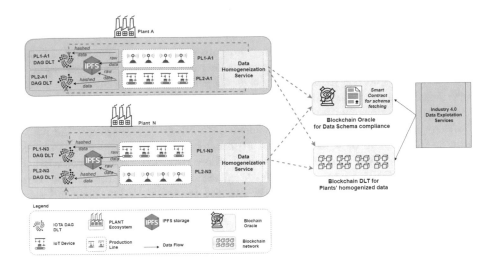

**Fig. 2.** The proposed interoperable plant blockchain and data homogenization via Smart Oracles

interoperable plant blockchain, which in turn stores it inside the IPFS storage system and keeps its references within the immutable ledger. Therefore, the main purpose of the interoperable plant blockchain is to securely store and manage the smart plant homogenized data references and provide access control to IPFS. This blockchain would also unify the data management of different industrial plants belonging to a same business conglomerate. Finally, this ledger would act as a bridge between the DAG DLTs that process the data from IIoT devices inside production lines, and other hypothetical DLT connections that might be addressed in a future work. Consequently, interoperability capabilities are required at this level. In order to connect the production lines DAGs and the plant blockchain, we propose to make use of a simple notary scheme that would interact with a smart contract from the destination blockchain in order to securely send the data.

## 4   Implementation

In this section we describe the implementation process of a prototype that we have implemented to prove the viability of our proposal. We employ real-world raw IIoT data to build our use case. We use IOTA as the production line DAGs to process the raw data, since IOTA is currently the most advanced DAG type DLT platform. ChainLink is an established oracle solution in the blockchain ecosystem. It enables simple deployment of oracle networks along with a great variety of connectors for DLT interoperability among other uses. Hence, due to the fact that at the moment ChainLink is the most advanced and suitable solution for our requirements, we selected it as our trustworthy data model provider service.

We use the JSON-based Eclipse Unide[2] standard as our data model. The Unide data model is specifically designed for manufacturing processes and it is trusted by several major parties such as SAP and Bosch. Finally, we use Hyperledger Fabric[3] as the interoperable plant blockchain. Fabric is a permissioned feeless blockchain with smart contracts support, which is ideal for an industrial plant ecosystem.

First, we have implemented a NodeJS IOTA client that periodically sends industrial raw data to an IPFS file system, and the resulting IPFS hash (e.g. QmZYVqj6qgdeHSj9TJSMDcX8e4n26PikkzfpqHB4N5nG92) to the IOTA DAG DLT. Afterwards, we implemented a NodeJS data homogenization service that performs the following operations:

1. Access the IPFS raw data using the hash that is stored in the production line IOTA DAG DLT.
2. Call a Solidity smart contract that retrieves the data model from a ChainLink decentralized oracle service.
3. Perform the data homogenization process. We defined the mapping between the raw data schema to the standard Eclipse Unide data model schema using the *jsonpath-object-transform*[4] NPM package.
4. Perform the validation of the resulting homogenized data using the Ajv JSON schema validator[5].
5. Add the used data model and the resulting homogenized data to IPFS.
6. Call a Hyperledger Fabric smart contract in order to store the IPFS hash of the homogenized data in the aforementioned blockchain.

The implementation of the IOTA client and of the data homogenization service has been made on top of an Ubuntu 20.04 LTS operating system with an i7 9th generation CPU, 16 GB of RAM and a SSD drive.

In Fig. 3 we show the raw data, the data model and the resulting homogenized data; In Fig. 4 we show the data model retrieval process inside the ChainLink oracle; Finally, in Fig. 5 we show the homogenized data hash from the destination Hyperledger Fabric plant blockchain.

## 4.1   Discussion

In this work we leverage decentralized oracles for data interoperability purposes, i.e. to securely gather the external IIoT data model and perform an homogenization process of the machine raw data. Decentralized oracle platforms such as ChainLink or Gravity [14] try to enable the development of fast, decentralized, and secure oracles for many applications. Gravity, however, is in a very early stage of development. On the other hand, ChainLink is a an established solution in the blockchain ecosystem. ChainLink enables simple deployment of

---

[2] https://www.eclipse.org/unide/specification/.
[3] https://www.hyperledger.org/use/fabric.
[4] https://www.npmjs.com/package/jsonpath-object-transform.
[5] https://ajv.js.org/.

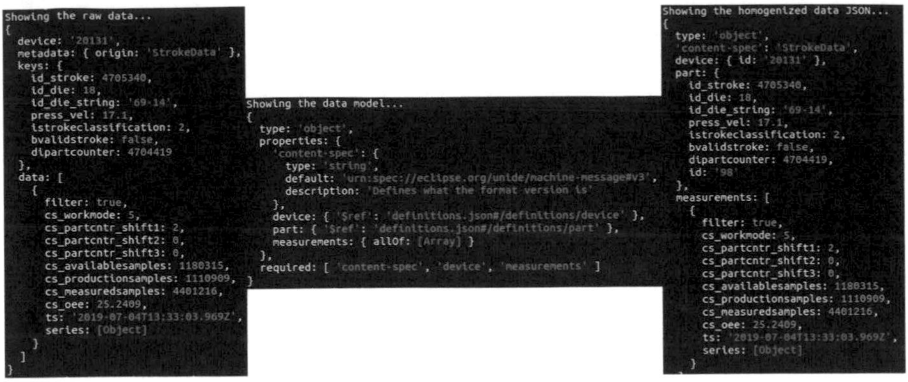

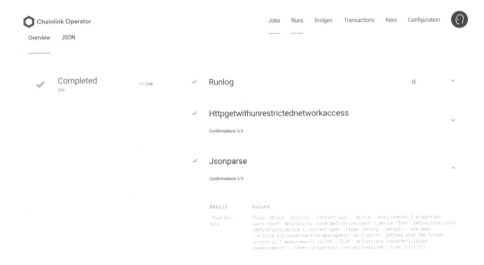

**Fig. 3.** From left to right: raw data, data model and the resulting data.

**Fig. 4.** ChainLink oracle data model retrieval tasks.

```
1 ▾ {
2     "_id": "2022-04-12T06:44:08.958Z-QmU14bMXWFLuMqeDtL6A86XD53K5X2DXmyaiaWxwTbnseK",
3     "_rev": "1-e23d902f922010a6cef207f083946869",
4     "datetime": "2022-04-12T06:44:08.958Z",
5     "docType": "hash",
6     "hash": "QmU14bMXWFLuMqeDtL6A86XD53K5X2DXmyaiaWxwTbnseK",
7     "id": "2022-04-12T06:44:08.958Z-QmU14bMXWFLuMqeDtL6A86XD53K5X2DXmyaiaWxwTbnseK",
8     "~version": "CgMBGAA="
9 }
```

**Fig. 5.** Homogenized data hash in Hyperledger Fabric.

oracle networks along with a great variety of connectors for DLT interoperability among other uses.

However, despite the significant amount of security (i.e. data integrity) that a decentralized oracle mechanism brings to an architecture when providing data, some delay may be introduced due to the complexity of an additional decentralized network in between. Nevertheless, ChainLink nodes take little time to process transactions. Indeed, our measurements show that ChainLink data model retrieving tasks are executed in less than 100 ms if the Ethereum blockchain is not used for the oracle data storage, which is our case. Therefore, we can conclude that using a decentralized oracle service for retrieving a JSON data model scheme does not incur a significant delay in the network when comparing to centralized service. Finally, according to our measurements, the whole process is executed in just 305 ms. However, this parameter depends on the employed hardware.

Regarding security, in the presented schema, the integrity of the data is ensured throughout the whole process, from when the data is generated in production lines up until it is homogenized, and finally accessed at the plant level. This is due to the employment of secure DLT technologies throughout the whole process (i.e., production lines DAG DLTs, decentralized blockchain oracles for data homogenization and plant processing blockchain), as shown in Fig. 6.

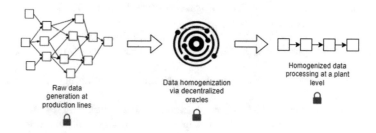

Raw data generation at production lines

Data homogenization via decentralized oracles

Homogenized data processing at a plant level

**Fig. 6.** Security of the data throughout the whole process.

## 5   Conclusion and Future Work

In this work we present an industrial data homogenization solution leveraging decentralized oracles, and an interoperable blockchain that manages the homogenized data across one or many industrial plants. We claim to achieve the integrity of the data throughout the whole process from when the data is generated and processed at a production line level through efficient DAG DLTs, up until it is homogenized and managed at a plant level. We also implement a prototype of the proposed solution based on IOTA, IPFS, ChainLink Oracle and Hyperledger Fabric. As future work, we plan to enable a more generic data homogenization process based on Model Driven Development (MDD) techniques, as well as, to conduct on-field experiments.

**Acknowledgements.** The European commission financially supported this work through Horizon Europe program under the RENAISSANCE project (grant agreement number 101021911). It was also partially supported by the Ayudas Cervera para Centros Tecnológicos grant of the Spanish Centre for the Development of Industrial Technology (CDTI) under the project EGIDA (CER-20191012), and by the Basque Country Government under the ELKARTEK program, project REMEDY - REal tiME control and embeddeD securitY (KK-2021/00091).

# References

1. Lasi, H., Fettke, P., Kemper, H.G., Feld, T., Hoffmann, M.: Industry 4.0. Business and information systems engineering **6**(4), 239–242 (2014)
2. Bhandary, M., Parmar, M., Ambawade, D.: A blockchain solution based on directed acyclic graph for IoT data security using IoTA tangle. In: 2020 5th International Conference on Communication and Electronics Systems (ICCES), pp. 827–832 (2020)
3. Stefanescu, D., Galán-García, P., Montalvillo, L., Unzilla, J., Urbieta, A.: Towards a holistic DLT architecture for IIoT: improved DAG for production lines. In: Prieto, J., Partida, A., Leitão, P., Pinto, A. (eds.) Blockchain Appl., pp. 179–188. Springer International Publishing, Cham (2022)
4. Wu, Y., Dai, H.N., Wang, H.: Convergence of blockchain and edge computing for secure and scalable IIoT critical infrastructures in industry 4.0. IEEE Internet Things J. **8**(4), 2300–2317 (2021)
5. Jirkovsky, V., Obitko, M., Marik, V.: Understanding data heterogeneity in the context of cyber-physical systems integration. IEEE Trans. Indus. Inf. **13**(2), 660–667 (2017)
6. Roman, R., Najera, P., Lopez, J.: Securing the Internet of Things. Computer **44**(9), 51–58 (2011)
7. Bodkhe, U., Tanwar, S., Parekh, K., Khanpara, P., Tyagi, S., Kumar, N., Alazab, M.: Blockchain for industry 4.0: a comprehensive review. IEEE Access **8**, 79764–79800 (2020)
8. Bellavista, P., Esposito, C., Foschini, L., Giannelli, C., Mazzocca, N., Montanari, R.: Interoperable blockchains for highly-integrated supply chains in collaborative manufacturing. Sensors **21**(15), 1–21 (2021)
9. Scheid, E.J., Hegnauer, T., Rodrigues, B., Stiller, B.: Bifröst: a Modular blockchain interoperability API. In: Proceedings—Conference on Local Computer Networks, LCN 2019-Octob, pp. 332–339 (2019)
10. Jiang, Y., Wang, C., Wang, Y., Gao, L.: A cross-chain solution to integrating multiple blockchains for IoT data management. Sensors (Switzerland) **19**(9), 1–18 (2019)
11. Gao, Z., Li, H., Xiao, K., Wang, Q.: Cross-chain oracle based data migration mechanism in heterogeneous blockchains. In: Proceedings—International Conference on Distributed Computing Systems 2020-Novem, pp. 1263–1268 (2020)
12. Wiraatmaja, C., Zhang, Y., Sasabe, M., Kasahara, S.: Cost-efficient blockchain-based access control for the internet of things. In: 2021 IEEE Global Communications Conference (GLOBECOM), pp. 1–6 (2021)

13. Al-Breiki, H., Rehman, M.H.U., Salah, K., Svetinovic, D.: Trustworthy blockchain oracles: review, comparison, and open research challenges. IEEE Access **8**, 85675–85685 (2020)
14. Pupyshev, A., Gubanov, D., Dzhafarov, E., Sapranidi, I., Kardanov, I., Zhuravlev, V., Khalilov, S., Jansen, M., Laureyssens, S., Pavlov, I., Ivanov, S.: Gravity: a blockchain-agnostic cross-chain communication and data oracles protocol. arXiv pp. 1–21 (2020)

# Towards Cost-Efficient Management for Power Purchase Agreements Using Blockchain Technology

Ivan Gutierrez-Aguero[1][(✉)] [ID], Yesnier Bravo[2] [ID], Daniel Landa[1] [ID], Oscar Lage[1] [ID], Iñaki Seco[1] [ID], and Aitor Castillo[2] [ID]

[1] TECNALIA, Basque Research and Technology Alliance (BRTA), Parque Científico y Tecnológico de Bizkaia #700, E-48160 Derio, Spain
{ivan.gutierrez,daniel.landa,oscar.lage,inaki.seco}@tecnalia.com
[2] Bettergy, Malaga Tech Park, Av. Juan López, Peñalver nº17, Floor 3a, 29590 Málaga, Spain
{ybravo,acastillo}@bettergy.es

**Abstract.** The difficulty for two parties to efficiently manage the contract conditions and clauses defined in a Power Purchase Agreement introduces several challenges. This manuscript addresses the current management limitations that arise during its lifecycle and offers a new vision adequate to the business implications, the need of transparency and the requirements of edge computing continuous monitoring. Considering the relevant actors, the main scenarios driven by the Power Purchase Agreements and their state-of-the-art contract designs, this manuscript proposes a novel method and procedure to enhance the cost-efficiency of the installations, granting the consumer self-management and the confidence and the guarantee of origin through a trustworthy distributed solution. The method proposed enables total traccability and integrity on the PPA performance in the long-term by applying blockchain technology. A new operational design is proposed, explaining the energy balance and energy cost formulas required by businesses. The results show that blockchain enhances the way of managing data, the data ownership, and the market opportunity.

**Keywords:** Power purchase agreement · Data management automation · Guarantee of origin · Blockchain · Smart contract

## 1 Introduction

A Power Purchase Agreement (PPA) is a long-term electricity supply contract between two parties. In this kind of contract, one of the parties represents a renewable energy producer, and the other one a buyer (trader or consumer) of that energy. The PPA helps to define the conditions of the agreement, like the minimum amount of energy to supply, the prices, or the penalties when the counterparty fails to comply.

Building a renewable plant entails a high initial investment. Energy producers or investors are willing to build the plant for a monthly recurrent payment. Here is where the

PPA takes place, to establish the agreements and clauses, in terms of energy production, consumption and costs.

The PPA by itself only stipulates different parties' obligations and consequences of breaking these obligations. To satisfy actual needs for the intervening parties, the PPA needs to monitor the plant performance and reduce errors concerning the cost-efficiency. To this extent, additional tools and devices are required for the PPA management, as well as support the relationship between the producer and consumer for a long period (10 + years).

This scenario brings a number of challenges [1, 2] for an efficient PPA management, and the PPA volumes increase each year, making management even more difficult. This mechanism is on the rise as the consumer intends to reduce the market price variation risks and the investment costs associated with operating renewable energy. At the same time, plant operators are focused on revenues for electricity generation from their renewable energy plants.

## 1.1 State-of-the-Art Limitations for PPA Management

Traditionally, the reliability of PPAs to optimize energy performance has been calculated to favour either the consumer or the producer. There are two research lines in the state of the art: One of them is focused on the financial attractiveness for producers [3, 4], while the other one is focused on the energy efficiency for consumers [5–7]. These studies often follow a profit-seeking approach, dealing with both energy and market prices uncertainties by optimizing the structure.

Other studies consider the performance and include predictive maintenance in estimating the most cost-effective schedules [3] and financial risks [4, 8]. While the state-of-the-art is focused on the predictability and optimization of PPA in terms of energy reliability and financial terms, our approach is to assume and manage uncertainties. The increment and diversity of recent studies on PPA applications [9–11] give a special relevance to this approach focused on the management enhancement.

To the best of the authors' knowledge, research combining blockchain and PPA is limited [12]. The studies are mostly focused on providing renewable energy guarantees of origin and energy tokenization. By contrast, our approach proposes to take advantage from blockchain for a cost-effective management.

However, there is still a lack of a guarantee in the traceability of the PPA clauses, there are no systems that provide an independent and decentralized measurement and verification process for the signatories of the PPA contract. Issues with the centralized approach include privacy concerns due to third party management of cloud servers, single points of failure, bottleneck in data flows, and others [13]. In addition, no reliable verification exists between the actual production and the corresponding invoices where the agreed price and clauses are reflected, during the contract term.

## 1.2 Cost-Efficiency in PPA Management

The cost analysis of a renewable energy installation is limited to its optimal sizing or to the reduction of the initial costs. As an example, [3] proposes avoiding expensive "corrective maintenance that may cause long downtimes". However, in the PPA there are other cost analysis factors related with operative costs, including the cost of the execution and data measurement and verification. Significant back-office work for its signatories exists due to a lack of trust in the long-term. Participants need additional security for the PPA management while automating control of what amount of energy is produced and consumed, or when contract clauses violated.

This could lead to greater operational efficiency, allowing employees to focus on tasks where they can provide greater value. As this cost usually represents part of the contract price, the new model could make renewable projects more affordable.

## 2    Materials and Methods

The blockchain-enabled PPA solution offers a secure communication flow, where data gathered from dataloggers is sent directly to blockchain. The dataloggers periodically collect information and send it signed through a blockchain decentralized application (DApp).

Once the data arrives to blockchain network, the related contractual clauses of the PPA are verified. This action is possible because the verification operations are implemented in the smart contracts and customized with each PPA's clauses that have been previously accepted by the contract signatories.

An alert will be raised if a clause is violated. The alerts are received by the authorized actors correctly subscribed to the topic of an event channel. As an alert means that there's a contract disturbance, the mediator business can intervene when needed, restoring the normal execution, and saving time and money for the signatories.

### 2.1    Addressing the Challenges in the PPA Management

PPAs can be very different due to its bilateral agreement nature. This manuscript presents a solution on the State-of-the-Art Limitations for PPA Management. Explained on Sect. 1.1, providing reliable ways to manage prices and clauses thus responding to traditional challenges [14]. A relevant solution has been modelled considering different situations of PPAs and actions derived from measurements and contract clauses. Also, it considers the datalogger or device in charge of measuring and registering the amounts of energy produced by the plant and consumed by the facility.

A key innovation in the energy transactions is the ability to get automatic balances based on the amount of energy generated by the energy production installation [15].

Moreover, to mutually reduce the risks and conflicts between parties, it is mandatory to monitor the violation of the agreed contract clauses. When a violation occurs during the life of a PPA, an automated and decentralized alert event registration ensures the execution of the contract agreements. Naturally, these alert events should be created from trusted and certified energy readings, needing dataloggers capable of ensuring the data origin. This way, the producer can provide evidence of the origin.

The consumers need a way of getting direct access to a reliable information source without intermediaries. This specific challenge can be met by using blockchain, following the line of decentralization. In summary, the PPA challenges tackled are:

1. Monitoring the violation of contract clauses.
2. Automating the billing process.
3. Certifying green energy origin.
4. Granting self-management to consumers.

## 2.2 Procedure

To address the challenges in the PPA management, a new operational design is presented. The proposed solution enables a blockchain-based scenario, where the consumers become autonomous, and the producers reduce management costs thanks to automatically settled agreements written in smart contracts.

Thanks to the blockchain network, a digital record of measurements allows to calculate the clauses of the PPA and back it at the distributed ledger. All the information is handled cryptographically to minimize the attack vectors.

The flow of the operation (see Fig. 1) begins with some offline negotiations between the producer and the consumer to agree on the PPA contract's clauses.

The producer uses a DApp to communicate with the smart contract and register the PPA (createPPA). Among other information the producer provides the generation limit values as well as the prices of the generated energy in each range bounded by those limits and registers the dataloggers (linkDataloggerToPPA) that are authorized to send their measurements into the PPA by providing their credentials. This is an in independent task from the creation of the PPA due to the possibility that the dataloggers change during the life of the PPA.

After a successful PPA creation the consumer receives informative event and proceeds to query the PPA details (getMyPPAs, getPPADetails) and, once validated, sign it to confirm the participation in the PPA (signPPA). The consumer uses a DApp to perform these actions in the smart contract.

Once the contract is registered, it is ready to accumulate measurements from the dataloggers. The dataloggers connected to the plant periodically collect measurements about the energy generation and consumption values. Then they use DApps to communicate with the smart contract and deliver the measurements (saveEnergy). With no information about the relationship of the dataloggers to the specific PPA, the smart contract gets that relationship from the credentials used by the datalogger to sign the measurements sent.

Once the PPA has received the measurements belonging to a full balance period, the balance is closed by the producer (closeBalancePeriod). At this stage, the smart contract will calculate the generation costs, notify the consumer by means of an informative event, and reset the balance period.

The producer can also validate the clauses established in the PPA by requesting the closing of a validation period (closeValidationPeriod). When requested, the smart contract will validate the clauses and issue clause violation alerts when applicable.

## 2.3 Formulas Addressed by Blockchain

The way to find a cost-efficient solution is to capture the advantage provided by blockchain for the automation of systems. The interactions performed between the relevant actors during the PPA contract lifecycle have some operation points where the smart contracts are a trusted source for the results needed.

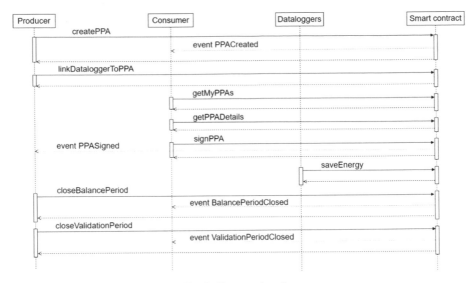

**Fig. 1.** Process data flow

To this extent, there are some PPA contract parameters that blockchain must manage to automate the calculations needed to leverage the power of the technology in benefit of the cost-efficiency. These parameters are:

– PPA contract reference
– Installed power
– Energy pricing
– Energy generation low threshold
– Energy generation high threshold
– Energy consumption low threshold
– Warranty expiration.

This solution provides four strategic operations that allow the actors to be autonomous, solving the issue of depending on a trusted third party. Due to the lack of formal research references for these formulas, the methodology to select them has been based on the direct business knowledge extracted from a set of energy suppliers.

**Energy balance.** In a PPA standard conditions, the consumer is intended to use the 100% of the generated energy. If the consumption is below the energy consumption low threshold, the remaining energy must be discarded or fed to the grid at a lower price. An automatic process in the blockchain implementation verifies the energy consumption and raises an event in case of energy consumption low threshold violation.

In the energy balance calculation, described in the Eq. 1, the total amount of energy generated $(X)$ and consumed $(Y)$ in a period $(k)$ is the summation (of size $n$) of all energy readings registered by a datalogger in that period. Then, the amount of energy discarded or fed to the grid is calculated by using the Eq. 2.

$$X_k = \sum_{i=0}^{n} x_i Y_k = \sum_{i=0}^{n} y_i \tag{1}$$

$$f_i = \frac{Y_i}{X_i} \tag{2}$$

An alert is triggered if the energy consumption result is below the threshold $F_{th} \in (0, 1)$, fulfilling the Eq. 3.

$$f_i < F_{th} * Y_i \tag{3}$$

**Total energy cost.** The total energy cost ($C$) is calculated using the Eq. 4, using the datalogger readings and the prices. The cost at the end of a period $i$ depends on the total amount of energy generated ($X_i$) compared to the *Minimum yearly generation* committed in the PPA. Thus, if $X_i$ is below the under-production threshold ($X_{min}$), the under-production price ($p_l$) is applied. Else, if $X_i$ is between $X_{min}$ and the overproduction threshold ($X_{max}$), the regular price ($p_r$) is applied. And finally, if $X_i$ is greater than $X_{max}$, then $p_r$ is applied up to $X_{max}$ and the overproduction price ($p_h$) from that value.

$$C(X_i, p_l, p_r, p_h) = \begin{cases} p_l * X_i, & X_i < X_{min} \\ p_r * X_i, & X_{min} < X_i < X_{max} \\ p_r * X_{max} + p_h * (X_i - X_{max}), & X_i > X_{max} \end{cases} \tag{4}$$

## 3   Results

A new solution for the management of the PPA contracts with a cost-efficient perspective has been proposed, based on the digitalization allowed by the usage of blockchain. The cost efficiency comes mainly from the trusted automation of the operations and the self-management that is a novelty from the user point of view.

The software components and the functionality are key points for this solution, since they help to cover the blind points existing on the PPA management [16, 17]. This section explains those materials that helped to give shape to this coverage, defining the stages where a consumer can become autonomous on its own management; designing the event broadcasting for the most important energy production installation alerts; and studying the main formulas of a PPA contract and selecting those where the automation with blockchain could give better results.

### 3.1   Self-managed Consumers

The proposal of enhancing the management of the PPA contracts by using blockchain proves a new concept of interactions between actors.

The producers can digitalize the settlement of the PPA contracts and give autonomy to their dataloggers. And the dataloggers can recollect measurements and deliver them in a direct way, without the need of trusting in a centralized service platform which would work as a black box. Due to the nature and the technologies used in the blockchain platform, every participant actor has full confidence of the integrity and non-repudiation of any information exchanged. Granting self-management to consumers means that they manage their own account instead of trusting in a third party.

In terms of cybersecurity, every user account managed by a producer, a consumer or a datalogger is independent from the PPA cloud platform, protected by a password and recoverable with a unique set of words, called mnemonic. This way, blockchain ensures the uniqueness and independency of each user. In case of industrialization of this proposal, the developers should consider that dataloggers are connected IoT devices and cybersecurity principles must be met from the design phase.

Regarding the managed data, the decentralization of the system has privacy implications that open two future lines: (i) sensitive data must be kept off chain and blockchain is an integrity guarantee or (ii) the network is consortiated or private, in a way that data privacy can be ensured in another way. The work performed in [12] can be the starting point for this matter.

## 3.2   A Distributed Solution

Given the nature of the described procedure, blockchain data will be directly managed by the corresponding actor. To this extent, the actors need to trust in the smart contracts deployed in each network node to manage the PPA contract lifecycle.

Ethereum, for instance, provides a trust-minimized platform where no trusted third party is needed and nevertheless, the whole system is trusted. So, this could be a good starting point for the development of a fully distributed solution.

Comparing the costs of the solution maintenance, the use of a public network has a cost attached to the transactions executed, whereas using a private network has the cost of the maintenance of the servers that host the network itself. Which kind of network to use is dependent to the needs of the platform involved and the relevant actors.

A formal research methodology has been followed to know the maintenance cost in a public based solution. The first step has been to take each function explained in the procedure (see Table 1) and calculate the average gas consumption, using the formula from Eq. 5, measured in *gas units*. Then, three public gas price calculators (ETH Gas Station, Blocknative and Gwei Gas Calculator) have been selected to calculate a median gas price per gas unit, with a value of 42 *gwei/unit* (reference date February 15, 2022), being 1 *gwei* 0.00000001 *ether*.

The transaction cost, measured in *gwei*, can be calculated applying the Eq. 6 for each transaction, multiplying it by the number of needed transactions per type for each time frame. Algorithm 1 shows how to calculate the total maintenance cost, applying the corresponding ether price to the summation of transaction costs. The ether price has a 52-week average value of 2,923.86 €/ether, with lower and higher bounds of 965.84 € and 4,636.43 € respectively (reference date June 15, 2022).

$$Total\ Gas = Initial\ Fee + Input\ Data\ Fee + Gas\ Used - Gas\ Refund \qquad (5)$$

$$GasFee = GasConsumption * GasPricePerUnit \qquad (6)$$

**Table 1.** Gas consumption per function in the procedure

| Function | Gas consumption |
| --- | --- |
| createPPA | 235,994 |
| linkDataloggerToPPA | 89,847 |
| signPPA | 78,423 |
| saveEnergy (first call) | 97,337 |
| saveEnergy | 52,349 |

**Algorithm** 1. Total maintenance cost calculation

**procedure** MaintenanceCalculation()
let *totalGwei* ← 0
**foreach** *type* in *Tx_type* **do**
*totalGwei* = *totalGwei* + count(*type*)*GasFee(*type*)
**end foreach**
**return** *etherPrice* ∗ *totalGwei* ∗ 0.00000001
**end procedure**

# 4   Discussion

## 4.1   PPA Data Management

The solution proposed by this manuscript presents seven (7) enhancements to the auton-omy of the actors. In the state-of-the-art solutions, the actors are forced to trust in a third party providing the service, the agreements, and the numbers. In addition, these services mean a bottleneck in PPA contract management, that is unlocked with this solution, adding smart contracts to provide automatic PPA executions and find a reliable way to manage prices and clauses from PPA contracts.

## 4.2   Data Ownership

Assuming that is important that certain parts of the data contained in a PPA contract can only be visualized by specific kinds of actor, this cost-efficient solution is built on the basis that the operations performed by some functions on the smart contract are only callable by specific roles (e.g., the owner of the contract, or a registered datalogger).

Each blockchain address (representing the user identifier) shall be assigned to one or more roles. When a user (origin) makes a call to a smart contract function generating a new transaction, the access allowance is validated by taking advantage of one of the major benefits offered by blockchain, that is non-repudiation.

To improve the user data protection, authors recommend using a DApp to manage one's own data while interacting with blockchain to operate PPA contract. Using cloud-hosted wallets from third party servers could incur on data and money loss [18].

### 4.3  Business Implications and Market Opportunity

Added to the enhancements to the autonomy of the actors, this manuscript presents demonstrable enhancements in cost-efficiency. The use of blockchain reduce PPA management costs due to eliminating, for example, personnel for operating the PPA. Common PPA operation tasks include periodic on-site monitoring, energy billing, contract clauses verification, and detection of anomalies. On the one hand, the proposed solution allows to manage PPA contracts even within trustless relationships. The data are saved in an immutable, reliable, and decentralized registry that can be verified at any time.

Also, smart contracts are implemented to automate frequent billing and clauses verification. In summary, we can manage a PPA contract with less personnel and resources in comparison with the conventional way. A renewable installation of 1 MWp can cost about 700,000 €, considering 0.7 €/Wp, with 20 years of operation. Usually, one operator can run up to four independent installations. Assuming a salary of 25,000 € per year, it represents an additional cost of 17.86% (125,000 €) over the project cost.

In accordance with Algorithm 1 depicted in 1.1 A Distributed Solution, the blockchain based solution has an initial cost of less than 14 € for operations needed to initialize the PPA contract. This operation occurs only once in the PPA contract lifecycle, so it can be disregarded. It also has an estimated maintenance cost of 1.6 € for the daily updates made by a datalogger to the ledger. This would make a total 11,650 € for the same 20 years of operation.

If an hourly update was required, the 37 €/day cost will turn into 273,000 € for the installation life. In this case, additional cost-efficiency actions shall be needed. Using a *layer 2* chain would significantly decrease the costs, becoming only hardware maintenance in case of consortium networks.

**Funding.** The research presented has been done in the context of SDN-microSENSE project. SDN-microSENSE has received funding from the European Union's Horizon 2020 research and innovation programme under grant agreement No 833955. The information contained in this publication reflects only the authors' view. EC is not responsible for any use that may be made of this information.

## References

1. di Castelnuovo, M., Dimitrova, M.: Corporate renewable energy procurement through PPAs in the United States. In: 3rd AIEE Energy Symposium Conference Proceedings, pp. 124–126 (2018)
2. Baines, S., Wrubell, S., Kennedy, J., Bohn, C., Richards, C.: HowToPPA: an examination of the regulatory and commercial challenges and opportunities arising in the context of private power purchase agreements for renewable energy. Review **57**, 389 (2019)
3. Lei, X., Sandborn, P.A.: Maintenance scheduling based on remaining useful life predictions for wind farms managed using power purchase agreements. Renew. Energy **116**, 188–198 (2018)
4. Das, P., Malakar, T.: A day-ahead sleeved power purchase agreement model for estimating the profit of wind farms in the Indian energy market. Int. Trans. Electr. Energy Syst. **31**(4), (2021)

5. Trivella. A.: Decision Making Under Uncertainty in Sustainable Energy Operations and Investments. Ph.D. thesis. Technical University of Denmark, Lyngby, Denmark (2018)
6. Ghiassi-Farrokhfal, Y., Ketter, W., Collins, J.: Making green power purchase agreements more predictable and reliable for companies. Decis. Support Syst. **144**, 113514 (2021)
7. Wu, O.Q., Babich, V.: Unit-contingent power purchase agreement and asymmetric information about plant outage. Manuf. Serv. Oper. Manag. **14**(2), 165–353 (2012)
8. Kaufmann, J., Kienscherf, P.A., Ketter, W.: Modeling and managing joint price and volumetric risk for volatile electricity portfolios. Energies (Basel) **13**(14), 3578 (2020)
9. Yaqub, M., Sarkni, S., Mazzuchi, T.: Feasibility analysis of solar photovoltaic commercial power generation in California. Eng. Manag. J. **24**(4), 36–49 (2012)
10. Vimpari, J.: Financing energy transition with real estate wealth. Energies (Basel) **13**(17), 4289 (2020)
11. Bruck, M.A.: Levelized cost of energy model for wind farms that includes power purchase agreements (PPAS). ProQuest Dissertations Publishing (2018)
12. Mollah, M.B., et al.: Blockchain for future smart grid: a comprehensive survey. IEEE Internet Things J. **8**(1), 18–43 (2020)
13. Uddin, M.A., Stranieri, A., Gondal, I., Balasubramanian, V.: A survey on the adoption of blockchain in IoT: challenges and solutions, Blockchain: Res. Appl. **2**(2), 100–006 (2021). ISSN 2096-7209
14. Huneke, F., Göß, S., Österreicher, J., Dahroug, O.: Power purchase agreements: financial model for renewable energies. Energy Brainpool White Paper (2018)
15. Zhang, W., Wang, X., Wu, X., Yao, L.: An analysis model of power system with large-scale wind power and transaction mode of direct power purchase by large consumers involved in system scheduling. Proc. CSEE **35**(12), 2927–2935 (2015)
16. Ueckerdt, F., Hirth, L., Luderer, G., Edenhofer, O.: System LCOE: what are the costs of variable renewables? Energy **63**, 61–75 (2013)
17. Bruck, M., Sandborn, P., Goudarzi, N.: A levelized cost of energy (LCOE) model for wind farms that include power purchase agreements (PPAs). Renew. Energy **122**, 131–139 (2018)
18. Saad, M., Spaulding, J., Njilla, L., Kamhoua, C., Shetty, S., Nyang, D., Mohaisen, A.: Exploring the attack surface of blockchain: a systematic overview, 2019. arXiv preprint arXiv:1904. 03487

# Tokenizing the Portuguese Accounting Standards System

Paulo Vieira[✉][iD] and Helena Saraiva[iD]

Guarda Polytechnic Institute, Avenida Dr. Francisco Sá Carneiro, n.o 50, 6300-559
Guarda, Portugal
vieirapaulo927@gmail.com,{pavieira,helenasaraiva}@ipg.pt

**Abstract.** In this paper we show through examples how it is possible
to tokenize the Portuguese Accounting Standards System. Using these
examples we design a smart contract for Ethereum as concept proof. This
paper it is the kick off of a project to design an ERC (Ethereum Request
Commment) for the Portuguese Accounting Standards System.

**Keywords:** Ethereum · Tokenization · ERC · Accounting standards
systems

## 1 Introduction

In Portugal, the reference system for carrying out Accounting records in the
business sector it is called Accounting Standards System (SNC, it is the Portuguese abbreviation). This system consists of a set of Accounting Standards for
Financial Reporting (NCRF, it is the portuguese abbreviation), and Interpreta-
tive Standards (NI, it is the Portuguese abbreviation). The system consists on
an addaptation of the International Accounting Standards (IAS- International
Accounting Standards and IFRS - International Financial Reporting Standards)
issued by the International Accounting Standards Board (IASB), and aims to
harmonize national accounting with the European Union (EU) standards, which
is based on this set of international standards.

In fact, in Portugal, as in most European countries, due to the application of
European Union legislation [1,2], a hybrid system has been implemented with
regard to accounting standardization, in which private companies listed on the
stock exchange report under IAS and IFRS adopted by the EU, and unlisted
private companies report by the SNC, in its various subsets of standards, in
accordance with the size and nature of the entities. The Public Sector entities
report in accordance with the European norms by the Accounting Standard
System for Public Administrations (the Portuguese abbreviation it is SNC-AP),
based on the International Public Sector Accounting Standards (IPSAS).

Accounting systems with these characteristics result from the harmonization
effort that has been carried out in most countries, at a global level [3]. The

Supported by organization UDI - Unit for Inland Development.

concept of accounting harmonization corresponds to a process that allows to increase the comparability of accounting practices, limiting their degree of variation [4] and it has been developed under the initiatives of the European Union in coordination with the IASB.

Several application proposals have recently emerged regarding the Blockchain concept, in order to create opportunities for business models and to improve the performance of some solutions [5–8]. Through the use of Blockchain is possible to conceive a system in which contractual positions between entities are available in shared and accessible on-chain, in an environment where agreements, processes, tasks and payments are recorded and accessible to all actors, through a record and digital identification, making possible its validation, storage and sharing [9]. In this sense, the use of Blockchain presents similarity to the functions that accounting, in any entity, should assume and ensure: making a record that is verifiable, that can be shared with all stakeholders and interested bodies in the records of operations and being that records secure and traceable.

A Smart Contract is code running on the Blockchain, with an address on the ledger. Each blockchain and all its nodes have the same ledger, and all transactions are recorded in this ledger. The main platform for running contracts on Blockchain is Ethereum. In Ethereum the process of tokenizing Standards is described through Ethereum Request Commments (ERCs).

Thus, this paper serves as a Proof of Concept (PoC) to show that it is possible to Tokenize the Portuguese Accounting Standardisation System using Smart Contracts. Tokenizing the Portuguese System also acts to demonstrate that it is possible to do the same for the accounting systems of other European countries.

## 2   Accounting Movements

In this section we present two examples of some of the most usual accounting movements and the registration in the respective accounts. The Portuguese account denomination were translated to English.

First example: registration of a purchase of Goods, in the General Ledger, subject to VAT at 23%, with payment made after the purchase, in the normal payment period (without interest). The movements of inventories (warehouse accounts) have not been considered:

At the moment of the purchase:

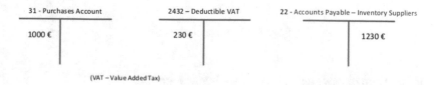

Fig. 1. Moment of the purchase

At the time of payment of the purchase invoice amount:

**Fig. 2.** Moment of payment of the purchase - invoice amount

The registration on the General diary-ledger.

**Table 1.** Movement on the diary-ledger

| Date | Description | Debit | Credit | Value (Euro) |
|---|---|---|---|---|
| Purchase date | Purchase with payments | 31 | | 1000 |
| | | 2432 | | 230 |
| | | | 22 | 1230 |
| Payment date | Payment of the purchase invoice amount | 22 | 12 | 11230 |

Second example: registration of a sale of Goods, in the General Ledger, with output VAT at 23%, with payment received after the sale, in the normal payment period (without interest). The movements of inventories (warehouse accounts) have not been considered:

At the moment of the sale:

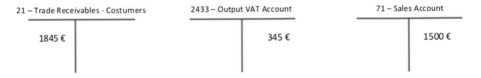

**Fig. 3.** Moment of the sale

At the time of payment of the sales invoice amount:
The registration on the General diary-ledger

**Fig. 4.** Receipt of the sales value - invoice amount

**Table 2.** Movement on the diary-ledger

| Date | Description | Debit | Credit | Value (Euro) |
|------|-------------|-------|--------|--------------|
| Sale date | Sale of goods with subsequent receipt | 21 | | 1845 |
| | | | 2433 | 345 |
| | | | 71 | 1500 |
| Date of receipt | Invoice amount received from sales | 12 | 21 | 1845 |

## 3   Smart Contracts

In this section we present a Smart Contract about the examples referred in the previous section. This contract should be seen as a PoC on the possibility of the Accounting Standardisation System becoming Tokenized, and is the start of a project under development in Portugal. The code of the contract it is described bellow. An account in the SNC it is designed by <account> is declared as <number_account> <account_name> but in the contract it is identified as <account_name>_< account _number>, example 21-Compras in the SNC is in the contract Compras_21. For each account it is defined in the contract two variables D<account> and C<account>; exemplifying with the account 21-Compras, it is associated DCompras_21 and CCompras_21, relative to respectively debit (D<account>) and credit (C<account>) on the respective bookkeeping account. In order to set calls to the contract, we have implemented functions, one for each account. The functions have the following syntax template declaration,

function set<account>(_D<account,_C<account>) public.... Following we present an exemplification to the 21-Compras account

```
function setCompras_21(uint _DCompras_21,
uint _CCompras_21) public {
        DCompras_21 += _DCompras_21;
        CCompras_21 += _CCompras_21;
}
```

The next lines shows the code of the contract in solidity

```solidity
pragma solidity ^0.8.1;
contract IntegerExample {
    uint public DDepositosOrdem_12;
    uint public CDepositosOrdem_12;
    uint public DCompras_21;
    uint public CCompras_21;
    uint public DClientes_22;
    uint public CClientes_22;
    uint public DIvaDedutivel_2432;
    uint public CIvaDedutivel_2432;
    uint public DIvaLiquidado_2433;
    uint public CIvaLiquidado_2433;
    uint public DVendas_71;
    uint public CVendas_71;

    function setDepositosOrdem_12(uint _DDepositosOrdem_12,
    uint _CDepositosOrdem_12)
    public {
        DDepositosOrdem_12 += _DDepositosOrdem_12;
        CDepositosOrdem_12 += _CDepositosOrdem_12;
    }

    function setCompras_21(uint _DCompras_21,
    uint _CCompras_21) public {
        DCompras_21 += _DCompras_21;
        CCompras_21 += _CCompras_21;
    }

    function setClientes_22(uint _DClientes_22,
    uint _CClientes_22)
    public {
        DClientes_22 += _DClientes_22;
        CClientes_22 += _CClientes_22;
    }

    function setIvaDedutivel_2432(uint _DIvaDedutivel_2432,
    uint _CIvaDedutivel_2432) public {
        DIvaDedutivel_2432 += _DIvaDedutivel_2432;
        CIvaDedutivel_2432 += _CIvaDedutivel_2432;
    }

    function setIvaLiquidado_2433(uint _DIvaLiquidado_2433,
    uint _CIvaLiquidado_2433) public {
        DIvaLiquidado_2433 += _DIvaLiquidado_2433;
        CIvaLiquidado_2433 += _CIvaLiquidado_2433;
    }

    function setVendas_71(uint _DVendas_71,
    uint _CVendas_71) public {
        DVendas_71 += _DVendas_71;
        CVendas_71 += _CVendas_71;
    } }
```

The contract is written and verified on the remix. The remix is also used to deploy to the Ethereum Ropsten Test Network using the MetaMask wallet. The creator of the contract is the account

0xc2612AB954A4Abda6D37dD4cC9b3F85FDfF52eD8

The contract it is inserted in the block 12198142. The contract address in the blockchain it is

0xe95d160609c6b81aaec8e6d4fa4666a1e8769cb2

this transaction have the hash

0 × 7fa9e0006d96d86c999743f64d76e2dcdd6aa5795d3b3f16990e363243 d8ad6d.

For this transaction, the creation of the contract, it was necessary to pay fees in Ether, 0.00099976565003199248 Ether, which at the current exchange rate is 3$ USA, or 2,85 EUROS. This amount is charged by the node that registers the block transaction.

**Fig. 5.** Remix IDE, http://remix.ethereum.org

The deployment of the contract can be seen in the Fig. 6. The calls of the contract allows to register the account movements and can be seen in the Meta-Mask as contract interactions. These contract calls correspond in the code the use of set functions named *function set<account_name>*

These movements were registered in the Blockchain as transactions. Each of these transactions cost 0.00011672 ETH, or $0.35 USD but in the features of the model, the contract calls are made with a gas price of 0 GWei and their cost will be zero. The compensation of the nodes, through the non-zero gas price (with value), will be done only by the registration of the contract and eventually by function calls associated with certain special accounting movements yet to be

**Fig. 6.** MetaMask Wallet

**Fig. 7.** Ethereum explorer relative of the wallet

defined in the ERC specification. These special movements may be, for example, capital increases, issue of shares, etc.

All these transactions, done through a Wallet (Fig. 6), can be seen on the wallet explorer demonstrated in the Fig. 7.

All contract transactions, its creation and calls, can be seen in the contract account showed here in Fig. 8.

**Fig. 8.** Transactions relative to the contract

The interactions with the user, the account user, and the wallet are showed in Fig. 9, that briefly will be an Dapp. There are three types of contract operations: contract creation, contract calls (setters) and variable value getters (getters). Ethereum nodes charge gas fees for the first two, but the last is fee-free. The getters are not registered operations in the Blokchain (in Fig. 9, blue buttons), but the creation and calls are (in Fig. 9,, orange buttons). (in the Fig. 9, blue buttons), but the creation and the calls they are (in the Fig. 9, orange buttons).

## 4  Conclusion and Future Work

This paper it is the start of a new project, for Tokenizing the portuguese Accounting Standards System, consisting on a PoC, that shows the project it is feasible. The development of the project is being done, for now, in a private git repository and as soon as possible will be available a first version for test, being the ERC still in preparation. Thus, it seems possible and advantageous to us that any operation carried out through a Blockchain can be simultaneously recorded, in accounting terms, on the Blockchain. This will contribute to benefit the transparency of operations, allowing real-time access; transparent usage; user identification; as well as ensure trust in this type of transactions, since one of the functions of accounting is to ensure trust by maintaining the public interest that accounting records allow.

To this end, we have started by developing the registration process in Blockchain of two very simple operations: a purchase and a sale of goods,

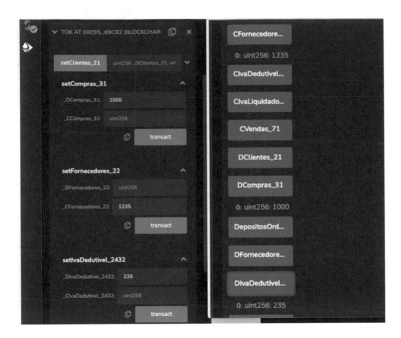

**Fig. 9.** Interface accountant-wallet

both with postponed payments. This implied four accounting records, performed simultaneously with the respective operations: the first being the purchase or sale commitment (using a smart contract), followed by the movements and records related to payments and receipts. The accounting entries were made in the traditional double-entry bookkeeping system, which proves that accounting in a Blockchain environment is not only cash accounting, instead it can be a true accounting support system because its technology allows it, due to its main features which are security, anonymity and data integrity. The issue of trust is also, in this field, essential, as accounting is the basis and underpinning of the entire financial system. Here, Blockchain technology can ensure the necessary trust due to its technical characteristics [10].

In our perspective, after this PoC, the application of Blockchain to accounting reveals that it is compatible with the traditional double-entry system, and may eventually include the possibility mentioned by [11]).

# References

1. Guerreiro, M.S., Rodrigues, L.L., Craig, R.: Institutional change of accounting systems: the adoption of a regime of adapted international financial reporting standards. Eur. Acc. Rev. **24**(2), 379–409 (2015). https://doi.org/10.1080/09638180. 2014.887477

2. Saraiva, H.I.B., Alves, M.D.C.G., Gabriel, V.: As raízes do processo formal de harmonização contabilística, a sua evolução e influência em Portugal. De Computis: Revista Española de Historia de la Contabilidad **12**(22), 172–204 (2015)
3. Nobes, C.: A half-century of accounting and business research: the impact on the study of international financial reporting. Acc. Bus. Res. **50**(7), 693–701 (2020). https://doi.org/10.1080/00014788.2020.1742446
4. Nobes, C., Parker, R.: Comparative International Accounting, 14th edn. Pearson. (2020)
5. Mohanty, D., Anand, D., Aljahdali, H.M., Villar, S.G.: Blockchain interoperability: towards a sustainable payment system. Sustainability **14**(2), 913 (2022). https://doi.org/10.3390/su14020913
6. Bracci, E., Tallaki, M., Ievoli, R., Diplotti, S.: Knowledge, diffusion and interest in blockchain technology in SMEs. J. Knowl. Manage. **26**(5), 1386–1407 (2022). https://doi.org/10.1108/JKM-02-2021-0099
7. Benchoufi, M., Ravaud, P.: Blockchain technology for improving clinical research quality. Trials **18**(1), 335 (2017). https://doi.org/10.1186/s13063-017-2035-z
8. Iansiti, M., Lakhani, K.R.: The truth about blockchain. Harvard Bus. Rev. **95**(1), 119–127 (2017)
9. Islam, A.K.M.N., Mäntymäki, M., Turunen, M.: Why do blockchains split? An actor-network perspective on bitcoin splits. Technol. Forecast. Soc. Change **148**, 119743 (2019). https://doi.org/10.1016/j.techfore.2019.119743
10. Pascual Pedreño, E., Gelashvili, V., Pascual Nebreda, L.: Blockchain and its application to accounting. Intangible Capital **17**(1), 1–16 (2021). https://doi.org/10.3926/ic.1522
11. Grigg, I.: Triple entry accounting. Syst. Inc. 1–10 (2005)

# Bibliometric Analysis on the Convergence of Artificial Intelligence and Blockchain

Maryam Hajizadeh[1](✉) , Morteza Alaeddini[2] , and Paul Reaidy[2]

[1] Université Grenoble Alpes, Grenoble IAE, 38000 Grenoble, France
Maryam.Hajizadeh@etu.univ-grenoble-alpes.fr
[2] Université Grenoble Alpes, Grenoble INP, CERAG, 38000 Grenoble, France
{Morteza.Alaeddini,Paul.Reaidy}@univ-grenoble-alpes.fr

**Abstract.** Arising from distributed artificial intelligence (AI) and blockchain (BC), the recently emerging notion of decentralized AI refers to transferring information and learning to various peer-to-peer connected machines which learn based on their locally available data and make decisions individually. Using a decentralized consensus mechanism without the need for trustworthy third parties or intermediaries, decentralized AI provides its users with processes, analytics, and decisions based on trusted, digitally signed, and securely shared data that have been transacted and stored on the BC in a decentralized manner. In order to identify core research focusing on both AI and BC and to find avenues for future research, this study employs thematic analysis of keywords of 1538 academic publications retrieved from WoS and Scopus databases, as well as bibliometric analysis of authors, affiliations, and sources to examine productivity, citation metrics, and bibliographic coupling. By highlighting the areas of digital transformation, environment/society, decentralized AI, DeFi, and cyber security as the focal points of the BC-AI integration, this paper intends to give researchers a comprehensive view of this convergence and can be used in the industry.

**Keywords:** Machine learning · Distributed ledger · Smart contract · Internet of Things (IoT) · Systematic review · Scientometric analysis

## 1 Introduction

Due to unending breakthroughs in cutting-edge information technology, perhaps the two growing technologies of artificial intelligence (AI) and blockchain (BC) have played a significant role in driving tremendous changes in most business models in recent years [1]. BC is an immutable, public digital ledger distributed among networked peers [2]. In order to reach consensus, every transaction in the BC is cryptographically signed and validated by all nodes [3], meaning that the nodes keep a copy of the complete ledger consisting of chained blocks of all transactions [4]. Another prominent field of great interest is AI, which allows a computing machine to learn, infer, and adapt depending on its data [4]. The ability of computer-aided machines to perform human behaviors requires skills such as understanding, generalizing, producing solutions, learning, and inferring from past experiences, which are defined as AI [5, 6].

J. Prieto et al. (Eds.): BLOCKCHAIN 2022, LNNS 595, pp. 334–344, 2023.
https://doi.org/10.1007/978-3-031-21229-1_31

On the one hand, the development of AI has been facilitated by the vast production and creation of data through sensing systems, Internet of things (IoT) devices, social media, and web programs [7]. On the other hand, the ability to configure the BC to manage transactions among people involved in decision-making or data generation and access is a characteristic of BC smart contracts [8]. Like autonomous systems, technologies built based on smart contracts seem to be able to learn and adapt to changes over time and make reliable decisions confirmed and certified by all BC nodes [4]. AI approaches using BC have assisted decentralized learning in analyzing and deciding on vast volumes of securely shared data stored and signed on BC [9].

The research on the combinations of AI and BC is still in the early stages [4]. The idea of recently developing decentralized AI is fundamentally a combination of the two technologies [10]. Because AI works with massive amounts of information, BC now seems to be a secure way to store this data. Focusing on this new approach and aiming to identify potentials and areas of interest for the convergence of two types of digital technologies, BC and AI, our motivation for conducting this research is to depict the status of articles on these two technologies through scientometric visualization.

This paper presents a comprehensive review of the research on BC and AI. Section 2 describes the method we have used to answer the following research questions [11]: (i) Thematically, research in this field can be classified into how many clusters? (ii) Who are the most prominent authors and sources in this field? (iii) Which countries have done the most research in this area? (iv) What are the promising avenues for future research in this field? In Sect. 3, we answer the questions by co-occurrence mapping of keywords, as well as by providing bibliometric analysis of other attributes of articles. By mapping periodic affiliations and keywords and discussing the research results, in Sect. 4, we determine several future directions. Section 5 concludes the paper.

## 2 Method and Data

This work uses a scientometric analysis method to find answers to research questions within the bibliometric attributes of journal articles, conference papers, and book chapters found in the Web of Science (WoS) and Scopus databases using search parameters linked to BC-AI themes. The query string includes the principal terms 'artificial intelligence' and 'blockchain,' along with their related keywords – computational intelligence, machine learning (ML), neural network, Bayesian network, deep learning, reinforcement learning, supervised learning, Q-learning, transfer learning, natural language processing, thinking computer system, expert system, evolutionary computation, intelligent agent, swarm intelligence, fuzzy logic, hidden Markov model, computer vision, simulated annealing, support vector machine, random forest, decision tree, learning algorithm, genetic algorithm, inductive logic program, intelligent tutoring, autonomous robot, distributed ledger, and smart contract. Also, we only focus on articles in English.

The results include 564 studies from WoS and 1468 records from Scopus. Finally, we obtained 1538 studies by eliminating duplicates, merging results, and meeting the criteria (i.e., language and document types). Documents used in this study are from 2008 onwards. The final data set is analyzed using the VOSviewer software tool and the Bibliometrix library in R. Table 1. Summarizes the data retrieval resulting from our search strategy.

**Table 1.** Characteristics of the selected publications.

| Data | Documents | Authors |
|---|---|---|
| • Retrieval date: 1/26/2022<br>• Databases: WoS; Scopus<br>• Timespan: 2008–2022<br>• Sources: 727<br>• Documents: 1,538<br>• Average years from publication: 1.95<br>• Average citations per document: 6.13<br>• Average citations per year per document: 1.55<br>• References: 47,118 | • Article: 607<br>• Book chapter: 48<br>• Conference paper: 883<br>• Keywords Plus[a]: 7,352<br>• Author's Keywords: 3,345 | • Authors: 4,278<br>• Author Appearances: 5,919<br>• Authors of single-authored documents: 122<br>• Authors of multi-authored documents: 4,156<br>• Single-authored documents: 132<br>• Documents per author: 0.359<br>• Authors per document: 2.78<br>• Co-authors per document: 3.85<br>• Collaboration Index: 2.96 |

[a]Terms appeared in the titles of an article's references but not appeared in the title of the article itself

## 3  Results

### 3.1  Time Evolution of the Studies

Figure 1 presents the number of papers in the field of BC-AI published and the mean of total citations between 2008 and early 2022. The number of total publications (TP) over time can be used to track the evolution of research interests.

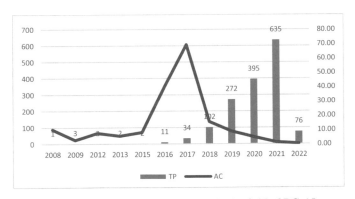

**Fig. 1.** Annual scientific production in the field of BC-AI.

The annual publication distribution shows an increasing trend. It is observed that BC and AI have become a field of research interest among academics and practitioners in the recent past – only 635 papers were published in 2021. Research in these fields was limited from 2008 through 2015 as publications are at a stable rate of 1, 2 articles per year. The number of papers rises gradually to 11 and 34 in 2016 and 2017, then increases sharply from 102 in 2018 to 635 in 2021. Based on the trend of recent years and the actual number of publications in the first month of 2022, approximately 1,066 publications are projected for 2022. Also, the average citations (AC) rate per year in Fig. 1 shows that the articles published in 2016 and 2017 have been cited more than others.

## 3.2 Thematics of the Studies

In order to identify important research topics in BC and AI, we first analyze authors' keywords that represent the vital content of the studies. The results confirm that smart contract, security, and IoT are at the top of the BC-AI keywords with 459, 355 and 321 occurrences, respectively. In order to remove general and merge similar terms, we build a thesaurus and use it in keyword mapping. To calculate the relatedness of the keywords for building a co-occurrence map, we employ the fractional-counting method, in which the weight of a co-occurrence link is fractionalized based on the number of other keywords in the publication [12]. We set at least 30 occurrences of keywords, meaning we cluster keywords that occur in more than 2% of documents. With these settings, 42 keywords meet the threshold.

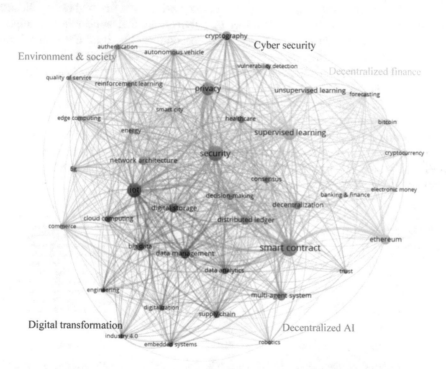

**Fig. 2.** Clusters resulting from the keywords co-occurrence analysis.

The network depicted in Fig. 2 is composed of five clusters of keywords. We name clusters according to the meaning and nature of most of the keywords belonging to them.

The red cluster, digital transformation, includes keywords of digital technologies that are primarily used to meet changing business and market requirements. This cluster reflects the authors' focus on the combined use of AI and BC in data, cloud, and IoT as well as their application in industry and the supply chain. The green cluster, environment & society, highlights the application of AI and BC to topics such as energy, smart cities and vehicles, commerce, and other fields of interest to society and the environment. The exciting thing about this cluster is that researchers use reinforcement learning algorithms to develop intelligent agents who perceive and interpret their environment, take actions and learn through trial and error. The other three clusters in Fig. 2 refer to the use of blockchain to provide reliable data for robots and intelligent agents to improve the quality of their decisions (blue), predictions in the field of decentralized financial (DeFi) using both supervised and unsupervised learning algorithms (yellow), and concepts regarding the protection of blockchain network against cyber attacks. (purple). The nodes or keywords that did not have a network with other keywords have the potential to become new research topics in the future [13] (see Sect. 4).

### 3.3   The Most Prolific Authors

Figure 3 shows in three ways the top ten authors who have the highest publication productivity in BC-AI. First, the number of publications that appear in the author's name (as the lead author or co-author) has been counted. The total citations of each author are presented in the second measure, and the third one indicates the authors' *h*-index, showing that authors have authored *h* papers, each of which has been referenced at least *h* times. From Fig. 3(i), Wang Y and Wang X are ranked first with 22 articles, followed by Li Y and Li J with 20 papers and Wang Z and Zhang Y with 18 articles. Although some authors are present in both Figs. 3(i) and 3(iii), in Fig. 3(ii), we come across several names that do not exist in those two figures. For instance, Wang X, Wang Y, Li J, Tanwar S, and Wang Z, who have high TP ratings, also have the highest *h*-index. In Fig. 3(ii), Atzei N, Bartoletti M, and Cimoli T are co-authors of an article with 649 citations which put them in the first citation rank.

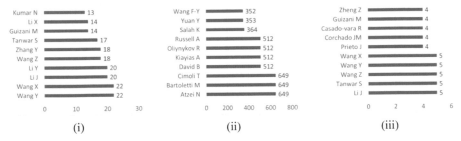

**Fig. 3.** Authors' productivity: (i) TP–total publications, (ii) TC–total citations, (iii) *h*-index.

To visualize the collaboration of authors with each other, we make a co-authorship map when the minimum numbers of publications and citations are 5 and 0, respectively (Fig. 4). Intensely collaborative authors from the red cluster include Li J, Li X, and Zhang

J, who study BC, ML, Ethereum, and multi-agent systems. The green cluster involves the authors studying BC, ML, IoT, and federated learning. Wang Y and Wang X are the most productive authors in this cluster. Tanwar S, Kumar N, and Gupta R emerge as solid contributors in AI and distributed ledgers in the blue cluster. In the yellow clusters, Chen Y and Wang S are the most collaborative authors whose main fields of interest are BC and privacy issues. Topics proposed in the violet cluster are BC and security, in which Guizani M and Chamola V have the maximum productivity.

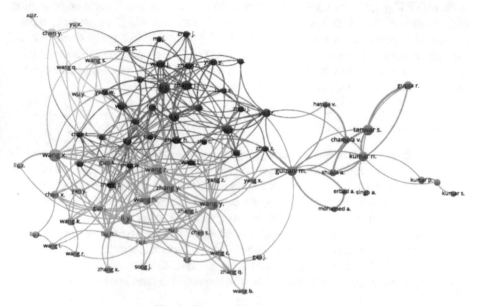

**Fig. 4.** The co-authorship network.

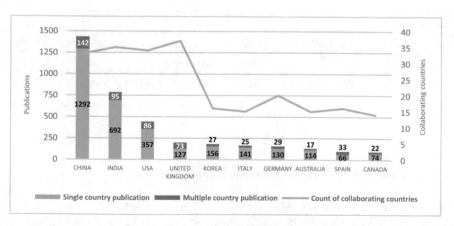

**Fig. 5.** Corresponding authors' countries.

### 3.4  Geographic Collaborations Perspective

The authors of the documents reviewed in this study are from 81 different countries. Figure 5 Shows 10 countries with the highest number of the identified papers according to the geographical locations of the authors. On top of all countries, China is the highest producer of research in this field with 1,434 studies, of which 1,292 documents have been published by Chinese researchers, and 142 publications have been prepared in collaboration with researchers from 33 other countries. In this regard, China is followed by India with 787 publications (692 publications alone and 95 publications in collaboration with 35 countries) and the United States with 443 documents (357 single publications and 86 joint publications with 34 countries). The results of our study show that the term blockchain is the main keyword in about half of China's research, followed by the term smart contract. The situation is similar in India and the United States, although the weight of focusing on these keywords may be somewhat different.

### 3.5  The Most Influential Sources

In clustering by coupling sources measuring and cluster labeling by authors' keywords with 500 units and 5% minimum frequency, we have two clusters: the first consists of smart contracts and ML, and the second covers IoT, security, and privacy. The top three publication venues in terms of total papers and total citations are shown in Table 2. By publishing 114 articles, Lecture Notes in Computer Science is the most popular series of distinguished conference proceedings in the field of BC-AI. To measure the impact of each source on this research community, we also evaluate those shortlisted publication venues based on the total number of citations; the results show that Lecture Notes in Computer Science has outperformed others in terms of the total number of citations.

**Table 2.**  Source impact: TP–total publications, TC–total citations.

| Source | TP | Source | TC |
|---|---|---|---|
| Lecture notes in computer science | 114 | Lecture notes in computer science | 1,518 |
| IEEE access | 50 | IEEE access | 936 |
| Advances in intelligent systems and computing | 46 | IEEE internet of things journal | 540 |

## 4  Discussion and Future Research Direction

Figure 6 presents the thematic evolution of BC-AI research based on the authors' use of keywords over time. From 2008 to 2015, most topics revolve around cryptography and the use of supervised learning. Gradually, other themes evolve in the 2016–2019 period due to spreading technology – AI methods (e.g., fuzzy logic and unsupervised learning), BC topics (e.g., smart contracts, distributed ledger, and consensus), IoT, privacy, and e-commerce. Interestingly, research focused on cryptography and supervised learning in the first era are mainly used in IoT field in the second period. The period of 2020–2022 witnesses research in IoT, security, and BC-related topics. Moreover, smart city and reinforcement learning are the new topics that have been addressed in this period. It seems that these topics, as well as cryptography and supervised learning methods that have returned to the field of interest of scholars, remain for future development.

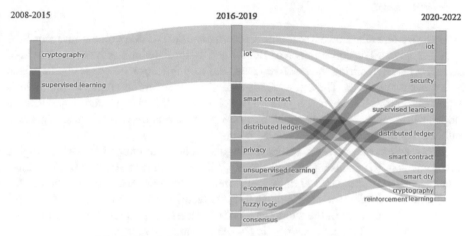

**Fig. 6.** Multi-fields plot between affiliation and keywords.

The co-occurrence map and the most frequent word analysis suggest that extant literature has made significant strides in understanding the BC and AI concepts and their application in the fields of IoT, security, privacy, supply chain, healthcare, and finance. Although research on these sectors seems mature, there are several gaps and dimensions that should be explored in the future. Managing and controlling localized data by fog nodes equipped with AI, ML and BC is an example.

Although BC and AI have advanced considerably in recent years, there remains a significant gap in terms of combining these two concepts. While empirical investigations of this combination have already highlighted a few possible benefits of this integration, future research could help further converge these technologies. Both technologies have distinct advantages that, if combined, will promise more reinforced benefits that would cause the applications of both fields to grow.

Research into different literature under this bibliographic analysis reveals several common avenues for future research. This includes research on their vulnerability, how they help develop autonomous renewable energy, their application in the circular economy, investigating the result of the combination of federated learning protocols with BC, or using game-theoretic mechanisms to improve cryptographic technologies for secure computation. Work on privacy-enhancing technologies (PETs) is one of the fields that has recently prompted scholars to study this integration. Recent changes to PETs using blockchain and smart contracts that focus on differential privacy (e.g., PETchain by Javed et al. [14]), federated learning (e.g., FLchain by Nguyen et al. [15]), and privacy-preserving (e.g., Deebak and Fadi [16]) are examples confirming garnering interest in this convergence. Establishing a mechanism for distributed trust among nodes in a BC network who use AI to decide whether to endorse other nodes seems to be another direction for future research in this area. Moreover, BC technologies have shown high adherence in operations and supply chain management, while studies contemplating the interplay of BC, AI, and supply chain are in the early stages.

## 5  Conclusion

In this paper, an evaluation of the global research trends in BC-AI publications from 2008 to 2022 is given. The convergence of BC-AI and their related topics has been a field of extensive research during the last five years, and its publication output is characterized by exponential growth. The latest trends indicate that scientific output in this field is expected to grow. The next few years will enlighten if this rising trend in publication output will remain. The study includes 1,538 publications on BC-AI, covering 4,278 authors, 727 publication venues, and 81 countries.

Five leading research areas can be distinguished in the domain of BC-AI: (i) digital transformation, (ii) environment and society, (iii) decentralized AI, (iv) DeFi, and (v) cyber security. The research area of DeFi seems to be more theoretical, whereas the research areas of digital transformation and cyber security have a more practical-oriented emphasis. The research in the subdomains of digital transformation, environment, and society stands in a dominant position in BC-AI research nowadays.

Some favorable aspects derived from the bibliometric analysis are as follows: (i) Since multi-authored publications make up more than 90% of all publications, it can be concluded that there is much collaborative research in the field of BC-AI; (ii) The fact that many different sources with a wide range of subject categories are devoted to BC-AI publications indicates the wide variety of research themes and the multidisciplinary character of the research in this domain.

China, India, and the United States are the most influential countries in this field. Based on the highest number of publications, Beijing University has the most heightened cooperation and activities. Moreover, according to the number of citations and papers, Lecture Notes in Computer Science have the highest impact among sources. Furthermore, although Wang Y and Wang X rank first in publishing, Atzei N, with "A survey of attacks on Ethereum smart contracts," is the most influential among authors. Also, our study discovers the potential future research in BC and AI, including a combination of federated learning protocols with BC, a distributed trust mechanism using BC and AI, and practical research to use BC and AI in supply chain management and in the fields of security and privacy.

Some limitations of this scientometric study are as follows. First of all, the search has been limited to publications listed on WoS and Scopus. Although they are among the most prominent global databases, they do not contain all publications in the field of BC-AI research. Other international databases such as ProQuest could have been used. It should also be noted that the integration of Scopus and WoS databases and the analysis of data within the final unique repository with VOSviewer and R are associated with problems and differences in their reports. The second limitation is the predominance of quantitative methods in bibliometric analysis, making the content or quality of publications uninterpretable [17]. This could lead to some publications being included in the analyses despite addressing a different topic than BC-AI convergence.

Finally, we hope that this bibliometric analysis of the confluence of AI and BC can give crucial information to the researchers and help them better understand the evolution of technical ecosystems and the growing literature in this cross-disciplinary field. Keeping up with the rapidly developing body of BC-AI research allows academics and practitioners to take advantage of the opportunities created by the confluence of the two domains to promote the use of AI for BC and BC for AI.

# References

1. Singh, S.K., Rathore, S., Park, J.H.: Blockiotintelligence: a blockchain-enabled intelligent IoT architecture with artificial intelligence. Futur. Gener. Comput. Syst. **110**, 721–743 (2020)
2. Nakamoto, S.: Bitcoin: a peer-to-peer electronic cash system. Decentralized Bus. Rev. 21260 (2008)
3. Alaeddini, M., Dugdale, J., Reaidy, P., Madiès, P., Gürcan, Ö.: An agent-oriented, blockchain-based design of the interbank money market trading system. In: Agents and Multi-Agent Systems: Technologies and Applications 2021, pp. 3–16. Springer (2021).https://doi.org/10.1007/978-981-16-2994-5_1
4. Salah, K., Rehman, M.H.U., Nizamuddin, N., Al-Fuqaha, A.: Blockchain for AI: review and open research challenges. IEEE Access **7**, 10127–10149 (2019)
5. Bello, O., Holzmann, J., Yaqoob, T., Teodoriu, C.: Application of artificial intelligence methods in drilling system design and operations: a review of the state of the art. J. Artif. Intell. Soft Comput. Res. **5** (2015)
6. Kalogirou, S.A.: Artificial intelligence for the modeling and control of combustion processes: a review. PrECS **29**(6), 515–566 (2003)
7. Schmidhuber, J.: Deep learning in neural networks: an overview. NN **61**, 85–117 (2015)
8. Wood, G.: Ethereum: a secure decentralised generalised transaction ledger. Ethereum Proj. Yellow Pap. **151**(2014), 1–32 (2014)
9. McMahan, B., Moore, E., Ramage, D., Hampson, S., Arcas, B.A.: Communication-efficient learning of deep networks from decentralized data. In: Artificial Intelligence and Statistics, pp. 1273–1282. PMLR (2017)
10. Hussain, A.A., Al-Turjman, F.: Artificial intelligence and blockchain: a review. Trans. Emerg. Telecommun. Technol. **32**(9), e4268 (2021)
11. Mejia, C., Wu, M., Zhang, Y., Kajikawa, Y.: Exploring topics in bibliometric research through citation networks and semantic analysis. Front. Res. Metrics Analytics **6** (2021)
12. van Eck, N.J., Waltman, L.: Visualizing bibliometric networks. In: Measuring Scholarly Impact, pp. 285–320. Springer (2014)
13. Effendi, D.N., Anggraini, W., Jatmiko, A., Rahmayanti, H., Ichsan, I.Z., Rahman, M.M.: Bibliometric analysis of scientific literacy using VOS viewer: Analysis of science education. J. Phys. Conf. Ser. **1** (2021). IOP Publishing, p. 012096
14. Javed, I.T., Alharbi, F., Margaria, T., Crespi, N., Qureshi, K.N.: PETchain: a blockchain-based privacy enhancing technology. IEEE Access **9**, 41129–41143 (2021)
15. Nguyen, D.C., Pathirana, P.N., Ding, M., Seneviratne, A.: Integration of blockchain and cloud of things: architecture, applications and challenges. IEEE Commun. Surv. Tutorials **22**(4), 2521–2549 (2020)

16. Deebak, B.D., Fadi, A.-T.: Privacy-preserving in smart contracts using blockchain and artificial intelligence for cyber risk measurements. J. Inf. Secur. Appl. **58**, 102749 (2021)
17. Pajo, A.T., Espiritu, A.I., Jamora, R.D.G.: Scientific impact of movement disorders research from Southeast Asia: a bibliometric analysis. Parkinsonism Relat. Disord. **81**, 205–212 (2020)

# Towards an Infrastructure Cost Model for Blockchain-Based Applications

Miguel Pincheira$^{(\boxtimes)}$, Elena Donini, Massimo Vecchio, and Raffaele Giaffreda

Fondazione Bruno Kessler, Trento, Italy
{mpincheiracaro,edonini,mvecchio,rgiaffreda}@fbk.eu

**Abstract.** Blockchain has become a core technology for developing decentralized applications allowing the safe interaction of untrusted actors. Given the early stage of blockchain research, the current literature has not addressed several inherent challenges related to system and software engineering, including the infrastructure costs and benefits. These are critical to evaluate the feasibility of any system and guide architectural choices during its design, implementation, and exploitation. This work presents a model for evaluating the infrastructure costs and benefits in blockchain applications. First, we propose a taxonomy to classify the application related transactions. Then, we propose a model to evaluate the infrastructure costs and benefits in applications using either public or private blockchains. The model is based on parameters that describe the systems and can be easily identified at any stage of the application life cycle. To illustrate the model validity, we quantitatively analyze a real use case. The results highlight the inherent complexity of characterizing the costs and benefits of blockchain applications, while the proposed models facilitate several aspects of this process. Moreover, it provides a solid ground to evaluate a blockchain-based system from the early stages of the development, eventually supporting the entire life cycle.

**Keywords:** Blockchain · Smart contracts · Dapps · Costs

## 1 Introduction

In the landscape of software systems, blockchain has become a core technology for developing new types of decentralized applications. Early adopters [1] embraced private blockchains as a seamless transition from traditional centralized systems to more decentralized ones. In recent years, an increasing number of blockchain-based systems have been thriving in public networks [2,3], embracing the possibility of interacting with unknown actors without intermediaries. Even if blockchain technology is attracting considerable interest, its adoption is still in an early stage [4]. From a system and software-engineering perspective, blockchain applications present inherent challenges [5] given by the specific life cycles that introduce unique constraints and characteristics in the system [6].

© The Author(s): under exclusive license to Springer Nature Switzerland AG 2023
J. Prieto et al. (Eds.): BLOCKCHAIN 2022, LNNS 595, pp. 345–355, 2023.
https://doi.org/10.1007/978-3-031-21229-1_32

These challenges are partially addressed in the literature, currently focusing on characteristics such as scalability, security, and performance [5]. However, very little is known about other characteristics (e.g., the infrastructure cost) that support the application development, deployment, and evaluation. The infrastructure cost is critical to evaluate the potential monetary value in terms of costs and benefits at an early and more advanced application development stage. A detailed and comprehensive infrastructure cost analysis can greatly increase the adoption of blockchain [6]. Although critical, the infrastructure cost is marginally addressed in the literature. Authors in [7] included a monetary analysis presented when quantifying the current scalability limits of Bitcoin and its decentralized components. Authors in [8] presented a cost comparison between Ethereum-based and cloud-based applications focusing on business process execution. Similarly, authors in [9] presented a financial evaluation framework that identifies factors influencing a blockchain application from a financial perspective (i.e., cost savings and benefits). Finally, authors in [10] presented an overview of smart contracts in the Ethereum network and analyzed several application metrics, including the cost in terms of GAS usage. Summarizing, the cost analyses in the literature consider metrics based only on the actual gas usage in a specific application scenario (e.g., public considering only a defined number of users and in a fixed time frame). As a result, literature analyses fail to provide enough generalization to evaluate different scenarios and evaluate the applications' entire life cycle. Furthermore, the literature studies focus only on one type of blockchain network (i.e., public or private) and do not provide a comparative analysis between them.

In this work, we present an infrastructure cost model for applications that integrates either public or private blockchains. We provide a simple conceptual tool to evaluate application costs and benefits from the early stages of development to the exploitation phase. First, we propose a taxonomy model to classify and generalize typical transactions in the life cycle of blockchain applications. Then, we define an infrastructure cost model based on the system parameters describing the application. These parameters are simple to identify and quantitatively analyze the monetary costs and benefits of using a public or a private blockchain network. The proposed model can evaluate the system cost and benefits in many different scenarios over the entire application's life cycle. Finally, we consider an application from the current literature and evaluate its costs and benefits using our model. The evaluation highlights the flexibility of the model to work for public and private blockchains, the simplicity of identifying the model parameters, and the ability to analyze the entire life cycle of the application. The contribution of this paper is three-fold. i) We propose a simple transaction taxonomy for blockchain applications. ii) We define an infrastructure cost model for private and public blockchains. iii) We propose a model that can evaluate the costs and benefits over the application's entire life cycle.

The rest of this paper is structured as follows: Sect. 2 formalizes the problem. Section 3 describes the proposed infrastructure cost model and transaction taxonomy. Section 4 evaluates the model in real use-case. We finalize the paper with our conclusion and plans for future work.

## 2   Problem Definition

The main objective of our model is to provide a simple tool for actors and stakeholders to evaluate the economic feasibility (i.e., costs and monetary benefits) of an application based on a private or public blockchain during its entire life cycle. Given a group of actors $A$ and a group of stakeholders $S$, the cost of the blockchain system $C(m)$ is calculated as the infrastructure needed to support the interactions $I$ between the actors. These interactions should generate a value unit $K$, which provides the monetary benefits $B(m)$ for the stakeholders. Our goal is to model $C(m)$ and $B(m)$ for actors and stakeholders. The model should be general enough to evaluate applications based on private and public blockchains. Moreover, the model should be able to evaluate the feasibility along the entire application life-cycle, i.e., from the early stages of development to a more advanced stage of exploitation.

$C(m)$ and $B(m)$ depend on the characteristics of the application and the blockchain, including the nodes running the network. The node number and configuration depend on the blockchain technical implementation (e.g., consensus model) and type (i.e., public or private). The configuration follows the application requirements (e.g., latency and throughput) and the existing trust between stakeholders [11]. On public blockchains, transactions that create new information (i.e., modify the state of the blockchain) have a monetary cost. Conversely, the transaction number does not directly affect the infrastructure cost in private blockchains. Private networks can reach a large transaction throughput with the same cost as a smaller one [12].

In this paper, we consider the following definition of a blockchain application and assumptions. We consider blockchain application software to support the interactions of a group of unknown actors identified only by their private/public keys. Actors have limited types of interactions to create and transfer information and value. Interactions are characterized by a maximum lifespan, which is an upper limit for the transaction processing time. Therefore, the lifespan of the transaction is the first factor in evaluating the feasibility of an application. We consider that the entire application logic is in the blockchain so that all information in the application is immutable, auditable, and accessible by anybody. This assumption is generally valid when there is no access restriction and facilitates comparing public and private networks. Finally, we consider that the application has a two-phase life cycle: Bootstrap and Operation, similarly to [8]. Note that off-chain computations and side-chains are beyond the scope of this paper. Moreover, we do not consider the cost of developing or updating software.

# 3   Proposed Cost and Benefit Model

This section describes the proposed model to evaluate the infrastructure costs and benefits for the actors and stakeholders at different stage of the application development. First, we define a transaction taxonomy, called $CRIV$, that classifies the interactions between the actors in the framework of blockchain applications during all the development and exploitation phases. This is because our model considers interactions among actors and stakeholders (i.e., transactions) as the functional units for the Life-Cycle Assessment (LCA) of an application. Then, we present a cost and benefit model based on the system parameters and the proposed taxonomy. We model the costs of public and private blockchain separately given that the transaction fees depend on the network type. Moreover, we model the monetary benefits for the stakeholders associated to a generic blockchain that can be public or private. Table 1 lists and describes the mathematical symbols used as the parameters for the model.

**Table 1.** Parameters for the proposed cost model.

| Parameter | Description | Parameter | Description |
|-----------|-------------|-----------|-------------|
| $A_0$ | Number of initial actors | $S$ | Number of stakeholders |
| $K$ | Value Unit | $L_k$ | $K$ lifespan |
| $P_C(m)$ | Cryptocurrency Price | $P_K$ | Value unit Price |
| $P_{node}$ | Node Price | $F_I$ | Interaction factor |
| $\mu$ | Time factor for transactions | $F_g$ | Growth factor |
| $O_C$ | Computational cost of $T_C$ | $F_V$ | Value factor |
| $O_R$ | Computational cost of $T_R$ | $F_t$ | Trust factor |
| $O_I$ | Computational cost of $T_I$ | $F_o$ | Operation factor |
| $O_V$ | Computational cost of $T_V$ | $F_k$ | Benefit factor |

## 3.1   Proposed Transaction Taxonomy for Blockchain Applications

In blockchain-based applications, interactions among actors are transactions. From the set $\mathcal{I}$ of all possible transactions, we focus only on a subset $\mathcal{T} = \{T_i, i = [C, R, I, V]\} \subset \mathcal{I}$ of transactions that creates new information for the application and the actors. Considering that these transactions vary greatly from one application to another, we propose the $CRIV$ taxonomy to easily identify core transactions and link them to the application's life cycle. $CRIV$ categorizes the interactions into four types of transactions: creation $T_C$, registration $T_R$, interaction $T_I$, and value $T_V$. Figure 1 shows the four transactions in our taxonomy along with the general life cycle of the application. Each type of transaction is defined as follows:

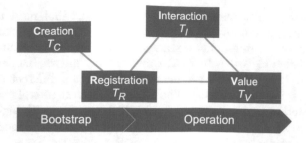

**Fig. 1.** Life cycle of a blockchain-based application.

- **Creation Transaction** $T_C$ deploys the application into the blockchain. $T_C$ may include one or more transactions happening during the Bootstrap phase.
- **Registration Transaction** $T_R$ is required at the first interaction of an actor with the system, to make the actor part of the application.
- **Interaction Transaction** $T_I$ are the most common interactions between actors. They produce information to be stored in the blockchain, without including any value transfer.
- **Value Transaction** $T_V$ is the most important transaction as it includes the value transfer between unknown actors.

## 3.2 Proposed Public Blockchain Cost Model

Given the life cycle of an application, we divide the cost model into two components, i.e., the bootstrap $C_B$ and operation $C_O$ costs. $C_B$ considers the transactions needed to deploy the logic of the application ($T_C$) and the transactions to register the initial actors ($T_R$). $C_O$ considers the transactions for the registration of new actors ($T_R$), the interaction transactions between actors ($T_I$), and the transactions that transfer value ($T_V$). Here, considering the general life cycle of an application, $C_B$ and $C_O$ are evaluated in a given month $m$, which is defined as the minimal time window for assessing the systems. This is to have an easier comparison with other types of monetary evaluations (e.g., budget planning). However, the cost model can be easily adapted to shorter and longer windows. Hence, the initial month ($m = 0$) corresponds to the bootstrap phase, and any other month ($m > 0$) indicates the operation phase. We define the costs of a public blockchain infrastructure $C_{pub}(m)$ as :

$$C_{pub}(m) = \begin{cases} C_B(m) = CT_C(m) + CT_R(m), & \text{if } m = 0 \\ C_O(m) = CT_R(m) + CT_I(m) + CT_V(m) & m > 0 \end{cases} \quad (1)$$

where $CT_i(m)$ is the total monetary costs paid in a month $m$, for all the transactions of type $i$, where $i = \{C, R, I, V\}$. The monetary cost of a single transaction of type $i$ links the computational cost $O_i$ of the transaction , with the price of the cryptocurrency $P_C(m)$ and a processing time factor $\mu$. Finally, the total number of transactions of type $i$ in a given month $m$ is give by $Q_i(m)$.

The price of the cryptocurrency $P_C(m)$ is a function that must be defined by the user by taking into account the high volatility of the cryptocurrencies price and the scenario to evaluate. For instance, using historic cryptocurrency prices, the user can define $P_C(m)$ with fixed value for any month (i.e., an average for all months). Similarly, the user can define $P_C(m)$ with a different monthly value (i.e., an average for each month). The factor $\mu$ is used to scale the price paid for each transaction (i.e., a transaction fee). Transactions with higher prices are more attractive for the node operators and typically are processed faster, since the the node that process the transaction will receive the fee as a reward. The total cost $CT_i(m)$ for each transaction $i$ is given by:

$$CT_i(m) = O_i\, P_C(m)\, \mu\, Q_i(m), \quad \text{with } i \in \{C, R, I, V\} \tag{2}$$

In the operational phase, the total number of transactions of each type $i$ in a given month $m$ is directly related to the number of actors $A(m)$ in the system defined as:

$$A(m) = \begin{cases} A_0, & \text{if } m = 1 \\ A(m-1)\, F_g & m > 1 \end{cases} \tag{3}$$

where $A_0$ is the initial number of actors in the system, and $F_g$ is the actor growth factor that describes the monthly growth of the system in terms of the actor number. Finally, the number of each type of transaction $Q_i(m)$ in the operation phase ($m > 0$) with respect to the actor number is defined as:

$$Q_i(m) = \begin{cases} 0 & \text{for } i = T_C, m > 0 \\ A(m) - A(m-1) & \text{for } i - T_R, m > 0 \\ A(m)\, F_I & \text{for } i = T_I, m > 0 \\ A(m)\, F_V & \text{for } i = T_V, m > 0 \end{cases} \tag{4}$$

$Q_C(m)$ is equal to 0 after the bootstrap phase ($m > 0$). $Q_R(m)$ is given by the number of new actors in that month. $Q_I(m)$ links the total number of actors in the system $A(m)$ using an interaction factor $F_I$. $F_I$ relates to the expected number of interaction transactions $T_I$ of each actor. For example, if each actor is expected to have at least two $T_I$ in the time unit $m$, the factor is set to $F_I = 2$. Lastly, $Q_V(m)$ links the total number of actors with factor $F_V$, which represents the value transfer transactions $T_V$ of each actor. For instance, when actors are expected to have at least one $T_V$ every two months, the factor is set to $F_V = 0.5$. The user can estimate the values of $F_I$ and $F_V$ at an early stage of development. In a more advanced stage, the factors can be estimated from the current activity in the system.

## 3.3    Proposed Private Blockchain Cost Model

For applications based on a private blockchain, we define the infrastructure cost $C_{Pri}(m)$ in a given month $m$ as divided into two components $C_B$ and $C_O$, indicating the bootstrap and the operation phase respectively. $C_B$ is the initial investment to acquire $N$ nodes (i.e., computers) with a price $P_{node}$ to create the network infrastructure. $C_O$ is the expense of running and operating the nodes. Similar to the traditional software systems, we estimate the operating costs $C_O$ as a percentage of $P_{node}$ using a scale factor $F_o$. $F_o$ is estimated by taking into account the system characteristics in terms of the hardware and software required to run the nodes. The infrastructure cost $C_{Pri}(m)$ in a private blockchain is defined as:

$$C_{Pri}(m) = \begin{cases} C_B(m) = N\,P_{node} & \text{if } m = 0 \\ C_O(m) = N\,F_o\,P_{node} & m > 0, \end{cases} \tag{5}$$

In our model, the node number $N$ is related to the number of stakeholders $S$ by a trust factor $F_t$:

$$N = S\,(1 - F_t). \tag{6}$$

where $F_t$ is the relation between the $N$ and the total number of stakeholders. For instance, if 100 stakeholders agree that only 30 different nodes are required to support the infrastructure, a trust factor of 70%. Similarly, a 100% trust factor will translate into a centralized system. Here, for simplicity, we consider that one node represents one stakeholder.

### 3.4   Monetary Benefits

For applications based on both public and private blockchains, the monetary benefits $B(m)$ for the stakeholders are derived from the **value units** $K$ transacted in the application and are given by:

$$B(m) = F_k\,Q_V(m)\,P_K \tag{7}$$

where $F_k$ is the benefit factor that indicates the expected value units for each value transfer transaction $T_V$. $Q_V(m)$ is the number of value transfer transactions in the month $m$, and $P_K$ is the price (i.e., total monetary value) of the value unit $K$. $P_K$ is the sum of all the monetary values assigned to each stakeholder $S$ (i.e., the benefit for each stakeholder). In a public network, this value may also be linked to the price of the cryptocurrency, such as $P_K(m) = 0.4\,P_C(m)$.

## 4   System Setup and the Model Evaluation

To evaluate the goodness of the proposed cost and benefit model, we consider a blockchain application existing in the literature [2]. This section also shows how to apply the proposed model by describing the rationale used for the model parameters.

## 4.1  Description of the Blockchain Setup

Here, we adopted the Ethereum blockchain, as it is considered a reference implementation for smart contracts and can be used in both public and private scenarios [13]. Furthermore, several private blockchains have taken Ethereum as the reference for their implementation [13]. Therefore, the benefit and cost model can be easily migrated to another blockchain type without any major issue.

For the price of the cryptocurrency $P_C(m)$, we used historical values that are available on several online websites, such as Etherscan.[1] The computational cost $O_i$ of the transactions i set equal to the gas required for their execution, and $\mu$ is the gas price on Ethereum. For the cost of the node $P_{node}$, we consider the minimum hardware requirements for an Ethereum node.[2] At the time of writing, this translates into a computer of 300 USD. We also consider the operation factor for the node $F_o$, as 40% of this cost.

## 4.2  Description of the Application

This section describes the application [2] and the parameters set up for the model. Some parameters are clearly stated in the paper, and others require our estimation. From [2], the group of unknown actors $A$ is composed of farmers using IoT devices to measure water consumption (i.e., a valve). The stakeholders $S$ are those interested in fostering the water savings. The value unit $K$ is a cubic meter of saved water. Furthermore, the lifespan unit is a day, as the water usage is reported daily. Hence, the application can also be executed in a public network. The application in [2] has four types of transactions, which directly map to our taxonomy and provides the values $O_C, O_R, O_I$, and $O_V$. For the price $P_C(m)$, we consider the historical values of Ethereum in 2020 and $\mu = 30$ gwei. [2] also considers an initial number of actors of $A_0 = 100$ users with a monthly growth of $F_g = 5\%$. Moreover, all the actors have to update their valve once a week, which translates into $F_I = 4$.

Here, we define that all actors will receive their reward once a month $F_V = 1$. For the benefits, we consider that each actor saves on average 4m³ in 100 ha farm as described in similar studies [14]. Hence, the benefit factor is set to $F_k = 4$.

The most complex part of estimating the values for this use case is defining a monetary value $P_K$ for a saved of m³. For this scenario, we consider an NGO that grants a value to the savings equal to the cost of irrigation m³ at 0.8 USD (based on the values presented in [14]). We also assume that an energy company offers a discount of 0.2 USD for saved m³. Finally, let us assume that a "eco-friendly" label will translate into 10 USD of additional benefits per month (a very conservative value). With only these three stakeholders ($S = 3$), the total monetary value of our value unit (saved m³ of water) is 11 USD. Table 2 summarizes the value of the parameters for the application.

---

[1] https://etherscan.io/chart/etherprice.
[2] https://docs.ethhub.io/using-ethereum/ethereum-clients/geth/.

**Table 2.** Parameters for the cost model of the water-management system from [2].

| Parameter | Value | Parameter | Value |
|-----------|-------|-----------|-------|
| $A_0$ | 100 | $S$ | 3 |
| $K$ | Saved m3 of water | $L_k$ | 1 day |
| $P_C(m)$ | ETH in 2020 | $P_K$ | 11 USD |
| $P_{node}$ | 300 USD | $F_I$ | 4 |
| $\mu$ | 30 gwei | $F_g$ | 005 |
| $O_C$ | 3343572 | $F_V$ | 1 |
| $O_R$ | 143947 | $F_t$ | 0.01 |
| $O_I$ | 26821 | $F_o$ | 0.4 |
| $O_V$ | 156580 | $F_k$ | 1 |

### 4.3 Evaluation of the Costs and Benefits

We evaluate the cost and benefits of the application for a period of 12 months with a step of one month. When using a private network, the monthly cost $C_{pri}(m)$ is fixed at 360 USD for all months. Conversely, when using a public blockchain, the cost $C_{pub}(m)$ varies with the number of transactions and the cryptocurrency's price $P_C(m)$. Here, we evaluate two different scenarios for the public blockchain: i) $P_C(m)$ is the same every month, defined from the yearly average cryptocurrency price, and ii) $P_C(m)$ varies every month considering the monthly average price Fig. 2 shows the costs and the benefits with a monthly step for one year for the private blockchain and public blockchains for both scenarios. As we can see in Fig. 2, the benefits of the application outweigh the cost in all three scenarios. The cost in a private scenario decreases as the number of actors increases, and it is lower than in a public network around the ninth month. Conversely, the cost per transaction on a public blockchain remains the same when considering the fixed year average price for ETH. However, when considering cryptocurrency price volatility, the costs can change drastically. In this evaluation, the costs are lower than the benefits in all cases. However, there are scenarios where the application will not be economically feasible (e.g., a lower value of $P_K$, or higher values for $P_C$). These scenarios will be easily identified by using our model and defining different parameters for the application.

## 5 Conclusions and Future Works

This work presented a cost model for evaluating the costs and benefits of blockchain-based applications from the early stages of development to the exploitation phase. First, we provide a taxonomy model to classify and generalize typical transactions in the life cycle of blockchain applications. Then, we define an infrastructure cost model based on the system parameters that describe the application. Finally, we quantitatively analyze the cost st and benefits of a

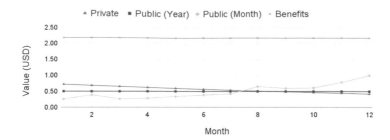

**Fig. 2.** Average monthly costs and benefits of water management system considering different scenarios.

real use case from current literature. The evaluation highlights the flexibility of the model to work for public and private blockchains, the simplicity of identifying the model parameters, and the ability to analyze the entire life cycle of the application. Hence, this study's rationale and practical evaluation can contribute to the growing knowledge of blockchain-based software engineering.

Future works include evaluating other applications to fine-tune the model's parameters better and provide reference values for the transactions in the taxonomy. Another interesting research path is creating a hybrid model that considers architectures and applications combining private and public networks.

# References

1. Ishan, M., et al.: Blockchain for 5G-enabled IoT for industrial automation: a systematic review, solutions, and challenges. Mech. Syst. Sig. Process. **135**, 106382 (2020)
2. Pincheira, M., et al.: Exploiting constrained IoT devices in a trustless blockchain-based water management system. In: 2020 IEEE International Conference on Blockchain and Cryptocurrency (ICBC) (2020)
3. Pincheira, M., et al.: A blockchain-based approach to enable remote sensing trusted data. In: 2020 IEEE Latin American GRSS ISPRS Remote Sensing Conference (LAGIRS) (2020)
4. Maximilian, W., Zdun, U.: Architectural design decisions for blockchain-based applications. In: The 3rd IEEE International Conference on Blockchain and Cryptocurrency (ICBC), May 2021
5. Destefanis, G., et al.: Smart contracts vulnerabilities: a call for blockchain software engineering? In: 2018 International Workshop on Blockchain Oriented Software Engineering (IWBOSE)
6. Vacca, A., et al.: A systematic literature review of blockchain and smart contract development: techniques, tools, and open challenges. J. Syst. Softw. **174**, 110891 (2021)
7. Croman, K., et al.: On scaling decentralized blockchains. In: Financial Cryptography and Data Security, pp. 106–125. Springer Berlin Heidelberg (2016)
8. Rimba, P., et al.: Quantifying the cost of distrust: comparing blockchain and cloud services for business process execution. Inf. Syst. Front. **22**(2), 489–507 (2020)

9. Demir , M., et al.: A financial evaluation framework for blockchain implementa-tions. In: 2019 IEEE 10th Annual Information Technology, Electronics and Mobile Communication Conference (IEMCON), pp. 0715–0722 (2019)
10. Wu, K. et al.: A first look at blockchain-based decentralized applications. Softw. Pract. Experience **51**(10), 2033–2050 (2021)
11. Schäffer, M., et al.: Performance and scalability of private ethereum blockchains. In: International Conference on Business Process Management, pp. 103–118. Springer (2019)
12. Ampel, B. et al.: Performance modeling of hyperledger sawtooth blockchain. In: 2019 IEEE International Conference on Intelligence and Security Informatics (ISI), pp. 59–61 (2019)
13. Kondo, M., et al.: Code cloning in smart contracts: a case study on verified con-tracts from the ethereum blockchain platform. Empirical Softw. Eng. **25**(6), 4617–4675 (2020)
14. Giannakis, E., et al.: Water pricing and irrigation across Europe: opportunities and constraints for adopting irrigation scheduling decision support systems. Water Supply **16**(1), 245–252, 09 (2015)

# A Benchmarking Study of Blockchain-Based Technology Implementation Cost: Public and Private Blockchain for Enterprise Level Organization Using Benchmarking Model

Richard(✉) ⓘ, Vera Angelina, Felix, and Michael Wangsa Mulia

Information Systems Department, School of Information Systems, Bina Nusantara University, Jakarta 11840, Indonesia
richard-slc@binus.edu, {vera.angelina,felix012, michael.mulia}@binus.ac.id

**Abstract.** The goal of this research is to benchmark two types of Blockchain, Public Blockchain and Private Blockchain, which can be implemented by organizations by considering the implementation cost. The cost of implementation is benchmarked using the Benchmarking methodology. The Benchmarking model is modified from the one that has been developed by G. Anand and R. Kodali [12]. There are 2 platforms of each type of Blockchain which are being benchmarked. Data were collected by using literature review methodology by analysing the indicators of each type that are gathered through the platforms' official websites, whitepapers, journals, and articles related to Blockchain. During the calculating process, there were assumptions added based on the collected data, and some were adapted from previous research. The result of this study is a template that can be used by organizations to self-calculate the cost of implementing Blockchain from various platforms but still follow the developed indicators.

**Keywords:** Benchmarking study · Blockchain · Public blockchain · Private blockchain · Implementation cost

## 1 Introduction

Nowadays, Blockchain technology has gained popularity, especially in Indonesia's tech community with the emergence of many new tokens and blockchain projects. Because of its huge potentials and benefits, enterprises are starting to adopt this technology into their business process. Blockchain technology can be utilized in many aspects of enterprises' business models and across industries, e.g., finance, insurance, human resources, supply chain management, and many more. A survey conducted in 2018 found out that more than 52% of 225 respondents' organizations were considering to apply blockchain solutions [1]. In 2021, the investment in blockchain technology around the world is predicted to reach $6.6B and will reach $19B by 2024 [2]. When it comes to implementing new technology, the cost is one of the important variables to consider. However, the cost of implementing a blockchain solution can be a little bit tricky to estimate due to its fluctuating gas fee. Therefore, in this paper, we tried to solve this issue by developing a template to calculate the implementation cost of public and private blockchain. The implementation of private blockchain is meant to be implemented by 1 organisation and not a consortium organisation.

## 2 Literature Review

### 2.1 Blockchain

Blockchain is a decentralised distributed ledger of data, which are shared and validated amongst the peers within a network [3]. There are three types of blockchain, public blockchain, private blockchain, and consortium blockchain [4]. Public blockchain is the most decentralized and transparent as anyone can read, write, and join the network [5]. Meanwhile, private blockchain to be part of the network or view the transaction history one need permission from the network owner. Hence, it is a very closed ecosystem and each member has their own role [5]. Consosrtium blockchain is similar to private blockchain which only invited participants can join the network. However, consosrtium blockchain is more suitable for multiple organizations. It provides more security in case of information alteration and hacking than private blockchain [4].

Enterprise needs to choose the right platform for their business. Some features need to be considered when choosing blockchain platform are the technical-related features (performance, security, consensus, etc.), business-related features (cost, reputation, use cases, etc.), and implementation-related features (ease of development, tokens and cryptocurrencies, Application Programming Interfaces (API) and Software Development Kits (SDKs)) [3]. Currently, there are approximately 1000 blockchain platforms developed and used for various use cases, such as supply chain management, cryptocurrencies, smart contracts, central bank money, and more [7]. In this paper, the authors focus on using the Ethereum network-compatible platforms as this platform has been supported by industry-leading companies such as Intel, Microsoft, J.P. Morgan, and Accenture [8]. For public blockchain, the chosen platforms are Ethereum and Polygon. On the other hand, the private blockchain platforms used are Hyperledger Fabric and Quorum.

### 2.2 Benchmarking

Benchmarking is a method to measure and evaluate the performance of several models with existing standards. According to Boychev [9], benchmarking is a system that aims to find, evaluate, and research best practices that can be implemented to increase efficiency [9]. According to Bhutta and Huq [10], there are various steps that can be taken when using the Benchmarking method and there are no definite steps that must be followed [10] Various kinds of benchmarking models have been developed and studied by many researchers to date, such as R. Camp, D. Garvin, C. Drury, T. Pilcher, R. Reider, S. Gunnar, K. Stefan, H. Belokorovina, N. Voevodina, A. Kulagina, O. Loginova, L. Kuznetsova, and others [11]. The models that have been developed are continuously modified and adapted to various relevant situations [12]. Therefore, in this study, we adapted the model that had been developed by G. Anand and Rambabu Kodali [12] and modified it again. There are 12 phases and 54 steps that have been identified by Anand and Kodali as best practice in Benchmarking.

# 3   Methodology

In conducting this study, the benchmarking methodology is used according to the benchmarking model developed by G. Anand and Rambabu Kodali [12]. However, the model was modified due to some steps that were not applicable at all. From 12 phases and 54 steps in the original model, only 6 phases and 9 steps are being applied.

## 3.1   Benchmarking Steps

**Subject Identification**. This phase consists of identifying the subject to be benchmarked. In this study, the subject is two types of blockchain technology, public blockchain and private blockchain. These types will be benchmarked based on their implementation costs.

**Subject Analysis**. There are 2 steps in this phase. The first one is understanding the subject's current condition based on the existing information. Blockchain technology has been developed to satisfy the requirements in various industries today. For example, the insurance industry exploits it to validate whether the claim has met with the agreed policy or not using a smart contract. Another example is in the health care system to track the supply chain management of the medicine. This technology is very suitable for enterprises because of its decentralized and immutable concept and thus the data is very transparent and easy to audit (Table 1).

The second step is characterizing the subject to comprehend its capability. The very distinct characteristics of the public blockchain and private blockchain are how the data can be accessed and the anonymity trait. In a public blockchain, anyone can join the network and therefore can read/write data within the network without anyone knowing who has done what (anonymous). On the other hand, in the private blockchain network, only those who have the permission can join and do what they have been permitted within the network. In addition, there are some different variables that impact the implementation cost of these types. The variables which were used in the calculation process were based on the paper published by Ernst & Young LLP in 2019 [13]. The details of the variables can be seen in Table 2.

**Selecting The Benchmarked Platforms**. There are many blockchain platforms that have been developed and used. However, in this study, we only choose 2 platforms from each type, Ethereum and Polygon for public blockchain and Hyperledger Fabric and Quorum for private blockchain. The considerations in selecting the public blockchain platforms are based on the market capitalisation, transaction per second, programming language, and the consensus protocol. A different approach was used to determine which private blockchain platform to be used in this study. The platforms were chosen based on their popularity in utilization by enterprises and also their applicability across the industry.

**Pre-benchmarking**. In this phase, we decided on how to collect the data. The method we used in this study is a literature review method. We gathered journals and reliable articles that were related to blockchain technology to be able to further understand this technology. In addition, to get the information related to the platforms we read the published whitepapers and visit the official websites. For the prices data, we collected from official websites that have historical data of the transaction fee such as etherscan.io,[1] polygonscan.com,[2] and messari.io.[3]

**Benchmarking**. There were three steps conducted simultaneously, such as doing the benchmarking study, data collecting, and sorting the collected data in Microsoft Excel. In the benchmarking process, we need data related to the implementation cost. We found out that there in public and private blockchain there were similar yet distinct indicators. We then categorized the cost into 2 in Microsoft Excel thus we can calculate the cost easily by exploiting the built-in function. The indicators were onboarding cost and cloud cost. For public blockchain, there is an additional cost which is the transaction fee or gas fee. The details of the cost can be seen in Table 1.

**Continuous Improvement**. After the benchmarking study is done, the result needs to be evaluated and further improved to keep its accuracy and stay updated.

## 4 Results

### 4.1 Define Assumptions and Calculations

The assumptions made in this study were created using the prior collected data. The costs required to implement a Blockchain technology can be categorized into two categories, Onboarding Cost and Cloud Cost. There are additional fees for Public Blockchain, Transaction Fees. The two types of blockchain have different indicators of implementation cost (Table 1). There are several variables that influence the implementation cost of both types of blockchain (Table 2). The consensus protocol variable will change the total of full nodes (Table 3) and cloud VM cost (Table 4). In addition, the cloud VM cost is also affected by the transaction size (Table 4). These assumptions of the impacted percentage were made according to Erns & Young LLP calculation [13].

The infrastructure service that is being used in this study is from Infura because this service can be used by both Ethereum and Polygon. Table 5 is showing the cost from Infura[4] service which was gathered directly from the official website on Dec 17th, 2021.

Cloud cost also can also be considered as server cost. In this study, we calculate the cost from 2 cloud server providers, Amazon Web Services[5] and Google Cloud.[6] The costs displayed below are for 1 full node.

---

[1] https://etherscan.io/chart/avg-txfee-usd, last accessed 2022/01/09.
[2] https://polygonscan.com/chart/gasprice, last accessed 2022/01/09.
[3] https://messari.io/, last accessed 2022/01/09.
[4] https://infura.io/pricing, last accessed 2022/01/09.
[5] https://calculator.aws/#/, last accessed 2022/01/10.
[6] https://cloud.google.com/products/calculator, last accessed 2022/01/09.

**Table 1.** Cost indicators for blockchain implementation.

| Cost | Public blockchain | Private blockchain |
|------|-------------------|--------------------|
| Onboarding cost | | |
| Full nodes | | ✓ |
| Infrastructure cost | ✓ | |
| Transaction fee | ✓ | |
| Cloud Cost | | |
| Cloud VM per full nodes | | ✓ |
| Annual cloud VM cost | | ✓ |
| Annual storage capacity | | ✓ |

**Table 2.** Blockchain implementation cost variables.

| Variables | Public blockchain | Private blockchain |
|-----------|-------------------|--------------------|
| Cloud service | | ✓ |
| Daily transaction volume | ✓ | ✓ |
| Transaction size | ✓ | ✓ |
| Consensus protocol | ✓ | ✓ |
| Payment method | | ✓ |
| Committed usage | | ✓ |

**Table 3.** Full nodes estimated impact percentage

| Consensus protocol | Estimated impact (%) |
|--------------------|----------------------|
| Proof of work | + 100 |
| Proof of stake | + 20 |
| Byzantine fault tolerance | 0 |
| Proof of authority | 0 |

In addition, there are equations for calculating the annual cloud VM cost and annual storage capacity cost. The annual cloud VM cost is calculated by adding the cloud service prices to the percentages of the estimated impact in Table 4.

$$Annual\ Cloud\ VM\ Cost = a + b + c \qquad (1)$$

where:

$a$ = service price.

**Table 4.** Cloud cost estimated impact percentage

| Variable | | Estimated impact (%) |
|---|---|---|
| Consensus protocol | Proof of work | + 100 |
| | Proof of stake | + 20 |
| | Byzantine fault tolerance | 0 |
| | Proof of authority | 0 |
| Transaction size | Small–150 bytes | − 5 |
| | Medium–250 bytes | 0 |
| | Large–500 bytes | + 10 |

**Table 5.** Infura price list

| Type | Price |
|---|---|
| Core | Free |
| Developer | $50 |
| Team | $225 |
| Growth | $1000 |

b = consensus estimated impact percentage.

c = transaction size estimated impact percentage.

On the other hand, the annual storage capacity cost is calculated by using the arithmetic formula by finding the first month's total storage (U1) and the total storage for n month (Un) in gigabytes. The results of U1 will be rounded up to the nearest multiple of 5 to anticipate excessive transaction (U1p). After obtaining the results of U1 and Un, then calculate the arithmetic formula.

$$U1 = \frac{x \times y \times 30}{z} \tag{2}$$

$$Un = U1p \times n \tag{3}$$

$$Annual\ Storage\ Capacity = \left(\frac{n(U1p + Un)}{2}\right) \times a \tag{4}$$

where:

x = daily transaction volume.

y = transaction size.

z = 1 GB in Bytes (1073741824).

n = n month.

a = service price.

**Table 6.** Cloud cost price list

| Variable | | Price | | |
|---|---|---|---|---|
| | | Amazon web services | | Google cloud |
| | | Monthly | All Upfront | |
| Committed usage | 1 year ($) | 1871.16 | 1746.74 | 1824.84 |
| | 3 years ($) | 1261.44 | 1097.63 | 1303.44 |
| Cloud storage (per Gb) ($) | | 0.12 | | 0.02 |

The transaction fee is affected by the size and volume of the transaction. In public blockchain the transaction fee is also known as gas fee. We have gathered the average daily transaction fee from etherscan.io, polygonscan.com, and messari.io since Sept 5th, 2021, to Jan 5th,2022. The data then processed to get the lowest price (small transaction), average price (medium transaction) and highest price (large transaction). The equation to find the transaction fee is by times the transaction fee with daily transaction volume and 365 days then add the transaction size estimated impact percentage.

$$Transaction\ Fee = j \times k \times 365 + c \tag{5}$$

where:

j = average transaction fee.

k = daily transaction volume.

c = transaction size estimated impact percentage.

## 4.2   Implementation Cost

To present the calculation results we simulated the implementation cost in several scenarios. The number in the scenarios' variables can be altered according to the needs (Tables 6, 7 and 8).

**Public Blockchain**. Table 9 shows the estimated implementation costs using Ethereum and Polygon platforms. The transaction fee is calculated based on Equation 5, while the infrastructure cost estimated from blockchain API provider cost to make smart contract interaction. The result shows Ethereum platform's cost is quite expensive compared to Polygon because Ethereum uses the Proof of Work consensus protocol which is high consumes energy and expensive. The transaction fee is greatly being impacted by the transaction volume dan transaction size. The higher/larger the volume and size are, the more expensive the cost of implementation will be.

**Private Blockchain**. Table 10 shows the estimation of the implementation costs using Hyperledger Fabric and Quorum platforms. The basic cost of cloud VM and storage are

**Table 7.** Average transaction fee

| Transaction | Ethereum ($) | Polygon ($) |
|---|---|---|
| Small–150 bytes | 4.9221 | 0.0059 |
| Medium–250 bytes | 19.9336 | 0.0218 |
| Large–500 bytes | 34.9450 | 0.1054 |

**Table 8.** Scenarios

| Variable | Scenario 1 | Scenario 2 |
|---|---|---|
| Daily transaction volume | 1000 | 2000 |

**Table 9.** Public blockchain implementation cost

| Transaction size | Variables | Ethereum | | Polygon | |
|---|---|---|---|---|---|
| | | S1 | S2 | S1 | S2 |
| Small–150 bytes | Infrastructure cost ($) | 600.00 | 600.00 | 600.00 | 600.00 |
| | Transaction fee ($) | 1,706,742.97 | 3,413,485.93 | 2,043.05 | 4,086.10 |
| | **Total cost ($)** | **1,707,343.0** | **3,414,085.9** | **2,643.1** | **4,686.1** |
| Medium–250 bytes | Infrastructure cost ($) | 600.00 | 600.00 | 600.00 | 600.00 |
| | Transaction fee ($) | 7,275,755.69 | 14,551,511.38 | 7,960.62 | 15,921.24 |
| | **Total cost ($)** | **7,276,355.7** | **14,552,111.4** | **8,560.6** | **16,521.2** |
| Large–500 bytes | Infrastructure cost ($) | 600.00 | 600.00 | 600.00 | 600.00 |
| | Transaction fee ($) | 14,030,433.82 | 28,060,867.64 | 42,298.43 | 84,596.85 |
| | **Total cost ($)** | **14,031,033.8** | **28,061,467.6** | **42,898.4** | **85,196.9** |

based on cloud basic price and calculated using Eqs. (1) and (4). . The implementation cost of private blockchain is quite similar because it depends on which server service provider we use. Moreover, the transaction volume has a little impact on the cost overall. In the results, all annual cloud storage cost has the same value due to the round-up assumption before (all storage used are 5 GB) and the daily transaction volume does not impact the annual cloud VM cost significantly.

**Table 10.** Private blockchain implementation cost

| Transaction size | Variables | Amazon web services | | Google cloud | |
|---|---|---|---|---|---|
| | | S1 | S2 | S1 | S2 |
| Small–150 bytes | Cloud VM ($) | 17,776.00 | 17,776.00 | 17,336.00 | 17,336.00 |
| | Cloud storage ($) | 46.80 | 46.80 | 7.80 | 7.80 |
| | **Total cost ($)** | **17,822.80** | **17,822.80** | **17,343.80** | **17,343.80** |
| Medium–250 bytes | Cloud VM ($) | 18,711.60 | 18,711.60 | 18,248.40 | 18,248.40 |
| | Cloud storage ($) | 46.80 | 46.80 | 7.80 | 7.80 |
| | **Total cost ($)** | **18,758.40** | **18,758.40** | **18,256.20** | **18,256.20** |
| Large–500 bytes | Cloud VM ($) | 20,582.80 | 20,582.76 | 20,073.20 | 20,073.20 |
| | Cloud storage ($) | 46.80 | 46.80 | 7.80 | 7.80 |
| | **Total cost ($)** | **20,629.60** | 20,629.60 | **20,081.00** | **20,081.00** |

## 5   Conclusion

From the obtained results we can conclude that the implementation cost of public and private blockchain is varied especially the public blockchain. The cheapest implementation cost is when using Polygon platform with a small transaction size and low daily transaction volume. However, it is recommended for enterprises to choose a private blockchain instead of a public one due to the implementation cost and data security. Moreover, the use cases in private blockchain are more suitable to satisfy the business requirements. The findings of this study can be further researched and modified according to the newest information available.

## References

1. Lacity, M.C.: Addressing key challenges to making enterprise blockchain applications a reality. MIS Q. Exec. **17**, 201–222 (2018)
2. Worldwide spending on blockchain solutions from 2017 to 2024. https://www.statista.com/statistics/800426/worldwide-blockchain-solutions-spending/. Accessed 07 June 2022
3. Nanayakkara, S., Rodrigo, M.N.N., Perera, S., Weerasuriya, G.T., Hijazi, A.A.: A methodology for selection of a blockchain platform to develop an enterprise system. J. Ind. Inf. Integr. **23** (2021). https://doi.org/10.1016/j.jii.2021.100215
4. Bhutta, M.N.M., et al.: A survey on blockchain technology: evolution, architecture and security. IEEE Access **9**, 61048–61073 (2021). https://doi.org/10.1109/ACCESS.2021.3072849
5. Jo, M., Hu, K., Yu, R., Sun, L., Conti, M., Du, Q.: Private blockchain in industrial IoT. IEEE Network **34**, 76–77 (2020). https://doi.org/10.1109/MNET.2020.9199796
6. Dattani, J., Sheth, H.: Overview of blockchain technology. Asian J. Convergence Technol. **5**, 1–3 (2019)
7. McGovern, T.: How many blockchains are there in 2022? https://earthweb.com/how-many-blockchains-are-there/. Accessed 06 June 2022

8. Oliva, G.A., Hassan, A.E., Jiang, Z.M.(: An exploratory study of smart contracts in the ethereum blockchain platform. Empir. Softw. Eng. **25**(3), 1864–1904 (2020). https://doi.org/10.1007/s10664-019-09796-5

9. Boychev, B.: Application of the benchmarking approach in the agricultural sector. In: Agribusiness and Rural Areas-Economy, Innovation and Growth 2021 Conference Proceedings, pp. 181–191. University Publishing House 'Science and Economics', University of Economics-Varna (2021). https://doi.org/10.36997/ARA2021.181

10. Bhutta, K.S., Huq, F.: Benchmarking–best practices: an integrated approach. Benchmarking: Int. J. **6**, 254–268 (1999). https://doi.org/10.1108/14635779910289261

11. Akimova, L.A., Osadcha, O.V., Akimov, O.O.: Improving accounting management via benchmarking technology. Financ. Credit Act.: Probl. Theory Pract. **1**, 64–70 (2018). https://doi.org/10.18371/fcaptp.v1i24.128340

12. Anand, G., Kodali, R.: Benchmarking the benchmarking models. Benchmarking: Int. J. **15**, 257–291 (2008). https://doi.org/10.1108/14635770810876593

13. Brody, P., Holmes, A., Wolfsohn, E., Frechette, J.: Total cost of ownership for blockchain solutions (2019)

# SSI4Web: A Self-sovereign Identity (SSI) Framework for the Web

Md Sadek Ferdous[1,2(✉)], Andrei Ionita[1], and Wolfgang Prinz[1]

[1] Fraunhofer Institute for Applied Information Technology, Schloss Birlinghoven,
53757 Sankt Augustin, Germany
md.sadek.ferdous@fit-extern.fraunhofer.de
{andrei.ionita,wolfgang.prinz}@fit.fraunhofer.de
[2] BRAC University, 66 Mohakhali, 1212 Dhaka, Bangladesh

**Abstract.** The traditional protected web services rely on a user authentication process. The utilisation of an identifier (e.g. username, email address and so on) and credential (e.g. password) still remains the most widely deployed user authentication process, even though such an authentication process is one of the major sources of security breaches. Moreover, in this traditional setting, the management and sharing of user identity information is cumbersome with limited user controls over their identity data. In recent times, SSI has emerged as a new mechanism for managing and exchanging identity information in a more user-centric and privacy-friendly way. There are many explorations of SSI in different application domains, however, its utility for the web mostly remains unexplored. In this work, we present *SSI4Web*, a framework for integrating Self-sovereign Identity (SSI) for providing web services in a secure passwordless manner with much more user control and greater flexibility. We provide its architecture, discuss its implementation details, sketch out its use-case with an analysis of its advantages and limitations.

**Keywords:** Self-sovereign identity · SSI · Blockchain · Web · Hyperledger Indy · Hyperledger Aries

## 1 Introduction

With the growing popularity of web services, we have experienced a plethora of web services offering a wide range of online services such as social networking, online banking, e-commerce and so on. Most of these services require a user authentication process which traditionally relies on a knowledge based mechanism utilising an identifier (e.g. username email address) and a corresponding credential (e.g. a password) [1]. Unfortunately, such a password based mechanism is regarded as a major source of numerous security breaches and has strong usability issues [2]. There have been numerous efforts to replace such a mechanism, however, none of these efforts have been successful enough to replace passwords yet on a large scale [2].

J. Prieto et al. (Eds.): BLOCKCHAIN 2022, LNNS 595, pp. 366–379, 2023.
https://doi.org/10.1007/978-3-031-21229-1_34

Introduced with Bitcoin [3], the concept of blockchain has received significant attention from all over the world. A blockchain is an example of a distributed ledger (an ordered data structure) which is shared and maintained using a distributed consensus algorithm by a number of Peer to Peer nodes which could be scattered across different geographical locations of the world [4]. Similar to blockchain, Self-sovereign identity (SSI) is recently emerging as a new paradigm to manage one's digital identity, that the user creates and controls throughout its life-cycle [5,6]. It is in contrast to the traditional setting of identity management which is mostly controlled by different service providers (SPs) [7]. SSI aims to disrupt this notion by giving more controls to the users to handle their identity life-cycle. Understandably, SSI has generated a great deal of interest among enthusiasts from industries, academics and the Governments. Therefore, there are a number of researches exploring how SSI could be deployed in different application domains such as healthcare [8], IoT [9] and so on. There are other researches which have explored how SSI can be integrated with OAuth [10], a widely deployed access delegation method and with SAML (Security Assertion Markup Language) [11], a widely used standard for creating and managing identity federations. However, none of the existing works considered how SSI could be utilised for the web in such a way which could eventually replace the traditional method of the password-based authentication. Hence, the utility of SSI for the web mostly remains unexplored.

In this work, we present **SSI4Web**, a novel framework for accessing restricted web services using SSI with the motivation to facilitate a passwordless, secure user authentication mechanism for the web with more privacy controls.

**Contributions:** The main contributions of the paper are presented below:

- We present SSI4Web, a framework for integrating SSI in order to access restricted web services.
- We explain the motivation behind SSI4Web and discuss its architecture and design choices which are based on a rigorous threat modelling and requirements analysis.
- We also discuss its implementation details and sketch out a use-case with a protocol flow.
- Finally, we analyse how SSI4Web satisfies the formulated requirements as well as its advantages, limitations and future work.

**Structure:** Section 2 outlines the threats and the corresponding functional, security and privacy requirements for SSI4Web. We present the architecture, implementation details and a use-case protocol flow in Sect. 3. We analyse how SSI4Web satisfies different formulated requirements in Sect. 4 along with a highlight of its advantages, limitations and possible future work. Finally, we conclude in Sect. 5.

## 2 Threat Modelling and Requirement Analysis

In this section, we present a threat model (Sect. 2.1) and analyse a number of functional, security and privacy requirements (Sect. 2.2) for SSI4Web.

## 2.1   Threat Modelling

To model threats with respect to this work, we have chosen a well established threat model called STRIDE [12], which encapsulates different security threats as presented below.

T1. **Spoofing Identity:** This threat implies that an attacker can access an online service by spoofing someone else's identity.
T2. **Tampering with Data:** An attacker can modify crucial information, e.g. claims in a Verifiable Credential (VC, explained later) for malicious purposes.
T3. **Repudiation:** This threat implies that a corresponding entity can repudiate certain invalid and illegal actions while accessing online services.
T4. **Information Disclosure:** Sensitive data are revealed to an attacker unintentionally.
T5. **Denial of Service (DoS):** The online service can be the target of a DoS attack.
T6. **Elevation of Privilege:** An attacker might use other attack vectors to elevate their access privilege within the online services.

In addition to these, we have considered an additional threat which is crucial for accessing restricted web services securely.

T7. **Replay:** An attacker might capture an request/response and submit it afterwards, thus launching a replay attack.

The privacy threats mostly emerge from the lack of any privacy control by any user. Based on this assumption, the identified threats are as follows.

T8. **Lack of Consent:** A VC is released without the consent of a user.
T9. **Lack of control:** Users have little control over how individual claims are released to an SP.

## 2.2   Requirement Analysis

In this section, we present a set of functional, security and privacy requirements. The functional requirements capture the core functionalities of the system while security and privacy requirements ensure that they mitigate the identified threats.

**Functional Requirements (FR):** The requirements are presented below.

F1. The system must integrate SSI mechanisms into its online services so that a user can access online services following the SSI steps.
F2. The system must be accompanied by a corresponding SSI wallet so that users can manage their self-sovereign identities using the wallet.
F3. Users should be able to access the services provided by an SP using different devices in a seamless manner.

**Security Requirements (SR):** Next, we present a set of security requirements to address the identified security threats.

S1. The system must ensure that users can be authenticated using a VC based SSI and utilise the VC to securely access a service. This will mitigate the *T1* threat.

S2. The system must guard against any unauthorised modification of VC data to mitigate the *T2* threat.

S3. The system must utilise digital signature in order to mitigate the *T3* threat.

S4. Any crucial data must be transmitted in encrypted format via networks so as to ensure the confidentiality of the data related to a VC. This can mitigate the *T4*.

S5. The system must take measures against any DoS attack so as to mitigate the *T5* threat.

S6. The system should utilise an access control mechanism (e.g. RBAC [13]) to ensure that an attacker cannot elevate their access privilege and thereby mitigating the *T6* threat.

S7. The system must take protective measures against any replay attack in order to mitigate the *T7* threat.

**Privacy Requirements (PR):** Privacy requirements are important to mostly mitigate the privacy threats. Only one privacy requirement is identified as presented below:

P1. The system must ensure that every VC is released only after the user has checked and consented. This mitigates threats *T8* and *T9*.

## 3    SSI4Web Framework

In this section, we summarise the concept of SSI (Sect. 3.1) and then discuss the architecture of the SSI4Web framework & its implementation details (Sect. 3.2) and protocol flow (Sect. 3.3).

### 3.1    SSI Concept

Plans for the next generation of digital identity concentrated on introducing user autonomy over the digital identity and the possibility of sharing it across any number of services. This gave rise to the notion of Self-sovereign Identity (SSI). There are three major entities in SSI (Fig. 1): issuer, holder and verifier.

In realisation of SSI, the W3C community developed *Decentralized Identities* (DIDs) and *Verifiable Credentials* (VCs) [14] as central constructs for SSI. DIDs are user-generated strings linked to an SSI entity's cryptographic public key. DID Documents (DID Docs in short) are JSON objects containing the DID, its linked cryptographic public keys and further metadata. On the other hand, *Verifiable Credentials* (VCs) are cryptographically-signed claims (attribute values) made by an issuer over a subject. Since the VCs are signed by the issuers, they can

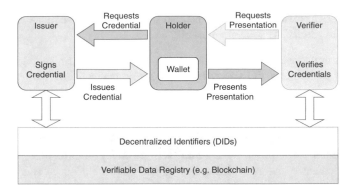

**Fig. 1.** SSI entities and their relations (add reference).

be verified by any other party (Verifier) that resolves their DID and checks the signature with the linked public key. The holder of the VCs stores them in a digital wallet and is able to present these, either individually or as a combination, to a verifier that requires the information in exchange for an action in the holder's favour (e.g. accessing a service). Within this setting, blockchain is used as a verifiable registry for storing DIDs or DID Docs because of its immutability, persistence and decentralisation properties.

### 3.2   Architecture and Implementation

The core motivation of the SSI4Web Framework (or simply SSI4Web in short) is to facilitate mechanisms for integrating SSI for offering protected web services by a Service Provider (SP). Since protected web services require, among other things, the authentication of the users, one of the principal use-cases within this setting is to use SSI for a passwordless user authentication mechanism for accessing protected web services, which is also the primary focus of this work.

In SSI4Web, a user utilises an SSI wallet to interact with the SP. In contrast to the SSI setting where different entities play different roles of issuers and verifiers, the SP within SSI4Web plays a dual role of an issuer and verifier. Within this setting, the SSI4Web must accommodate the following four crucial SSI functionalities.

- **Establishing the connection:** An SSI connection needs to be established between the wallet of the user and the SP.
- **Providing VC:** The SP can release a VC, when requested, to the user using the previously established connection. This process is analogous to the user registration process in the traditional web setting.
- **VC Storage:** The received VC must be stored securely in the user wallet.
- **Requesting Proof:** The SP requests a proof of the previously released VC from the user using the same established connection. This process is analogous to the authentication process in the traditional web setting.

Fig. 2 illustrates the architecture of SSI4Web. The architecture has been designed in such a way that it can facilitate the stated functionalities discussed previously. From Fig. 2, SSI4Web has five major entities: SP, user, wallet, mediator and blockchain. The SP is responsible for offering SSI based web services and utilises an SSI agent for managing SSI interactions. The user utilises the wallet to interact, via a mediator, with the SP and to store received VCs. The responsibility of the mediator is to forward messages back and forth between the wallet and SP. In addition, a mediator could potentially store messages destined for the wallet in case the wallet is offline and supply it when the wallet becomes online. A blockchain is used as a verifiable data registry. Both the wallet and the SP Agent interact with the blockchain as part of the SSI protocol flow (discussed later). The solid arrows in Fig. 2 represent a non-SSI direct communications between two entities whereas the dotted arrows represent the SSI communications.

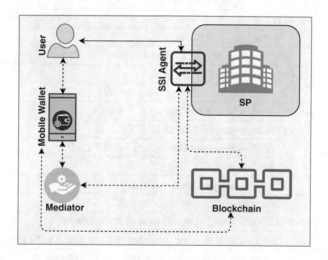

**Fig. 2.** Architecture of SSI4Web.

We have developed a Proof of Concept (PoC) of our proposal using a number of tools and frameworks. The core tools utilised to develop the PoC are Hyperledger Aries [15] and Hyperledger Indy [16]. Hyperledger Aries provides a set of libraries to develop SSI applications which are available in different programming languages. For our PoC, we have utilised the Hyperledger Aries Cloud Agent Python (ACA-Py) library [17]. The SSI agent has been developed using NodeJS [18] which depends on the ACA-Py library to serve different SSI functionalities. As the Mobile Wallet we have utilised an open source Aries-compliant mobile App facilitated by the Aries Mobile Agent React Native project [19]. It relies on the Indicio Public Mediator [20], a public mediator based on ACA-Py. The web services have also been developed with NodeJS. Hyperledger Indy is the blockchain platform that is being increasingly used in SSI applications.

In our PoC we have utilised an Indy testbed called *BCovrin Dev* (http://dev.greenlight.bcovrin.vonx.io/).

### 3.3   Use-Case and Protocol Flow

Next, we present a use-case with protocol flows to show how SSI4Web could be utilised for accessing restricted web services. However, at first, we introduce mathematical notations and data model in Tables 1 and 2.

**Table 1.** Cryptographic Notations.

| Notations | Description |
|---|---|
| $SP$ | Service provider (Playing the role of an issuer & verifier) |
| $U$ | User (Playing the role of a holder) |
| $M$ | Mediator |
| $K_{SP}^{U}$ | Public key of $SP$ for $U$ |
| $K_{SP}^{-1|U}$ | Private key of $SP$ to be used for $U$ |
| $K_{U}^{SP}$ | Public key of $U$ for $SP$ |
| $K_{U}^{-1|SP}$ | Private key of $U$ to be used for $SP$ |
| $DID_{SP}^{U}$ | Decentralised identifier of $SP$ for $U$ |
| $DID_{U}^{SP}$ | Decentralised identifier of $U$ for $SP$ |
| $VC_{SP}^{U}$ | A verifiable credential issued by $SP$ to $U$ |
| $N_i$ | A fresh nonce |
| $\{\}_K$ | Encryption operation using a public key $K$ |
| $\{\}_{K^{-1}}$ | Signature using a private key $K^{-1}$ |
| $H(M)$ | SHA-512 hashing operation of message $M$ |
| $[...]_K$ | Communication over an channel encrypted with key $K$ |
| $[...]$ | Communication over an unencrypted channel |
| $[[...]]$ | Optional data |

**Data Model:** The developed PoC is designed as a request and response system (denoted with *req* and *resp* respectively in Table 2). A *req* can be of four types: *serviceReq*, *inviteReq*, *credReq* and *proofReq* which represent a service request, SSI invitation request, credential request and proof request respectively, where the last three requests correspond to different SSI functionalities. Similarly, there are four responses which correspond to the four requests.

The *serviceReq* contains the URL of the requested service and optionally an invite cookie (denoted with *inviteCookie*) and a service cookie (denoted with *serviceCookie*). The functionalities of these cookies will be explained when we present the protocol flow. On the other hand, an SSI invite request (*inviteReq*) consists of the URL of the SP SSI agent (*url$_{SP}$*) and a nonce. There are two corresponding responses, *inviteResp1* and *inviteResp2*. The former response contains

**Table 2.** Data Model.

| |
|---|
| $req \triangleq \langle serviceReq, inviteReq, credReq, proofReq \rangle$ |
| $resp \triangleq \langle serviceResp, inviteResp[1-2], credResp, proofResp \rangle$ |
| $serviceReq \triangleq \langle url_{s_i}, [[inviteCookie, serviceCookie]] \rangle$ |
| $inviteReq \triangleq \langle url_{sp}, N_i \rangle$ |
| $inviteResp1 \triangleq \langle N_i, DID_U^{SP} \rangle$ |
| $inviteResp2 \triangleq \langle N_i, DID_{SP}^{U} \rangle$ |
| $credReq \triangleq \langle \text{``Requesting Credential''} \rangle$ |
| $attrReq \triangleq \langle a_1, a_2, ..., a_n \rangle$ |
| $attrResp \triangleq \langle (a_1, av_1), (a_2, av_2), ..., (a_n, av_n) \rangle$ |
| $credResp \triangleq \langle VC_{SP}^{U} \rangle$ |
| $VC_{SP}^{U} \triangleq \langle \{ (a_1, av_1), (a_2, av_2), ..., (a_n, av_n) \}_{K_{SP}^{-1|U}} \rangle$ |
| $initiateProofReq \triangleq \langle \text{``Initiating Proof Request''} \rangle$ |
| $proofReq \triangleq \langle a_1, a_2, ..., a_n \rangle$ |
| $proofResp \triangleq \langle VC_{SP}^{U} \rangle$ |

the previously used nonce and a DID of the user to be used with $SP$, denoted with $DID_U^{SP}$. Similarly, the later response contains the same nonce and a DID of $SP$ to be used with $U$, denoted with $DID_{SP}^{U}$.

Next, a credential request is denoted with $credReq$ which just contains a string as shown in Table 2. The $attrReq$ contains the list of attribute names and $attrResp$ contains the attribute name-value pairs. A credential response ($credResp$) corresponds to a $credReq$ and consists of a VC provided by $SP$ to $U$ and hence denoted with $VC_{SP}^{U}$. Such a VC is simply represented with a digitally signed set of attribute name-value pairs and modelled as shown in Table 2. Finally, an invitation proof request is denoted with $inviteProofReq$ which just contains a string as shown in Table 2. A proof request ($proofReq$) contains the list of attribute names whose values must be present in the released VC (VP) and the corresponding proof response ($proofResp$) contains the released VC.

**Algorithm:** Now, we present the algorithm snippets of three important functions for the SP SSI Agent, namely handling an invitation response (`handleInvi- teResp` function in Algorithm 1), generating a VC (`handleInviteResp` function in Algorithm 2) and handling a proof response (`handleProof` function in Algorithm 2). The `handleInviteResp` function is responsible for handling $inviteResp1$. Within the function, the DID (e.g. $DID_U^{SP}$) is retrieved and then the DID is used to retrieve the corresponding DID Doc from the blockchain (line 4 and 5 in Algorithm 1). Then, $K_U^{SP}$ is retrieved from the DID doc. In response, the agent at first generates a new key pair ($K_{SP}^{U}$ and $K_{SP}^{-1|U}$) and then generates a new DID Doc (line 8 in Algorithm 1). This DID Doc is pushed to the blockchain (line 9). Finally, $inviteResp2$ is created using $K_{SP}^{U}$ which is then returned back (line 10–11).

**Algorithm 1.** Algorithm Snippet for `handleInviteResp` function

| | |
|---|---|
| 1 | **Start** |
| 2 | ... |
| 3 | **function** `handleInviteResp`(*inviteResp1*) |
| 4 | $DID_U^{SP} := inviteResp1.DID_U^{SP}$; |
| 5 | Retrieve the corresponding DID Doc using $DID_U^{SP}$ from the blockchain; |
| 6 | Retrieve $K_U^{SP}$ from the DID Doc; |
| 7 | Store $K_U^{SP}$ and $DID_U^{SP}$ in the agent Wallet; |
| 8 | Generate $K_{SP}^U$ and $K_{SP}^{-1|U}$, create a DID Doc using $K_{SP}^U$; |
| 9 | Push the newly created DID Doc in the blockchain; |
| 10 | Prepare *inviteResp2* using $K_{SP}^U$; |
| 11 | **return** *inviteResp2*; |
| 12 | ... |

The `generateVC` function is used to generate and return a new VC by using the retrieved attributes from *attrResp* (line 4 to 6 in Algorithm 2). Finally, the `handleProof` function is used to handle a proof response. For this, at first the user presented VC ($VC'^U_{SP}$) is retrieved (line 8) which is then used to compare the previously supplied VC to the user ($VC^U_{SP}$). If they match, a `TRUE` response is returned, otherwise a `FALSE` false is returned (line 10–14).

**Algorithm 2.** Algorithm snippet for `credResp` & `handleProof`

| | |
|---|---|
| 1 | **Start** |
| 2 | ... |
| 3 | **function** `generateVC`(*attrResp*) |
| 4 | *atValSet* := Retrieve attributes and their values from *attrResp*; |
| 5 | $VC_{SP}^U := createVC(atValSet)$; |
| 6 | **return** $VC_{SP}^U$; |
| 7 | **function** `handleProof`(*proofResp*) |
| 8 | $VC'^U_{SP} := proofResp.VC'^U_{SP}$; |
| 9 | Retrieve $VC_{SP}^U$ from the wallet for $U$; |
| 10 | **if** $VC'^U_{SP} == VC^U_{SP}$ **then** |
| 11 | **return** *TRUE*; |
| 12 | **else** |
| 13 | **return** *FALSE*; |
| 14 | **end** |
| 15 | ... |

**Protocol flow:** Now, we present a combined protocol flow for showcasing how SSI4Web facilitates the three SSI functionalities (creating a new connection, releasing a credential and releasing a proof). The flow is sketched in Table 3 and discussed next.

**Table 3.** Web-SSI protocol

| | | |
|---|---|---|
| $M1$ | $U \rightarrow SP$ : | $[N_1, serviceReq(url_{s_1})]_{HTTPS}$ |
| $M2$ | $SP \rightarrow U$ : | $[N_2, \text{SSI Option Page}]_{HTTPS}$ |
| $M3$ | $U \rightarrow SP$ : | $[N_2, \text{New Invitation Option}]_{HTTPS}$ |
| $M4$ | $SP \rightarrow U$ : | $[N_3, inviteReq]_{HTTPS}$ |
| $M5$ | $U \rightarrow M$ : | $[SP, N_3, inviteResp1]$ |
| $M6$ | $M \rightarrow SP$ : | $[N_3, inviteResp1]$ |
| $M7$ | $SP \rightarrow M$ : | $[U, N_3, inviteResp2]$ |
| $M8$ | $M \rightarrow U$ : | $[N_3, inviteResp2]$ |
| $M9$ | $SP \rightarrow U$ : | $[N_4, \text{Credential Option Page}, inviteCookie]_{HTTPS}$ |
| $M10$ | $U \rightarrow SP$ : | $[N_4, \text{Credential Reuqest}]_{HTTPS}$ |
| $M11$ | $SP \rightarrow U$ : | $[N_5, attrReq]_{HTTPS}$ |
| $M12$ | $U \rightarrow SP$ : | $[N_5, attrResp]_{HTTPS}$ |
| $M13$ | $SP \rightarrow M$ : | $[U, \{N_5, credResp\}_{K_U^{SP}}]$ |
| $M14$ | $M \rightarrow U$ : | $[\{N_5, credResp\}_{K_U^{SP}}]$ |
| $M15$ | $U \rightarrow SP$ : | $[N_6, \text{Initiate Proof Request}]_{HTTPS}$ |
| $M16$ | $SP \rightarrow M$ : | $[U, \{N_6, proofReq\}_{K_U^{SP}}]$ |
| $M17$ | $M \rightarrow U$ : | $[\{N_6, proofReq\}_{K_U^{SP}}]$ |
| $M18$ | $U \rightarrow M$ : | $[SP, \{N_6, proofResponse\}_{K_{SP}^U}]$ |
| $M19$ | $M \rightarrow SP$ : | $[\{N_6, proofResponse\}_{K_{SP}^U}]$ |
| $M20$ | $SP \rightarrow U$ : | $[N_1, url_{s_1}, serviceCookie]_{HTTPS}$ |

i The user submits a request to an SP to access one of its services at $url_{s_i}$.

ii The SP returns a web-page having different options for SSI (e.g. creating invitations, release credentials and request a proof).

ii The user choose the "New Invitation" option.

iv The SP (its SSI agent) generates a new invitation which is encoded as a QR code and is displayed to the user.

v The user utilises their wallet to scan the QR code. The wallet then generates a new key pair ($K_U^{SP}$ and $K_U^{-1|SP}$) and creates a new DID ($DID_U^{SP}$) and DID Doc utilising the generated key pair. The DID Doc is pushed to the blockchain. The wallet then prepares *inviteResp1* using $DID_U^{SP}$ and returns it back to the SP via the mediator $M$, steps M5 and M6 in Table 3.

vi The SP utilises its `handleInviteResp` function (Algorithm 1) to handle *inviteResp1* and prepare *inviteResp2* which is then returned to the user wallet via mediator $M$ (steps 7 and 8 in Table 3). The user wallet retrieves required data from *inviteResp2* and combines it with data from $DID_U^{SP}$ to save as a new connection to the wallet (*Test* in Fig. 3a).

vii Once the connection is established, the SP shows a page to the user to request for credentials along with an *"inviteCookie"* which holds metadata about the new invitation.

viii The user submits a request to release a credential.

ix The SP returns a page where the user can submit different values for the required attributes, steps M11 and M12 in Table 3.

x The SP prepares a VC ($VC_{SP}^{U}$) and a *credResp* with the VC which is then returned back to the user via M, steps M13 and M14 in Table 3. After receiving, the wallet validates the signature of the VC using the public key of the SP for this particular connection as stored in the wallet. If it is successful, the VC is stored in the wallet (Fig. 3b).

xi Next, the user is forwarded to the SSI option page again where the user chooses a proof request option.

xii The SP generates a *proofReq* which is then sent to the user wallet via M, steps 16 and 17 in Table 3.

xiii The wallet parses the requested attributes and any condition from the *proofReq* and matches with the stored credentials in the wallet. The matched VC is shown to the user for their approval (Fig. 3c). Once the user approves, the wallet prepares a *proofResponse* using the approved VC which is then returned back to the SP, steps 18 and 19 in Table 3.

xiv The SP validates the VC as before and once validated, the user is forwarded to the requested url ($url_{s_i}$) with a session cookie (*"serviceCookie"*). The user can then can access the service as long as they are not logged out.

(a) Established connection

(b) VC Attributes

(c) Proof Request

**Fig. 3.** Establishing a connection and receiving a VC

The information from the connection metadata in the wallet can be utilised to access SP services from any browser of any devices using SSI4Web. For example, if a user would like to access the services from the SP from a different device with the previously established connection, the user can do so using a secure two-factor method. At first, the user needs to collect the connection name from the wallet (*Test* in Fig. 3a) and click the *'Active Connection"* from the SSI Option Page. Then, the user will have the option to search for the connection using the connection name. When an active connection is found, an one time password will be sent to the previously saved email address (supplied while creating a new connection). Once the code from the email address is supplied, the user can utilise the previous connection in the browser of the new device and the similar flow can be used to access the requested service.

# 4 Discussion

In this section, we examine how our framework has satisfied its different requirements (Sect. 4.1), discuss its advantages and limitations (Sect. 4.2) and highlight possible future works (Sect. 4.3).

## 4.1 Analysing Requirements

**Functional Requirements:** The SSI4Web implementation effectively satisfies all functional requirements. For example, SSI4Web requires that the user can access restricted services by following the SSI steps as outlined in the protocol flow, thereby satisfying *F1*. The user needs to utilise the SSI Wallet to access any SP service and thus satisfying *F2*. The implementation allows a user to access restricted services from multiple devices in a seamless manner which satisfies *F3*.

**Security Requirements:** It is evident from the discussion of the use-case that SSI4Web satisfies *S1*. *S2* and *S3* are satisfied as well since each VC is digitally signed by the issuer (SP in our case). As evident from the protocol flow, every interactions with the SP and all SSI interactions between the user and SP where sensitive data are transmitted take place either via HTTPS or encrypted channels, thereby satisfying *S4*. Finally, we have utilised nonces extensively in all interactions to satisfy *S7*. Regarding *S5*, every single web service is prone to DoS attacks and SSI4Web is not an exception in this regard. Additional steps are required to mitigate these threats which is beyond the scope of the presented work. Similarly, satisfying *S6* would require to deploy an access control mechanism. In the current implementation of SSI4Web our goal was to showcase if SSI could be utilised for simple web authentication. That is why S6 has not been considered in the presented work.

**Privacy Requirements:** The wallet utilised in the SSI4Web enables a user to check what attributes are released to the SP (Role attribute in Fig. 3c). Only after the user's consent, the VC is released to the user. This satisfies *P1*.

## 4.2 Advantages and Limitations

The advantages of SSI4Web are summarised below:

- SSI4Web provides a fully passwordless way of accessing restricted online services. This is also advantageous for the service providers as no additional security measures are needed to ensure password security at their ends.
- Users can access such services from multiple devices in a seamless manner without compromising their security. This is much better than accessing services using passwords across multiple devices in different browsers.

– Users have the ultimate control to determine how they want to release their credentials to whom and only with their explicit consent.

The current implementation of SSI4Web has some limitations:

– The wallet plays a central role for the user in any SSI setting. As all the VCs and connections are stored in the wallet, a secure backup mechanism is crucial. The wallet used in our PoC does not have any backup functionality.
– Using SSI4Web to access restricted websites presents a novel way to access any website. There are multiple new steps involved in comparison to the traditional way of accessing restricted websites and it represents a new learning curve for any user. The wide-scale adoption of such SSI based mechanism will depend on the proper training of the user and their willingness to adopt this method.

### 4.3   Future Work

In future we would like to explore the following:

– Developing a secure backup and synchronisation mechanism for the SSI wallet so as to provide a seamless user experience for any user across multiple devices.
– Studying the usability of proposed approach to understand how difficult it is for the user to adopt this approach.
– Utilising SSI4Web to adopt in other forms of identity management models (e.g. Federated Identity Management [21]). In this current version, we have only explored the SILO model [21]. This will showcase that such a model can be utilised in advanced use-cases as well.

## 5   Conclusion

In this work, we have presented SSI4Web, a Self-sovereign Identity (SSI) based framework for accessing restricted web services. The main objective of the framework is to facilitate a passwordless, secure user authentication mechanism with better user control over their identity data. We have presented its architecture which is based on a threat model and requirement analysis, discussed its implementation details and highlighted a use-case to show its applicability. With these contributions, we strongly believe that SSI4Web will usher a new domain of research in this respective domain.

## References

1. Katsini, C. , Belk, M., Fidas, C., Avouris, N., Samaras, G.: Security and usability in knowledge-based user authentication: a review. In: Proceedings of the 20th Pan-Hellenic conference on informatics, pp. 1–6 (2016)

2. Bonneau, J., Herley, C., Van Oorschot, P.C., Stajano, F.: The quest to replace passwords: a framework for comparative evaluation of web authentication schemes. In: IEEE Symposium on Security and Privacy, pp. 553–567 (2012)
3. Nakamoto, S.: Bitcoin: a peer-to-peer electronic cash system. Tech. Rep. Manubot (2019)
4. Chowdhury, M.J.M., Ferdous, M.S., Biswas, K., Chowdhury, N., Kayes, A., Alazab, M., Watters, P.: A comparative analysis of distributed ledger technology platforms. IEEE Access **7**(1), 167 930–167 943 (2019)
5. Mühle, A., Grüner, A., Gayvoronskaya, T., Meinel, C.: A survey on essential components of a self-sovereign identity. Comput. Sci. Rev. **30**, 80–86 (2018)
6. Ferdous, M.S., Chowdhury, F., Alassafi, M.O.: In search of self-sovereign identity leveraging blockchain technology. IEEE Access **7**, 103 059–103 079 (2019)
7. Ferdous, M.S.: User-controlled identity management systems using mobile devices. PhD. Thesis. University of Glasgow (2015)
8. Shuaib, M., Alam, S., Alam, M.S., Nasir, M.S.: Self-sovereign identity for healthcare using blockchain. Mater. Today: Proc. (2021)
9. Kulabukhova, N., Ivashchenko, A., Tipikin, I., Minin, I.: Self-sovereign identity for iot devices. In: International Conference on Computational Science and Its Applications, pp. 472–484. Springer (2019)
10. Hong, S., Kim, H.: Vaultpoint: a blockchain-based ssi model that complies with oauth 2.0. Electronics **9**(8), 1231 (2020)
11. Yildiz, H., Ritter, C., Nguyen, L.T., Frech, B., Martinez, M.M., Küpper, A.: Connecting self-sovereign identity with federated and user-centric identities via saml integration. In: IEEE Symposium on Computers and Communications (ISCC), pp. 1–7. IEEE (2021)
12. Shostack, A.: Threat modeling: Designing for security. Wiley (2014)
13. Sandhu, R.S., Coyne, E.J., Feinstein, H.L., Youman, C.E.: Role-based access control models. Computer **29**(2), 38–47 (1996)
14. Verifiable Credentials Data Model 1.0. Accessed: 27 Apr 2022. [Online]. Available: https://www.w3.org/TR/vc-data-model/
15. Hyperledger Aries. Accessed: 10 Nov 2021. [Online]. Available: https://www.hyperledger.org/use/hyperledger-aries
16. Hyperledger Indy. Accessed: 10 Nov 2021. [Online]. Available: https://www.hyperledger.org/use/hyperledger-indy
17. Hyperledger Aries Cloud Agent—Python. Accessed 05 Dec 2021. [Online]. Available: https://github.com/hyperledger/aries-cloudagent-python
18. Node.js. Accessed: 10 Nov 2021. [Online]. Available: https://nodejs.org/en/
19. Aries Mobile Agent React Native. Accessed: 01 Nov 2021. [Online]. Available: https://github.com/hyperledger/aries-mobile-agent-react-native
20. Indicio Public Mediator. Accessed: 01 Nov 2021. [Online]. Available: https://indicio-tech.github.io/mediator/
21. Josang, A., AlZomai, M., Suriadi, S.: Usability and privacy in identity management architectures. ACSW Front. **2007**, 143–152 (2007)

# The Communicational Universe of Cryptocurrencies. An Approach to the Current Scientific Importance

Sergio Manzano[1,2], Javier Parra-Domínguez[1,2(✉)], Francisco Pinto[3], Alfonso González-Briones[1,2,3], and Guillermo Hernández[1,2]

[1] BISITE Research Group, University of Salamanca, Salamanca, Spain
{smanzano,Javierparra}@usal.es
[2] IoT Digital Innovation Hub, Valladolid, Spain
[3] AIR Institute–Deep Tech Lab, Valladolid, Spain

**Abstract.** At present, the universe of cryptocurrencies is highly important and proof of this is the commitment of countries and institutions to the advancement of cryptocurrencies. This bibliometric research aims to discover the true reality of scientific progress in publications concerning cryptocurrencies and their communication. In any science or technological development, communication is important; in fact, communication is a digital asset that can make cryptocurrencies advance fast enough and in a profitable way. The methodology used is based on a bibliometric analysis of the Scopus database, showing a positive result in terms of the increase in the number of publications, but without being too high in quartiles such as the first or second, which means that there are still no renowned scientific authors in this field, nor is there a clear link to other fields of study. The implication for the advancement of cryptocurrencies by communication is vital as a business development and scientific concern should focus on these aspects.

**Keywords:** Cryptocurrency · Communication · Digital assets

## 1 Introduction to Cryptocurrencies and the Importance of Their Communication

### 1.1 The Cryptocurrency Universe

Until not so long ago, cryptocurrencies were seen as unreliable assets or elements and, therefore, with a somewhat dubious future [1, 2]. However, the current progression means that, in some cases, they are becoming essential assets not only for inexperienced investors with no previous experience or knowledge but also for certain governments. Therefore, the uncertainty of Elon Musk's statements at the beginning of 2021,[1] the Chinese government's announcement of a new cryptocurrency-related ban, or the increase

---

[1] https://www.cnbc.com/2021/09/28/tesla-ceo-elon-musk-says-us-government-should-avoid-regulating-crypto.html.

J. Prieto et al. (Eds.): BLOCKCHAIN 2022, LNNS 595, pp. 380–387, 2023.
https://doi.org/10.1007/978-3-031-21229-1_35

in lawsuits and complaints about scams involving cryptocurrency investments [3] are now a thing the past.

Currently, the best-known cryptocurrencies are Bitcoin, Ethereum, Binance, Cardano, Ripple, Litecoin and Dogecoin [4], whose main characteristics, such as internationality, anonymity and the fact that they do not require a banking system, in addition to the development of these currencies is due to blockchain technology [5–9]. The application of the technology mentioned above gives cryptocurrencies the power to work in their primary applications, such as:

- The investment in the cryptocurrency itself [10–13].
- Carrying out operations and transactions [14–16].

All of the above helps us realise that several actors are involved in the crypto universe [17–19]. Each one has a criterion or a specific point of view due to their communication about the technology [20–22]. Understanding communication is a critical factor when it comes to sizing the technology in the social mind [23–25], even more so when it comes to technologies that are undoubtedly abstract at first; the primary motivation of this study is to observe the presence of cryptocurrencies in the texts published in the prior communication publications.

To achieve this objective, we will end the introduction with incorporating a subsection dedicated to communication, move on to a Sect. 2 where we will establish a bibliometric study methodology, analyse the results in Sect. 3, and end with the discussion and conclusions.

### 1.2   The Communication of Technology. The Case of Cryptocurrencies

As we indicated earlier, there are many people involved in the development of cryptocurrencies, from governments to anonymous citizens, who need the optimal dimension of the communicational message to end up trusting a particular technology or, on the contrary, to end up denigrating it [26]. Some of the aspects that involve the knowledge of those involved are extremely important, as follows:

- Legal nature of cryptocurrencies
- Whether or not to regulate cryptocurrencies [27]
- The economic implications [28, 29].

These should be added necessary messages such as those concerning the European Union's proposal for the MiCA[2] regulation.

All of the above crossroads of messages to one audience or another make communication work difficult, and, although at the level of dissemination, the power of communication is arbitrary in terms of cryptocurrencies, it is at the scientific level where it becomes essential to meet the needs of the public [30]. It is necessary to introduce the

---

[2] Proposal for a Regulation of the European Parliament and of the Council on Markets in Crypto-assets and amending Directive (EU) 2019/1937 (COM/2020/593 final), **MiCA**.

relevance that progress in scientific publications on cryptocurrencies has for developing the technology linked to them and the asset they represent.

In this sense, using a quantitative methodology to define the discursive approach given to the news coverage of cryptocurrencies can help to identify the messages that are configured around them, whether it is positive or negative, whether it informs about the risks involved or whether is approached from a sensationalist point of view. Theories such as framing or methodologies such as content analysis, both of which are of great importance in Communication Sciences [31, 32], could help in this regard. Moreover, these methodologies allow us to go a step further, and it is even possible to analyse the most widespread approaches to information about cryptocurrencies in social networks [33].

Therefore, analysing from a theoretical point of view the message, the approach, or the volume of information on the media coverage that cryptocurrencies are receiving is of utmost importance to understand how public opinion is shaped around this technology.

In this way, this research aims to determine whether this is a prolific field at present or whether, on the contrary, it is necessary to delve deeper into the study and wait for more excellent scientific production to be able to establish conclusive results.

## 2  Methodology

To determine the interest that cryptocurrency research has aroused among the scientific community in Communication and Information Sciences, bibliometric indicators were analysed using the Scopus platform. It was decided to use this database due to its importance among the scientific community and the volume of results it could provide. In this case, the sample was limited so that the search equation yielded, from among all Q1 publications in Communication in the SJR ranking, those texts in which the concept "Cryptocurrencies" appeared. The search was limited to 2021 to include full sample years and avoid inconsistencies.

It was decided to limit the sample to the Q1 publications of SJR because, although it is not a sample that can be extrapolated to the entire universe of scientific publications, it is beneficial for initial analysis to understand the importance of this type of study in the most influential journals.

After carrying out the corresponding search in SJR and extracting the Source ID of the 104 publications at Q1, the following equation was used in Scopus. For operational reasons, this publication has shortened it using ellipses.

ALL (CRYPTOCURRENCIES) AND [SRCID (18651) … OR SRCID (17730)] AND PUBYEAR < 2022

The search yielded 34 results since the first record in 2014. At the time of writing (May 2022), seven more articles have been published.

Through this analysis, the researchers aim to answer the following research questions:

- *RQ1: What is the trend in the annual volume of texts related to the study of cryptocurrencies among the prominent publications specialising in communication?*
- *RQ2: Which SJR Q1-indexed communication journals have published the most texts on cryptocurrencies?*

- *RQ3: Who are the most prolific researchers in these studies, and from which institutions and countries do these texts originate?*
- *RQ4: How do these studies relate communication and cryptocurrencies to other scientific disciplines?*

In this way, the aim is to find out the interest that this type of research awakens in academic authors and whether it has increased over time, within which fields of study there is more excellent scientific production in this respect, and what its presence is within the total scientific output of each field understood independently.

## 3  Results

The bibliometric parameters have been analysed based on quantity indicators: increasing, decreasing, emerging and trend fields to respond to RQ1. In addition, to answer RQ2 and RQ3, a second analysis of the data was carried out, broken down by publication, author, institution and country of origin. To clarify RQ4, a final set of data has been used to show other subjects, other than Communication, in which each text has been catalogued independently.

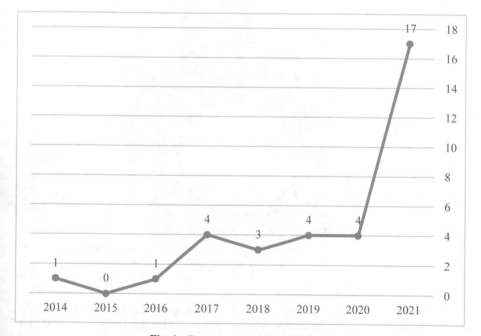

**Fig. 1.** Documents published by year

Figure 1 shows how the number of texts has increased considerably in the last year since the first publication in 2014: 17. Also, in 2014 and 2016, only one reader was published, and between 2017 and 2020, there was a practically stable volume, around 4 per year. However, after applying different models, it has not been possible to adjust the growth to any definite trend significantly.

The 34 texts are distributed among 15 publications, and 89 of the 104 that makeup Q1, 85.57% of the total, have not addressed cryptocurrencies in any of their articles. On the other hand, the journals indexed in Q1 in Communication that have edited the most texts are Internet Policy Review (10); Telematics and Informatics (6); Digital Communications And Networks (3); Games And Culture (2); International Journal Of Communication (2), and New Media And Society (2). The remaining nine articles have each been edited by one publication, respectively.

The most prolific authors in this field are Bodó, B. (3); Brekke, J.K. (3); De Filippi, P. (2); and Hoepman, J.H. (2). The authors are mainly from the United States (9), United Kingdom (6); China and Germany (4); Australia and the Netherlands (3); and Canada, Finland and Italy (2). The reference universities would be Amsterdam and Durhan (3) and Harvard, Radboud, Hamburg and Nanjing (2).

Finally, Fig. 2 shows how the texts published in the Q1 Communication journals in SJR also belong to other disciplines such as Computer Science and Social Sciences (24 out of 34 readers), Environmental Sciences (10 out of 34), Engineering (8), Arts and Humanities (4) and Psychology (3).

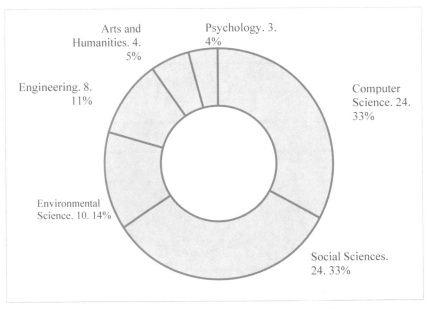

**Fig. 2.** Themes of texts: number of texts recorded in each discipline in addition to communication, and percentage of occasions where each reader is registered in another category out of the total (73 records for 34 texts).

## 4   Discussion

In light of the data reflected by the above analysis, and taking into account the methodological and sampling limitations of this study, it can be determined that, although the trend is indeed positive and, especially in the last year, the number of texts related to cryptocurrencies is increasing considerably in the publications specialised in Communication with the most significant impact according to the SJR index, the volume of articles is still too low to establish a solid conclusion (RQ1). However, 41.18% of the total number of articles published last year have been published in the previous four months may give a clue to the future.

Concerning RQ2, it is particularly noteworthy that 85.57% of the publications anlysed have not yet published anything on cryptocurrencies, and the only 15 journals that have done so are merely testimonial cases: 9 have only published one article, and the six main ones do not exceed ten articles in any case.

The low volume of publications means that RQ3 and RQ4 do not have a fascinating response today.

An important issue to include in the discussion is the necessary estimation of the impact on publications in the Q1 range, taking into account the real effects of Covid-19 on the evolution of cryptocurrencies themselves. Finally, it is essential to note the more significant future potential of the work by having a greater volume and range of data for analysis.

## 5   Conclusions

Whether it is because public interest in cryptocurrencies and their news coverage has not yet translated into a sufficiently significant scientific production, because academia has not yet focused on this issue, or because there is no interest in academia in developing this type of analysis, it is still too early to draw solid conclusions. However, given the importance of the schedule and the levels of attention, it has among the general population, the authors of this article consider it of great importance to make progress in this kind of research.

It is essential now that such a minimal part of the total universe as the Q1 of Communication in SJR has been studied. The general state has been intuited, to extend this first analysis employing a significant sample and to elaborate not only bibliometric studies but also content analysis and meta-analysis that help to understand the results that have been obtained in the few years of research.

In this way, future-and necessary-research will be able to start from a solid base that will facilitate the development of content analyses that help to understand the importance and impact that cryptocurrencies have on the messages issued not only by the mass media but also by influencers and opinion leaders on social networks.

In this sense, technologies such as data intelligence are essential.

**Acknowledgments.** This work has been partially supported by the Institute for Business Competitiveness of Castilla y León, and the European Regional Development Fund under grant CCTT3/20/SA/0002 (AIR-SCity project).

# References

1. Milutinovic, M.: Cryptocurrency. Економика-Часопис за економску теорију и праксу и друштвена питања **1**, 105–122 (2018)
2. Schmidt, W.C., González, A.: Fintech and tokenization: a legislative study in Argentina and Spain about the application of blockchain in the field of properties. ADCAIJ: Adv. Distrib. Comput. Artif. Intell. J. **9**(1), 51–59. Retrieved from https://revistas.usal.es/index.php/2255-2863/article/view/ADCAIJ2020915159 (2020)
3. Jiménez-Serranía, V., Parra-Domínguez, J., De la Prieta, F., Corchado, J.M.: Cryptocurrencies Impact on Financial Markets: Some Insights on Its Regulation and Economic and Accounting Implications. In: Prieto, J., Partida, A., Leitão, P., Pinto, A. (eds.) BLOCKCHAIN 2021. LNNS, vol. 320, pp. 292–299. Springer, Cham (2022). https://doi.org/10.1007/978-3-030-86162-9_29
4. Biernacki, K., Plechawska-Wójcik, M.: A comparative análisis of cryptocurrency wallet management tools. J. Comput. Sci. Inst. **21**, 373–377 (2021)
5. Valdeolmillos, D., Mezquita, Y., González-Briones, A., Prieto, J., Corchado, J.M.: Blockchain technology: a review of the current challenges of cryptocurrency. In: International Congress on Blockchain and Applications, pp. 153–160. Springer (2019)
6. Ahmad, P.: A review on blockchain's applications and implementations. ADCAIJ: Adv. Distrib. Comput. Artif. Intell. J. **10**(2), 197–208 (2021)
7. Domínguez, J., Roseiro, P.: Blockchain: a brief review of agrifood supply chain solutions and opportunities. ADCAIJ: Adv. Distrib. Comput. Artif. Intell. J. **9**(4), 95–106 (2020)
8. Srivastav, R.K., Agrawal, D., Shrivastava, A.: A survey on vulnerabilities and performance evaluation criteria in blockchain technology. ADCAIJ: Adv. Distrib. Comput. Artif. Intell. J. **9**(2), 91–105 (2020)
9. Casado-Vara, R., de la Prieta, F., Prieto, J., Corchado, J.M.: Blockchain framework for IoT data quality via edge computing. In: Proceedings of the 1st Workshop on Blockchain-Enabled Networked Sensor Systems, pp. 19–24 (2018)
10. Maiti, M., Vukovic, D., Krakovich, V., Pandey, M.K.: How integrated are cryptocurrencies. Int. J. Big Data Manage. **1**(1), 51–59 (2020)
11. Umer, M., Awais, M., Muzammul, M.: Stock market prediction using machine learning (ML) algorithms. ADCAIJ: Adv. Distrib. Comput. Artif. Intell. J. **8**(4), 97–116 (2019). https://doi.org/10.14201/ADCAIJ20198497116
12. Pérez-Pons, M.E., Parra, J., Hernández, G., González, J., Corchado, J.M.: Machine Learning and Financial Ratios as an Alternative to Altman's Z-Score Bankruptcy Model in Spanish Companies. In: Bucciarelli, E., Chen, S.-H., Corchado, J.M., Parra D., J. (eds.) DECON 2020. SCI, vol. 990, pp. 130–139. Springer, Cham (2021). https://doi.org/10.1007/978-3-030-75583-6_13
13. Schmidt, W.C., González, A.: Fintech and tokenization: a legislative study in Argentina and Spain about the application of blockchain in the field of properties. ADCAIJ: Adv. Distrib. Comput. Artif. Intell. J. **9**(1), 51–59 (2020)
14. Carlei, V., Adamo, G., Ustenko, O., Barybina, V.: Stacking Generalization via Machine Learning for Trend Detection in Financial Time Series. In: Bucciarelli, E., Chen, S.-H., Corchado, J.M., Parra D., J. (eds.) DECON 2020. SCI, vol. 990, pp. 159–166. Springer, Cham (2021). https://doi.org/10.1007/978-3-030-75583-6_16
15. Li, T., Chen, H., Sun, S., Corchado, J.: Joint Smoothing and tracking based on continuous-time target trajectory function fitting. IEEE Trans. Autom. Sci. Eng. **16**(3), 1476–1483 (2019)
16. Unzueta, M., Bartolomé, A., Hernández, G., Parra, J., Chamoso, P.: System for recommending financial products adapted to the user's profile. In: Advances in Intelligent Systems and Computing, AISC, vol. 1239, pp. 117–126 (2021)

17. Garcia, A.R., Garcia, P.H.R.: Cryptocurrencies: the communication inside blockchain technology and the cross-border tax law. Int. J. Blockchain Cryptocurrencies **1**(1), 22–41 (2019)
18. Burkert, R., van Hagen, J., Wehrmann, M., Jansen, M.: Protection Against Online Fraud Using Blockchain. In: Prieto, J., Partida, A., Leitão, P., Pinto, A. (eds.) BLOCKCHAIN 2021. LNNS, vol. 320, pp. 34–43. Springer, Cham (2022). https://doi.org/10.1007/978-3-030-86162-9_4
19. De Silva, A., Thakur, S., Breslin, J.: Towards micropayment for intermediary based trading. In: Lecture Notes in Networks and Systems, LNNS, vol. 320, pp. 222–232 (2022)
20. Grossman, A.O., Petrov, A.V.: Cryptocurrencies as a social phenomenon. Obs. Sreda Razvit **4**, 62–66 (2017)
21. Carare, A., Ciampoli, M., De Gasperis, G., Facchini, S.D.: Case Study: The Automation of an over the Counter Financial Derivatives Transaction Using the CORDA Blockchain. In: Prieto, J., Partida, A., Leitão, P., Pinto, A. (eds.) BLOCKCHAIN 2021. LNNS, vol. 320, pp. 128–137. Springer, Cham (2022). https://doi.org/10.1007/978-3-030-86162-9_13
22. Carnevale, A., Occhipinti, C.: Ethics and Decisions in Distributed Technologies: A Problem of Trust and Governance Advocating Substantive Democracy. In: Bucciarelli, E., Chen, S.-H., Corchado, J.M. (eds.) DECON 2019. AISC, vol. 1009, pp. 300–307. Springer, Cham (2020). https://doi.org/10.1007/978-3-030-38227-8_34
23. Wood, A.F., Smith, M.J.: Online Communication: LINKING Technology, Identity and Culture. Routledge (2004)
24. Choon, Y., Mohamad, M., Deris, S., Illias, R., Chong, C., Chai, L., Omatu, S., Corchado, J.: Differential bees flux balance analysis with optknock for in silico microbial strains optimization. PLoS ONE **9**(7) (2014)
25. Casado-Vara, R., González-Briones, A., Prieto, J., Corchado, J.: Smart contract for monitoring and control of logistics activities: pharmaceutical utilities case study. Adv. Intell. Syst. Comput. **771**, 509–517 (2019)
26. Lan, X., Wu, Y., Shi, Y., Chen, Q., Cao, N.: Negative emotions, positive outcomes? Exploring the communication of negativity in serious data stories. In: CHI Conference on Human Factors in Computing Systems, pp. 1–14 (2022)
27. Cuervo, C., Morozoya, A., Sugimoto, N.: Regulation of crypto assets. FinTech Notes 1. International Monetary Fund (2019)
28. Babkin Alexander, V., Burkaltseva Diana, D., Pshenichnikov Wladislav, W., Tyulin Andrei, S.: Cryptocurrency and blockchain-technology in digital economy: development genesis. St. Petersburg State Polytech. Univ. J. Econ. **67**(5), 9–22 (2017)
29. Birch, D.G.: What does cryptocurrency mean for the new economy? In: Handbook of Digital Currency, pp. 505–517. Academic Press (2015)
30. Davies, S.R., Halpern, M., Horst, M., Kirby, D.A., Lewenstein, B.: Science stories as culture: experience, identity, narrative and emotion in public communication of science. J. Sci. Commun. **18**(5) (2019)
31. Muñiz, C.: Framing as a research project: a review of concepts, fields, and methods of study. Profesional de la información **9**(6) (2020)
32. Ardèvol-Abreu, A.: Framing o teoría del encuadre en comunicación. Orígenes, desarrollo y panorama actual en España. Revista Latina de Comunicación Social **70**, 423–450 (2015)
33. Chow-White, P., Al-Rawi, A., Lusoli, A., Phan, V.: Social construction of blockchain on social media: framing public discourses on twitter. J. Commun. Technol. **4**(2), 1–31 (2021)

# Evaluation and Comparison of a Private and a Public Blockchain Solution for Use in Supply Chains of SMEs Based on a QOC Analysis

Vanessa Carls[✉], Lambert Schmidt, and Marc Jansen

Institute of Computer Science, University of Applied Sciences Ruhr West, 46236 Bottrop, Germany
vanessa.carls@hs-ruhrwest.de

**Abstract.** SMEs (Small and Medium Enterprises) are more and more interested in using blockchain technologies nowadays and one of the first questions that arises when dealing with the subject of blockchains is whether to choose a private or a public blockchain. This research utilizes requirements of a blockchain solution of three different SMEs for a QOC analysis (Questions, Options and Criteria) and consults the opinions of three blockchain experts to make a decision between a private and a public blockchain based on these evaluations and the corresponding weightings from the interviewed SMEs. In addition, the mentioned criteria also highlight IT requirements that are important for SMEs with regard to blockchain applications. Considering the specified requirements, their weightings, and the evaluation of the two given alternatives, it turns out that a public blockchain is preferable for this particular use case.

**Keywords:** Blockchain · QOC analysis · Supply chain · Private · Public · SME · IT requirements

## 1 Introduction

Today Blockchains are not only known for cryptocurrencies, but also for the other use cases which make use of the blockchains core features. The advantages of a distributed ledger technology include transparency, traceability and security of the data, that it stores [1]. All these features can serve to optimize and digitize a company and its business processes. The immutability of data builds trust between different actors in business relationships. Adopting new blockchain technologies in enterprises is still a challenge for many. The first problem companies face in such an implementation is deciding the type of blockchain. In addition to the initial public blockchains, there is also the possibility of integrating a private blockchain into the company. Both options have different advantages and disadvantages. For this paper three SMEs were asked to give their opinion on different

J. Prieto et al. (Eds.): BLOCKCHAIN 2022, LNNS 595, pp. 388–397, 2023.
https://doi.org/10.1007/978-3-031-21229-1_36

criteria to evaluate the decision between a public and a private approach and to answer the question which type of blockchain technology (private or public) is more suitable for use in an SMEs supply chain. A total of ten employees were interviewed. 22 Criteria were developed based on the aspects that companies identified as being of particular importance. The options (private and public) were evaluated based on the degree to which they met the criteria. To make the final decision between private and public, a score is calculated for both options.

## 2    State of the Art

Public blockchains can be used anonymously by all users and, in addition, the blockchain is distributed across many nodes. Every participant has the same rights and every participant has access to the entire transaction history. A public blockchain can be operated with low initial costs, since already existing infrastructure (computing and storage capacities of servers) is used, but it is often tied to a volatile cryptocurrency. Here, trust in a higher authority is not required; rather, trust is based on the security of the technology, the consensus mechanism and the entire network [2].

Private blockchains, on the other hand, have regulated access and sometimes participants know each other. In this case, trust is often placed again on a single instance for example on the company running the blockchain. A private blockchain is (usually) spread across fewer nodes and tailored to the needs of the business, which requires a larger initial investment, but does not incur cryptocurrency costs that are incurred in a public blockchain for each transaction [2,3].

In direct comparison, the setup of public blockchains is more straight forward, less costly and they are more resilient towards failures, since a large number of nodes would have to be affected to harm the blockchain. But private blockchains normally have a higher transaction speed because of the smaller number of nodes [2,4].

Blockchains are already being applied in various fields in supply chains.Hackius and Petersen[5] identified the ease of paperwork, identifying counterfeit products, origin tracking and operating the internet of things as potential use cases for blockchains. While Berneis et al. [6] identified Luxury Supply Chains, Food Supply Chains and Supply Chains in Healthcare as currently relevant application areas in their literature review.

## 3    Methods

The QOC Analysis (introduced in [7]) is an analysis that is used for decision making. To

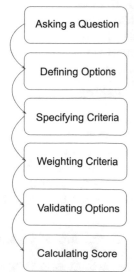

**Fig. 1.** Procedure of QOC analysis.

make reasonable decisions regarding entrepreneurial behaviour it is important to take different opinions and evaluations into account. The QOC analysis helps to figure out which option is the best one. QOC stands for question, options and criteria. There is always a question that needs to be answered. The possible answers represent the options and the criteria stand for significant features or characteristics that are relevant to decide the question. We have chosen the QOC method because we wanted to get an unaffected business view regarding the criteria, which are important to the different SMEs, that is not influenced by expert opinions. Figure 1 illustrates the procedure of the analysis. The first step is to determine the question, the corresponding options and the criteria. After that step the criteria are weighted ($importance_k$) regarding their importance to decide the question. In the next step the options are validated to decide which option fits the requirements best, for this purpose each option is evaluated regarding the different criteria ($compliance_{ik}$). In the case that several participants take part in the qoc analysis, the average of all given ratings is used for the calculation of the $importance_k$ and the $compliance_{ik}$. At the end of the analysis a score is calculated for each option:

$$score_i = \sum_{k=1}^{j} importance_k * compliance_{ik} \tag{1}$$

$i$ is the number of the option, $j$ is the number of criteria, k is the number of a single criterion. The option that reaches the highest score is the one that answers the question best.

## 4    Evaluation

To define the options and the criteria that are used in the QOC analysis three SMEs were consulted. These consist of a manufacturer of wooden materials, a technical wholesaler and a specialist in composite and fastening technology.

### 4.1    Asking the Question

Within the scope of a research project, these companies want to integrate the blockchain technology into their supply chains. To implement this, the first step is to figure out whether a private or a public blockchain would be better suited for the desired purposes. That point leads to the question that the QOC analysis tries to answer: Which type of blockchain technology (private or public) is more suitable for use in an SMEs supply chain?

### 4.2    Definition of Options

The research question results in the two options for the QOC analysis: a private blockchain and a public blockchain.

## 4.3  Definition of Criteria

The criteria were elaborated in small workshops with the different companies in multiple iterations. For this purpose, the companies defined several requirements that are the most important for them regarding the implementation of a blockchain solution. The resulting criteria are listed below (see Table 1). The table provides information about the criteria, the assigned category, as well as a description. The 22 criteria were divided into 11 categories by the SMEs: (data) security, energy demand, acceptance, transaction efficiency, anonymity, interoperability, scalability, transparency, costs, usability and IT-know-how.

## 4.4  Weighting of Criteria

Every SME weighted the collected criteria on a scale from zero to 100, where zero is "not important" and 100 is "very important". A total of ten employees were interviewed, who are either part of the research project or experts in the IT field. The weights of four employees were received from the company in the wood manufacturing sector, five from the technical wholesaler and one from the specialist in composite and fastening technology. The outcomes of these ratings are shown in Fig. 2.

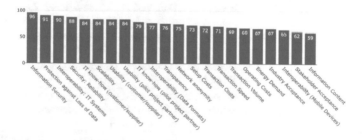

**Fig. 2.** Importance of Criteria evaluated by SMEs.

The weightings of the criteria show that the Information Security is by far the most important criterion for all of the SMEs with an average value of 96. The least important criterion is the information content per transaction (in the category transaction efficiency) with an importance score of 59.

Figure 3 shows the standard deviation in the respondents' importance ratings for each criterion. In this case, the standard deviation is an indicator of the consensus among the participants. The higher this standard deviation, the greater the differences in the ratings of the importance of the criteria. Here it can be seen that the criterion of information security is not only the criterion rated as most important, but the respondents also do agree on this point, as the standard deviation is lowest there. Figure 3 also shows that the transaction speed is the criterion on which the participants disagree the most with a standard deviation of 21.

**Table 1.** List of Criteria.

| Category | Criteria | Description |
|---|---|---|
| Security | Reliability | Security against a failure of the System |
| (Data) Security | Information security | Security through encryption i.e. the Information of the data can be decrypted only by selected persons |
| | Protection against loss of data | protect data from permanent loss |
| Energy demand | Carbon footprint | Energy demand (e.g. measured by Carbon Footprint) |
| Accept-ance | Industry acceptance | Acceptance in the industry (SMEs, supply chain) |
| | Stakeholder acceptance | Acceptance of the Technology by internal Stakeholders |
| Trans-action efficiency | Transaction speed | Speed of transactions to be performed (min. few seconds/max. up to 3 h) |
| | Transaction volume | Number of Transactions that can be performed |
| | Information content per transaction | Information content per transaction |
| Anonymity | Network anonymity | Anonymity in the blockchain for users |
| Interopera-bility | Mobile devices | Connection of the technology to mobile devices (e.g. smartphones, tablets, etc.) |
| | IT Systems | Connection of the technology to existing IT systems (e.g. ERP Systems) |
| | Flexibility towards different data formats | Use of data standards (e.g. JSON, XML etc.) |
| Scalability | Simplicity of connecting new participants | On-boarding of new participants to the system (in terms of Hardware and software requirements) |
| Transpar-ency | Insight into transaction history | Subsequent examining of executed transactions by (selected) participants |
| Costs | Operating costs | These include, for example, power consumption, maintenance costs, transaction costs etc |
| | Transaction costs | Costs of a single transaction |
| | Setup costs | Cost of building or acquiring the blockchain solution |
| Usability | Customer side/supplier side | User-friendliness of the system for customers/suppliers in daily use |
| | On the side of the pilot project partner | User-friendliness of the system for the own company in daily use |
| IT Know-How | Customer-side/supplier-side | Level of IT know-how required by the customer/supplier to operate, implement, maintain, and use the system |
| | On the side of pilot project partners | Level of IT know-how required by the company to operate, implement, maintain and use the system |

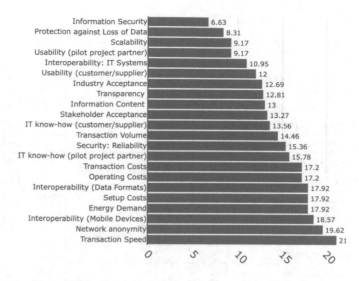

**Fig. 3.** Standard deviation of the weightings.

## 4.5   Validation of Options

Three blockchain experts evaluated the options regarding the compliance with the criteria (see Table 2). Therefore each option gets a score (from 0 to 100) for every criterion defining how good it fits the criterion. Some criteria are independent of the selected technology, which is why they were each rated at 50. Some other criteria were also evaluated similarly, as they have the same impact for both options. An example would be Information Security, because with both options, the data must be protected (for example by using encryption or hashing) from access by unauthorized persons. This is also important in a private blockchain, as there is also data that should only be able to be viewed by certain people with the appropriate authorization.

The following criteria show the biggest differences in compliance with a private blockchain versus a public blockchain. The following section is intended to explain why the compliance ratings vary so significantly [2,3,8].

- Security: Reliability: Since private blockchains operate with fewer nodes than public blockchains, the resilience is remarkably increased for public blockchains. The more nodes are used, the less likely it is that they will all be out of service at the same time.
- Security: Data Security: Protection against loss of data: The same applies to the data loss protection. Due to the higher number of nodes, there are on the one hand more backups of the data and on the other hand the data can only be changed with great difficulty by manipulation.
- Energy Demand: Carbon footprint: Considering the higher number of nodes in a public blockchain, it might be expected that the energy demand is higher in it. However, if each company implemented its own private blockchain,

**Table 2.** Evaluation of options.

| Criteria | Options | |
|---|---|---|
| | Private | Public |
| Security: reliability | 36.67 | 86.67 |
| Security: data security: information security | 96.67 | 96.67 |
| Security: data security: protection against loss of data | 30.00 | 96.67 |
| Energy demand: carbon footprint | 16.67 | 63.33 |
| Acceptance: industry acceptance | 53.33 | 60.00 |
| Acceptance: stakeholder acceptance | 50.00 | 53.33 |
| Transaction efficiency: transaction speed | 70.00 | 43.33 |
| Transaction efficiency: transaction volume | 70.00 | 43.33 |
| Transaction efficiency: information content per transaction | 50.00 | 50.00 |
| Network anonymity | 10.00 | 86.67 |
| Interoperability: mobile devices | 50.00 | 50.00 |
| Interoperability: IT systems | 43.33 | 66.67 |
| Interoperability: flexibility towards different data formats | 50.00 | 50.00 |
| Scalability: simplicity of connecting new participants | 26.67 | 73.33 |
| Transparency (insight into transaction history) | 96.67 | 96.67 |
| Costs: operating costs | 10.00 | 93.33 |
| Costs: transaction costs | 100.00 | 46.67 |
| Costs: setup costs | 3.33 | 86.67 |
| Usability: customer side/supplier side | 63.33 | 63.33 |
| Usability: on the side of the pilot project partner | 50.00 | 50.00 |
| IT know-how: customer-side/supplier-side | 40.00 | 60.00 |
| IT know-how: on the side of pilot project partners | 10.00 | 76.67 |

the energy consumption would be much higher than using a single public blockchain and its already existing infrastructure for all of them.

– Transaction efficiency: Transaction Speed: The transaction speed is mostly higher in a private blockchain than in a public blockchain because, on the one hand, the consensus mechanism works on a smaller scale and, on the other hand, the capacities are limited to a smaller number of users. Furthermore, the transactions can be adapted to the given use case.

– Network Anonymity: In a public blockchain, anyone can participate without revealing their identity. Since private blockchains are often regulated by access restrictions, it is usually necessary to identify oneself.

– Scalability: Since with a private blockchain, access authorizations usually have to be granted in order to connect new participants or even completely new nodes have to be set up and connected, with a public blockchain it is much easier to create your own account through the corresponding services offered.

- Costs: Operating Costs: The operational costs of a private solution are relatively high, because electricity and server maintenance have to be paid. With the public solution, a fee must only be paid when using the blockchain.
- Costs: Setup Costs: Building a private blockchain solution involves building an entire network and the blockchain implementation, whereas a public approach may just add a web server to the blockchain implementation to host the applications.
- Costs: Transaction Costs: Transaction costs are non-existent in the private solution, while the public solution is usually linked to a volatile cryptocurrency and every transactions has to be paid.
- IT know-how: on the side of the pilot project partners: In this criterion, the infrastructure of the private solution must also be maintained, whereas with the public solution only the connection to the infrastructure must be maintained.

### 4.6   Findings

Table 3 provides an overview of the calculated scores ($importance_k *$ $compliance_{ik}$) of the different criteria $k$ for the specific option $i$. The higher the score, the stronger the influence of the criterion on the final score of the option and, accordingly, on the decision between private and public blockchain. Here we can see that especially the criteria reliability, protection against loss of data, operating costs, setup costs, IT know-how on the customer/supplier side and network anonymity drive the score for the public blockchain up. The reasons for these higher scores were highlighted in the last section.

After evaluation and assessment of the QOC analysis by the expert group, the public blockchain scores with 116046.67 and the private blockchain with 78906.67. Based on this QOC analysis the public blockchain solution seems to be the better choice for the surveyed SMEs. These findings are consistent with Clohessy and Acton.

Clohessy and Acton [8] found that large companies tend to use private blockchains because they had control over "access privileges to modify or read the blockchain state of their applications is restricted to only a few authorized users". Whereas SMEs use public blockchains because of the "ease of accessibility, setup and access to information resources" and one SME rejected the private blockchain "option due to the complexity, cost and lack of business use cases for their industry.

## 5   Outlook

After analyzing whether a private or a public blockchain solution is more suitable for the use case, the next step is to investigate which specific public blockchain should be used by SMEs. In addition to the options of private and public blockchains, there are also different specifications such as permissoned or permissionless blockchains, which can also be considered. A subsequent analysis

**Table 3.** Scores of criteria for public and private blockchain.

| Criteria | Options | |
|---|---|---|
| | Private | Public |
| Security: Reliability | 3226.96 | 7626.96 |
| Security: Data security: Information security | 9280.32 | 9280.32 |
| Security: Data security: protection against loss of data | 2730 | 8796.97 |
| Energy demand: Carbon footprint | 1116.89 | 4243.11 |
| Acceptance: Industry acceptance | 3573.11 | 4020 |
| Acceptance: Stakeholder acceptance | 3100 | 3306.46 |
| Transaction efficiency: transaction speed | 4970 | 3076.43 |
| Transaction efficiency: transaction volume | 4830 | 2989.77 |
| Transaction efficiency: information content per transaction | 2950 | 2950 |
| Network anonymity | 750 | 6500.25 |
| Interoperability: mobile devices | 3250 | 3250 |
| Interoperability: IT systems | 3899.70 | 6000.3 |
| Interoperability: Flexibility towards different data formats | 3850 | 3850 |
| Scalability: simplicity of connecting new participants | 2240.28 | 6159.72 |
| Transparency (insight into transaction history) | 7346.92 | 7346.92 |
| Costs: Operating costs | 680 | 6346.44 |
| Costs: Transaction costs | 7200 | 3360.24 |
| Costs: Setup costs | 243.09 | 6326.91 |
| Usability: customer side/supplier side | 5319.72 | 5319.72 |
| Usability: On the side of the pilot project partner | 4200 | 4200 |
| IT know-how: customer-side/supplier-side | 3360 | 5040 |
| IT know-how: on the side of pilot project partners | 790 | 6056.93 |
| Score | 78906.67 | 116046.67 |

would be of interest here. Although the result of the survey was clear, it can be further investigated to what extent the result of the analysis changes when more companies are surveyed. It should also be investigated to what extent the results presented here can be applied or adapted to larger companies. The same analysis should also be carried out with larger companies in order to compare the results, as for example the evaluation of the infrastructure costs of a private blockchain solution might be evaluated differently for larger companies. With more data, it would also be possible to find out which factors play a role in the decision for a public or a private solution and to what extent the size of the company plays a role. Furthermore, the intended use case should be recorded in order to investigate the influence of different use cases on the choice of blockchain. On this basis, a process model could then be created for deciding on the appropriate blockchain solution. In addition, other use cases distanced from the supply chain could be explored regarding the use of a private or public blockchain solution.

# 6   Conclusion

In summary, based on the result of the QOC analysis, a public Blockchain is clearly preferable. The criteria of information security (96), protection against loss of data (91) and interoperability of IT systems (90) were the most important for the surveyed companies. According to the experts' evaluation, the following criteria showed the greatest difference in valuation: Operating costs (difference of 83.33), Network anonymity (difference of 76.67) and protection against loss of data (difference of 66.67). Transaction costs and transaction efficiency, specific transaction speed and transaction volume, were the only criteria where the experts decided in favor of the private solution. The other criteria were either rated equally or fell in favor of the public solution. For the SMEs surveyed and their supply chain use cases, a public blockchain solution better meets the defined requirements.

**Acknowledgements.** This research was supported by the European Regional Development Fund 2014–2020 within the BC4SC (Blockchain for Supply Chain) project (EFRE 0200617).

# References

1. Kuhn, R., Yaga, D., Voas, J.: Rethinking distributed ledger technology. Computer **52**(2), 68–72 (2019)
2. Monrat, A.A., Schelén, O., Andersson, K.: A survey of blockchain from the perspectives of applications, challenges, and opportunities. IEEE Access **7**, 117 134–117 151 (2019). [Online]. Available: https://doi.org/10.1109/ACCESS.2019.2936094
3. Jabbar, A., Dani, S.: Investigating the link between transaction and computational costs in a blockchain environment. Int. J. Prod. Res. **58**(11), 3423–3436 (2020) [Online]. Available: https://doi.org/10.1080/00207543.2020.1754487
4. Lai, R., Chuen, D.L.K.: Chapter 7—blockchain—from public to private. In: Chuen, D.L.K., Deng, R. (eds) Handbook of Blockchain, Digital Finance, and Inclusion, vol. 2. Academic Press, pp. 145–177 (2018) [Online]. Available: https://www.sciencedirect.com/science/article/pii/B9780128122822000073
5. Hackius, N., Petersen, M.: Blockchain in logistics and supply chain: trick or treat? In: Proceedings of the Hamburg International Conference of Logistics (HICL), vol. 23, pp. 3–18 (2017) [Online]. Available: https://doi.org/10.15480/882.1444
6. Moritz Berneis, D.B., Winkler, H.: Applications of blockchain technology in logistics and supply chain management-insights from a systematic literature review. Logistics **5**(43) (2021). [Online]. Available: https://doi.org/10.3390/logistics5030043
7. MacLean, A., Young, R., Bellotti V., Moran, T.: Questions, options, and criteria: Elements of design space analysis. Hum.-Comput. Interact. **6**, 201–250 (1991)
8. Clohessy, T., Acton, T.: Investigating the influence of organizational factors on blockchain adoption. Ind. Manage. Data Syst. **119**(7), 1457–1491 (2019) [Online]. Available: https://doi.org/10.1108/IMDS-08-2018-0365

# Identification of False Stealthy Data Injection Attacks in Smart Meters Using Machine Learning and Blockchain

Saurabh Shukla$^{(\boxtimes)}$ ⓘ, Subhasis Thakur ⓘ, Shahid Hussain ⓘ, John G. Breslin ⓘ, and Syed Muslim Jameel ⓘ

National University of Ireland Galway, Galway, Ireland
{saurabh.shukla,subhasis.thakur,shahid.hussain,john.breslin,
muslimjameel.syed}@nuigalway.ie

**Abstract.** The current challenging issue in the Advanced Metering Infrastructure (AMI) of the Smart Grid (SG) network is how to classify and identify smart meters under the effect of falsification attacks related to stealthy data, which are injected in such small numbers or percentages that it becomes hard to identify. The problem becomes challenging due to sudden variation in the pattern for the power consumption of electrical data, the existing approaches and techniques are making such stealthy data attacks unrecognizable. Therefore, to identify such a small portion of data attacks we proposed a novel solution using a combination of blockchain, Fog Computing (FC), and linear Support Vector Machine (SVM) with Principal Component Analysis (PCA). In this paper, we proposed a 3-tier blockchain-based architecture, an advanced system model and a classification algorithm in a blockchain-based FC environment. The blockchain system is used here to verify the electrical data transmission and transaction between the user and the utility centre. Whereas FC is used to provide real-time alert messages in case of any false attacks related to stealthy data. A detailed analysis of the generated results is conducted by benchmarking with the other state-of-the-art techniques. The algorithm and model show a marked improvement over other technologies and techniques. The simulation of the algorithm was conducted using iFogSim, Ganache, Truffle for compiling, Python editor tool, and ATOM IDE.

**Keywords:** Smart grid · Fog computing · Blockchain · Linear SVM · PCA · Security · Privacy · Cyber-physical system · False data attack · Cyber-attack · Machine learning

## 1 Introduction

The Advanced Metering Infrastructure (AMI) in a Smart Grid (SG) network consists of multiple smart meters (IoT devices) that communicate and collect the energy data in a two-way or bi-directional form from customers to utility. Architectural communication is used by smart meters for the transmission of energy data in a periodic manner, to various data management servers limited to the utility centres. The involved servers processed

J. Prieto et al. (Eds.): BLOCKCHAIN 2022, LNNS 595, pp. 398–409, 2023.
https://doi.org/10.1007/978-3-031-21229-1_37

the energy data received from smart meters for critical operations such as billing, load balancing, daily monitoring of threshold peaks, shift demand, and demand response. This integrity of the energy or electrical data from devices like smart meters plays a major role in the success of SG. However, cyber, or physical attacks on smart meters are increasing every day as the smart meters relate to the cyber-physical world or with web-based portals through which they are connected to the smart home network at the customer site. This type of cyber-physical attack consists of many smart meter devices connected to similar feeding data and then generates spoof false energy data readings. The false stealthy data injection leads to huge monetary losses to the consumers of smart cities. Identification and authentication of smart meters which are involved/engaged in the transmission or usage of stealthy false electrical data consumption are key security and privacy challenges. The quantifiable parametric value which distinguishes the false stealthy data from the original clean electrical data is known as the mean average value of false attack strength [1]. The lower marginal value of attack value/ strength is stealthy and hence it becomes difficult to detect/identify them to quantify the parametric value of such attacks comprising a large set of smart meters in the SG network is known as the "Scale of False Attack" [2]. Such minimal or lower kinds of stealthy attacks have a great significant impact on the utility bills/consumption readings when compromised to total cyberattacks cost. These attacks compromise the smart meter services. Such attacks are usually conducted by cybercriminals and organizational competitors to lower the marginal value of false data per smart meter by making hide them behind the random data generated from smart meter devices. This makes these smart meters hard to catch.

Therefore, providing an enhanced classification technique to identify smart meters which are involved in the transmission of stealthy false data is of utmost importance. This classification approach uses a hybrid of Supervised and Unsupervised learning algorithms i.e., Support Vector Machine (SVM) and 2-Principal Component Analysis (PCA) using blockchain technology for verification of transactions and use of Fog Computing (FC) to alert the end-users for real-time intruder's attack.

## 2 Contribution

P In this paper, we propose a novel anomaly detection model, for false stealthy data injection, consumption, and transmission in smart meters. The model will be able to identify and detect a wide range of cyber-attack from very low to a high margin of attack strength and its scales. The model will be using FC for alarm notification and detection of false alarms, and missed alarms, compared to the other state-of-the-art techniques. To accomplish this, the contributions of this paper are as follows:

1. We proposed a novel 3-tier blockchain-based machine learning architecture for false stealthy data identification in smart meters.
2. Next, we proposed and designed a novel and hybrid system model using blockchain and machine learning techniques in the FC environment.
3. At last, we designed and implement a hybrid blockchain-based machine learning algorithm for the identification of false stealthy data injection attacks on smart meters.

The paper is organized as follows: Sect. 3 discussed the related work. Section 4. Presents the 3-tier architecture. Whereas Sect. 5 presents the system model. Section 6 presents the hybrid algorithm. Section 7 discussed the experimental results and evaluation. And at last, Sect. 8 concludes the work.

## 3   Related Work

In this section, a complete detailed analysis of the current existing works related to false data attacks on smart metres and smart grids is conducted. Some of the techniques are highlighted in [3], the authors proposed a Scoring Framework (SF) for identifying stealthy false data attacks on smart meters. They highlighted the issue of low-margin false data injection in smart meters and smart grid networks. Their proposed method is based on a classification approach to identify the low strength of false data. Furthermore, they used the scoring technique with indexed-based solutions.

In [4], the authors designed a Zero parameter-information Data Injection Attack (DIA) class technique called ZDIA. They highlighted the problem of stealthy data attacks for tampering with the smart meter data without being noticed by the consumers. Through this, the attacker can extract the topology and branch information of the system parameter. They can easily predict the system states for various cyber-attacks. Next, they proposed countermeasures to address the vulnerability of various attacks conducted by cyber-criminals on smart grid networks using ZDIA. Furthermore, they presented a strategy and defined techniques against such DIA.

In [5], the authors proposed a rea-time-based-Tier Attack Identification (2-TAI) scheme to detect false stealthy data in smart meter infrastructure. In the first tier, the use of harmonics to the arithmetic mean ratio of total power consumption of electrical data is conducted to check whether it is under a safe margin or not. Whereas, in the second-tier detection scheme the monitoring of the sum of the residuals i.e., the difference between the proposed metric ratio and the safe margin is conducted over a longer period and days. If the sum of residuals exceeds the average margin range, then the presence of a data falsification attack is confirmed. They further used the technique of the system identification phase for safe margins and standards limits. A real-time AMI microgrid data set is used collected from two different centres.

In [6], the authors proposed a new technique based on the Dimensionality Reduction Method (DRM) for the identification of FDIA in the smart grid. Their proposed method consists of two phases: a dimensionality reduction phase-1 and Gaussian-based semi-supervised learning phase-2 to minimize the error and to increase the distinction between the normal value of data and the bad data value. This generates a reduced dimension based on a new orthogonal. At last, the threshold value is checked using the Gaussian mixture model to identify the FDIA. In [7], the authors proposed a Graph Neural Network (GNN) approach to detect and identify the presence of FDIA. They applied Graph filters using Auto Regressive Moving Average (ARMA). This technique automatically detects the localize FDIA in the SG network. They exploited the inherent graph topology of the SG network with spatial correlations of measurement data.

## 4    Blockchain-Based 3-Tier Architecture

In this section, we discussed the proposed blockchain-based 3-tier architecture in the FC environment. The architecture was designed to minimize and detect the FSDIA percentage for electrical data transmission, distribution and consumption using smart meters in an SG network system. See Fig. 1 for the architecture design.

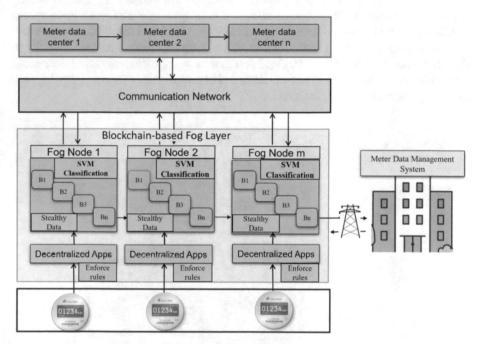

**Fig. 1.** 3-Tier architecture

Figure 1 illustrates the blockchain-based 3-tier architecture for the identification of false stealthy injection in smart meters. The proposed architecture utilizes the concept of machine learning technique called Linear SVM and 2-PCA classification for separating the outliers from the electrical data and readings collected from smart meters. Fog nodes are used here for running the machine learning algorithm inside the different processes and to send real-time notifications for any kind of cyberattacks to end-users. Whereas blockchain is used here to record and secure the various electrical data transactions in different chains of blocks called ledgers. The architecture consists of smart meters, data connectors, Meter Data Management System, (MDMS), and fog nodes. The electrical data is sent to meter data centres using a communication network.

## 5    Conventional and Advanced System Model

In this section, we discussed the difference between the conventional system model and the proposed advanced system model for FSDIA in smart meters and SG networks.

Furthermore, we discussed the advanced system for secure electrical data transmission communication in an SG network which enables the identification and detection of false stealthy data. The traditional model consists of smart meters, IoT devices, sensors, Remote Terminal Units (RTU) and a meter data acquisition system. The conventional model is prone to cyber-attacks when false measured data is injected through Supervisory Control and Data Acquisition (SCADA), and it disturbs the state estimation of the SG network system. The conventional system model lacks advanced security features such as blockchain techniques. The conventional model is an open network for outside intruders and CPS attackers to manipulate the smart meter readings and insert false data during electric data transmission. The model lacks the real-time decision-making capability to identify and detect the FSDIA. See Fig. 2 shows the novel system model for SG network communication, identification of false data, and classification of false stealthy injected data using fog nodes and blockchain.

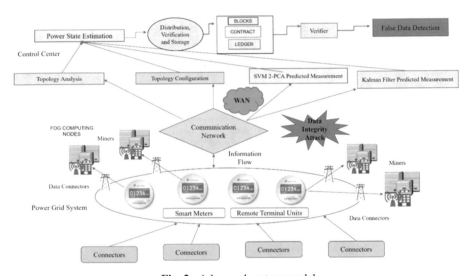

**Fig. 2.** Advanced system model

The classification of electrical data and variables is conducted using a supervised machine learning approach. Linear SVM and 2-PCA techniques are used for reducing the number of parametric values and severity levels involved in the data generation, transmission, distribution, and consumption and can easily classify by distinguishing the outliers and similar patterns. The FC nodes are used here as miners to process and mined the electrical data in a blockchain environment. Figure 2 shows the advanced system model for false stealthy data identification in the SG network. The novel Blockchain-based False Stealthy Data Identification Algorithm (BFSDIA) algorithm works in coordination with fog nodes where the classification of false data is conducted inside the fog nodes. The electrical data is classified using L-SVM which generated the predicted measured value which is then updated using the Kalman filter and these updated values are sent for state estimation where the electrical data transaction are verified using blockchain nodes.

# 6   Blockchain-Based False Stealthy Data Identification Algorithm (BFSDIA)

In this section, we discussed the proposed advanced hybrid BFSDI algorithm. The algorithm works to identify and detect the false stealthy data injected into an SG network during electrical data transmission. The novel algorithm utilizes the technique of machine learning algorithms called L-SVM and 2-PCA. This technique is used to classify the data along a hyper-plane to identify the outliers and false data. The data classified to the highest varied 2-PCA values using the ML algorithm to display not-under-risk and under-risk electrical data depending upon the false stealthy injected data presence. Furthermore, this section includes algorithm symbol notations, algorithm steps, and pseud-code.

**BFSDI Algorithm Symbol Notations**

| | |
|---|---|
| $S_e$: | State estimation |
| $FC_N$: | Fog computing nodes |
| $C_s$: | Current sample |
| $V$: | Voltage |
| $P_l$: | Parametric level |
| $S_l$: | Security level |
| $S_m$: | Smart meter |
| $E_{dt}$: | Electrical data |
| $F_d$: | False data |
| $L_{SVM}$: | Linear Support Vector Machine |
| $w_v$: | Weight vector |
| $D_t$: | Distance |
| $D_b$: | Decision boundary |
| $Mg_{w_d}$: | Marginal width |
| $Ce_{DS}$: | The list of classified electrical data |
| $A_l$: | Electrical data packets having anomalies |
| $C_V$: | Cross-validation |
| $V_M$: | Variable minimization |
| $2\text{-}PCA$: | 2-Principal Component Analysis |
| $R_d$: | Reduced dimension |
| $O_m$: | Original measurement |
| $T_v$: | Threshold value |
| $N_m$: | New measurement |
| $C_t$: | Coordinate transform |
| $FSDA$: | False Stealthy Data Attack |
| $P_v$: | Parametric value |
| $O_L$: | Outliers |
| $Rx$: | Communication channel |
| $Tx$: | Communication channel |
| $D_V$: | Data verification |
| $B_C$: | Blockchain |

## BFSDI Algorithm Steps

**Step 1:** Creation of an FC system.
**Step 2:** Collection of different state estimations.
**Step 3:** Reading the values of voltages and current samples.
**Step 4:** Checking of parameter and security levels.
**Step 5:** Selection of targeted smart meters.
**Step 6:** Injection of electrical data in smart meters.
**Step 7:** Identification of false data using Linear SVM.
**Step 8:** Classification of electrical data using Linear SVM.
**Step 9:** Next, training of the machine learning model.
**Step 10:** Perform variable minimization using PCA.
**Step 11:** Conduct cross-validation.
**Step 12:** Dimension reduction and removal of identified outliers.
**Step 13:** Obtain the parametric level and threshold value.
**Step 14:** Data verification using blockchain.
**Step 15:** Identify the False stealth data attack on smart meters.
**Step 16:** To check the different types of failures from the attack on the states of smart meters.
**Step 17:** Apply the Kalman filter for the updated value both for predicted values and measured values.
**Step 18:** Notify the users and meter management system.

## BFSDI Algorithm

**Input**: Smart meter state estimation, smart meter data, electrical readings, measured values, and injected electrical data.

**Output**: Notification of stable threshold marginal value for false stealthy data.

**Init**: Perform initialization tasks, such as the reading of electrical data, network filters and IP addresses.

1:   **START**
2:   *(FC-based blockchain system is created)*
3:   **While** *collect* $S_e$ *in* $FC_N$ **do**
4:       *Read V and* $C_s$
5:       *Check* $P_l$ *and* $S_l$ **end**
6:       *Select* $S_m$ *using* $S_e$
7:   $E_{dt}$ *injection in* $S_m$
8:   $F_d$ *identification using* $L_{SVM}$
9:   $E_{dt}$ *classification using* $L_{SVM}$
10:  **While** *(iter* $\leq$ maximum iteration) **do**
11:  **Function** *Def Distance (self, $w_d$, with_lagrange = True):*
12:  $D_t = self.y * (np.dot(self.X, w_v)) - 1$
13:      *get* $D_t$ *from* $D_b$ *from the current* $D_b$
14:      *if* $Mg_{w_d} == 1$

15:          get $D_t$ from $D_b$

16:          generate $Ce_{DS}$ and $A_l$

17:      **else if** $Mg_{w_d} == 0$

18:          then $D_t$ not retrieve

19:    **end if**

20: **end if**

21: **end Function**

22: **Function** Train (Model)

23:    do $C_V$

24:      $V_M$ using 2-PCA

25:      Construct $R_d$

26:      get $R_d$ and $O_m$ of $S_m$ data

27:      Obtain $T_v$ from $L_{SVM}$

28: **end Function**

29: **For** $N_m$

30:    do perform $C_t$

31:    Compute $P_v$

32:    $D_V$ using $B_C$

33: **end For**

34: **if** $P_l \geq T_v$

35:      Possible $FSDA$

36:   **else** No FSDA detection

37: **end if**

38: $LS_e$ in $FC_N$ using $H_C$ of $P_B$ and $C_B$

39: **if** $FSDA ==$ Minor Failure || Moderate Failure || Severe Failure || Catastrophic

40:    then do check $S_m$ states

41:      Remove identified $O_L$ and $V_B$ identified using SVM

42:    else

43:      System $==$ stable state

44: **end if**

45: **Function** KALMAN_STATE result

46: KALMAN_update ( )

47: Update Predict and Measure values

48: do notification using Rx/Tx

49: **end Function**

50: **End**

## 7 Results and Discussion

This section discusses the performance evaluation along with a simulation overview and settings for the proposed model and algorithm for generated results. In this section, the execution of the proposed Blockchain-based False Stealthy Data Identification Algorithm (BFSDIA) is analyzed. Next, to verify the proposed algorithm and system model a benchmarking of the existing techniques was conducted. This further helps in examining the robustness of the performance measures. The simulation of the BFSDI algorithm is conducted in the iFogSim Simulator. The algorithm uses the Linear SVM and 2-PCA techniques for the classification of false stealthy electrical data by minimizing the number of variables across a hyperplane. The missing values and outliers are removed and filled with a mean data value. These missing values are removed using a Kalman filter. To demonstrate the highest variation the 2-PCA values are used in the identified false stealthy electrical data. The algorithm is implemented using NetBeans and python with several main packages. The baseline for this simulation is the maximum false stealthy data detection accuracy percentage. See Fig. 3 for the topology configuration of deployed smart meter devices and fog nodes in the proposed system model using the iFogSim simulator.

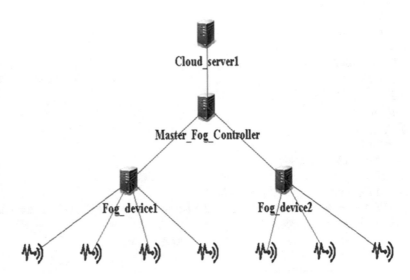

**Fig. 3.** Graphical user interface configuration

Figure 3 shows the graphical physical topology configurations built-in in the iFogSim simulator. The configuration in Fig. 5 is fully based on the concept of a proposed system model and BFSDI algorithm. The configuration consists of distributed smart meter devices placed at the edge of networks along with fog nodes that are further connected to cloud data centres. Whereas Fig. 4 shows the classified false stealthy injected electrical data using the hybrid machine learning approach called linear SVM and 2-PCA.

Figure 4 shows the not-under-risk electrical data which is classified as false stealthy free data shown by green colour dots and under-risk false stealthy electrical data shown by red colour dots. The X-axis shows the electrical data PCA in the 1st column, and the Y-axis shows the electrical PCA 2nd column.

Figure 5 shows the percentage of a node with false stealthy data injection in blockchain-based FC servers and cloud-based utility centres along with the false stealthy data detection accuracy. The minimum false stealthy data detection accuracy of the BFSDI algorithm in FC-based servers and cloud-based utility centres is 63 and 58%.

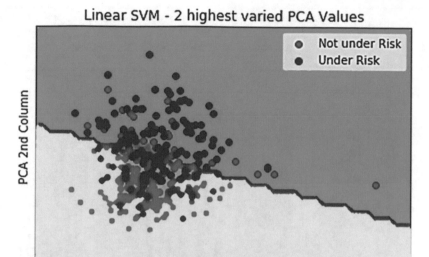

**Fig. 4.** Classification for false stealthy electrical data using L-SVM and PCA

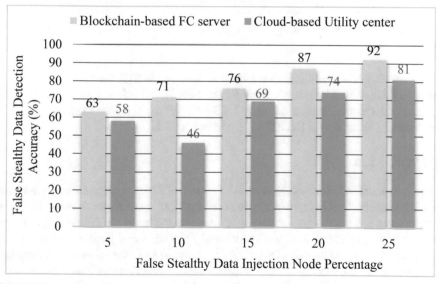

**Fig. 5.** False stealthy data detection accuracy versus node with false stealthy data injection percentage

Whereas the maximum detection accuracy of the BFSDI algorithm in fog servers and cloud environments is 92 and 81%.

Figure 6 shows the comparison of a node with false stealthy data percentage versus false data detection accuracy of the proposed BFSDI algorithm with the other existing

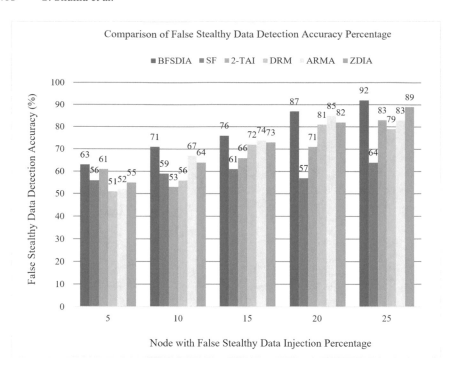

**Fig. 6.** Anomaly detection accuracy versus node with anomaly percentage performance evaluation

techniques. The minimum false data detection accuracy is 51% for DRM. Whereas the minimum false data detection accuracy for the BFSDI algorithm is 63%. However, the maximum false data detection accuracy is 89% for ZDIA. Whereas the maximum false data detection accuracy for the BFSDI algorithm is 92%.

## 8   Conclusion

The advancement in Industry 4.0 has evolved the role of AMI for smart cities and societies in an SG network. This SG network consists of multiple smart meter devices. The smart meter devices are the perfect examples of Internet-of-Things (IoT) that communicate the energy information flow in a bi-directional form from end-users to utility centres. Therefore, it is prone to various kinds of cyber-attacks from outside attackers or intruders. Sometimes a similar feeding of data is transmitted to smart meters that generate spoof false energy data readings. This false stealthy data is transmitted in such minimal or low forms that they are hard to distinguish from the main electrical data. The injection of, malicious data insertion, and false stealthy data inside the smart meter devices have generated a large risk to the SG network. Hence to overcome this issue we have proposed a 3-tier blockchain-based architecture, an advanced system model in an FC environment, and a novel hybrid machine learning algorithm that utilizes the concept and technique of both blockchain and FC. The algorithm uses the L-SVM and 2-PCA machine learning techniques for the classification and identification of FSDIA on smart meter devices.

**Acknowledgements.** This publication has emanated from research supported in part by a research grant from Cooperative Energy Trading System (CENTS) under Grant Number REI1633, and by a research grant from Science Foundation Ireland (SFI) under Grant Number SFI 12/RC/2289_P2 (Insight), co-funded by the European Regional Development Fund.

# References

1. Zhang, M., et al.: False data injection attacks against smart gird state estimation: construction, detection and defense. Sci. China Technol. Sci. **62**(12), 2077–2087 (2019). https://doi.org/10.1007/s11431-019-9544-7
2. Sengan, S., et al.: Detection of false data cyber-attacks for the assessment of security in smart grid using deep learning. Comput. Electr. Eng. **93**, 107211 (2021)
3. Bhattacharjee, S., Madhavarapu, P., Das, S.K.: A diversity index based scoring framework for identifying smart meters launching stealthy data falsification attacks. In: Proceedings of the 2021 ACM Asia Conference on Computer and Communications Security (2021)
4. Zhang, Z., et al.: Zero-parameter-information data integrity attacks and countermeasures in IoT-based smart grid. IEEE Internet Things J. **8**(8), 6608–6623 (2021)
5. Bhattacharjee, S., Das, S.K.: Detection and forensics against stealthy data falsification in smart metering infrastructure. IEEE Trans. Dependable Secure Comput. **18**(1), 356–371 (2018)
6. Shi, H., Xie, L., Peng, L.: Detection of false data injection attacks in smart grid based on a new dimensionality-reduction method. Comput. Electr. Eng. **91**, 107058 (2021)
7. Boyaci, O., et al.: Joint detection and localization of stealth false data injection attacks in smart grids using graph neural networks. IEEE Trans. Smart Grid **13**(1), 807–819 (2021)

# Workshop on Beyond the Promises of Web3.0: Foundations and Challenges of Trust Decentralization (WEB3-TRUST)

# Enhancing the Anonymity and Auditability of Whistleblowers Protection

Sergio Chica[1]([⊠]) [ID], Andrés Marín[1] [ID], David Arroyo[2] [ID], Jesús Díaz[3] [ID], Florina Almenares[1] [ID], and Daniel Díaz[1] [ID]

[1] Universidad Carlos III de Madrid, Leganés, Spain
`sergio.chica@uc3m.es`
[2] ITEFI/Consejo Superior de Investigaciones Científicas, Madrid, Spain
[3] Input | Output, Madrid, Spain
`https://iog.io/`

**Abstract.** In our democracy a trade-off between checks and balances is mandatory. To play the role of balances, it is necessary to have information that is often only obtainable through channels that ensure the anonymity of the source. Here we present a work in progress of a system that provides anonymity to sources in a open and auditable system, oriented to audit systems of critical infrastructure and built on our previous work AUTOAUDITOR [5].

**Keywords:** Permissioned blockchain · Anonymous whistleblowing · Group signatures · ECDHE

## 1 Introduction

In 2015, Mr. David Kaye, United Nations Special Rapporteur on the promotion and protection of the right to freedom of opinion and expression, submitted a report on the protection of sources of information and whistleblowers, highlighting the key elements of a framework for the protection of source and whistleblowers: a communication platform to disseminate that information; and a legal system and political culture that effectively protect both the source and the communication platform. Twenty-eight States responded to a questionnaire requesting information on national norms protecting sources and whistleblowers. Four years later, the Directive (EU) 2019/1937 of the European Parliament and of the Council addresses the protection of the persons who report breaches of Union law. This directive explains that the subject of such protection are

This work has been partially supported by COMPROMISE Project financed by (PID2020-113795RB-C31/AEI/10.13039/501100011033), by the SPIRS Project with Grant Agreement No. 952622 under the EU H2020 research and innovation programme and by Madrid regional CYNAMON project (P2018/TCS-4566), co-financed by European Structural Funds ESF and FEDER.

workers in public and private organizations, who play a crucial role in *exposing and preventing such breaches and in safeguarding the welfare of society.* That behavior is often discouraged for fear of retaliation. The Directive enforces the need to *introducing effective, confidential and secure reporting channels and by ensuring that whistleblowers are protected effectively against retaliation.*

Anonymous reporting is not new, it was already common in the roman law, and often has lead to malicious reporting. There are objective differences between anonymous reporting and whistleblowing. A whistleblower always has clearance to access confidential information about the law breach, which is not necessarily the case for malicious reporting. Before providing protection, a proof of identity as a worker of the organization should be required.

Directive [6] recognizes the importance of a balanced and effective protection to whistleblowers in different areas: in public procurement, in financial services, in manufacturing, importing and distribution of fire arms, defense and explosives, transportation security, environment protection, nuclear safety, food chain and animal health, public health, and in the security of networks and information systems. In this context [6] introduces: *a requirement to provide notification of incidents, including those that do not compromise personal data, and security requirements for entities providing essential services across many sectors, for example energy, ....*

New tools are being engineered in the protection of power supply companies. Tools aiming at the automation of the security auditing and pentesting of the vast number of devices, networks and networked systems involved in such companies. Proposal [10] is an example, that also works on the auditing of the process, presenting distributed ledger technology and specific smart contracts to store the audit reports of each organization. The permissioned blockchain is distributed in nodes from competing organization of energy supplying companies. The permission system allows for sharing some security information, while other is kept private within the organization. Security information sharing is important to improve the response when a security incident (intrusion, data breach, etc.) happens. If we know which tests have been correctly executed before the investigation, that will be a clue to identify which was the entry point of the cyberattack and will help other organization to protect from similar attacks.

We aim at providing a platform with the capabilities to support anonymous and confidential communications for potential whistleblowers. We expect that such support will be enough to prevent some EU law breaches. Our motivation is to provide transparency and support for potential disclosures. We are working on a system based on group signatures [3] and ephemeral Diffie-Hellman agreements that offers sufficient protection to potential whistleblowers.

## 2    State of the Art

Among the open-source projects to facilitate anonymous disclosure, we can cite GlobaLeaks y SecureDrop. GlobaLeaks complies with ISO 37002 and EU Directive 2019/1937. In addition, it allows GDPR data retention policies configuration, and is easily distributable to different nodes, each of them storing the documents of their own whistleblowers. SecureDrop is a more centralized system, designed to be installed on a newspaper owned site. Whistleblowers are expected to use pen drives with a Tails software distribution that routes all the traffic through an onion network, and makes no use of the computer disk or the operating system in order to leave no traces of the session. In both projects, journalists and whistleblowers communicate and exchange messages exclusively through the system so that, neither the communications company nor providers such as email services can trace the source of a whistleblower.

By default, both systems provide server encryption, and none of them allow whistleblowers to give a proof of their reliability as sources. To overcome this limitation, group signatures system can be used—being part of an organization is a requirement to join the group. Signing with a group credential proves belonging to an organization, therefore demonstrating legal access to the documentation being disclosed.

In [9] a comparison is proposed between different forms to protect anonymous disclosures: web forms, WikiLeaks, and both open-source systems GlobaLeaks and SecureDrop. Finally, the article proposes the use of blockchain and smart contracts along with IPFS (InterPlanetary File System), however, no details are given.

Research [13] proposes a system that combines blockchain and ring signatures to provide true anonymity to whistleblowers. Even more, only the signer has the option to revoke their anonymity. On the other hand, when using a ring signature scheme with RSA, the complexity and computation time to sign and send a disclosure are high. Verification of the signature is fast, but it gets more complicated and time consuming to verify the claims for credit of a signer who revokes their anonymity.

One of our goals is to implement a solution that improves the complexity of [13], and thus we leverage group signatures together with ephemeral agreements based on a public key pair generated by the whistleblower. This being the case, if the whistleblower wants to revoke their identity, they can present the ephemeral private key used to generate the symmetric key which decrypts the disclosure. A second requirement which is not in [13], is the confidentiality of the disclosures: the whistleblower should be able to deliver to a chosen person, and maintain a subsequent exchange with that person. This requirement originates from the draft of the future Spanish national law that will transpose the European directive [6]. The draft explicitly mentions a reference person, the person in charge of the internal information system. The system we propose and describe in the next section perfectly fits into the internal information system referred in the draft of the law.

A key piece to ensure anonymity of whistleblowers is the group signatures scheme PS16 [12], a scheme of type Sign-Randomize-Prove based on signatures of type Camenisch-Lysyanskaya [1]. In other words, it is based on a scheme of signatures using message vectors, compatible with efficient zero-knowledge proof protocols. Given a PS16 signature issued by the group manager (i.e., the group member credential), the additional zero-knowledge proof protocols permit the user to prove possession of said signature without revealing the signature—achieving the anonymity required in the schemes of group signatures. The group manager keeps a list with the registries of the group joining process for each member. In order to open a group signature, the manager must iterate over such list.

## 3    Architecture

Our proposal is based on HyperLedger Fabric, hereinafter fabric for short, a permissioned blockchain. Fabric is used to offer auditability to the technical reports done by organizations over time.

The architecture is divided in two sections as shown in Fig. 1: the first section, performed by the Whistleblower; and the second section, conducted by the Recipient.

The Whistleblower section is divided in two stages: the registration stage (Stage 1 W), required to obtain group credentials, is outlined in Sect. 3.2; and the disclosure publication stage (Stage 2 W), explained in Sect. 3.3. In a production system, a minimum of $k$-people registrations to obtain group credentials must be enforced in order to achieve a $k$-anonymity on each disclosure.

The Recipient section (Stage 1–2 R) is composed of a registration stage, mandatory in order to be marked as recipient of anonymous disclosures, enabling Whistleblowers to have access to their certificate, is described in Sect. 3.2; and a read stage. In an ideal setting, every member of the fabric network will be registered as Recipient. In the remaining cases, in accordance with the draft of the Spanish law, at least the person in charge of the internal information system must be registered.

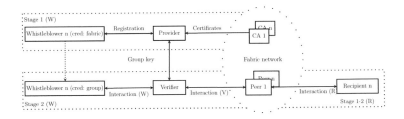

**Fig. 1.** Interaction between the components of the system

### 3.1    Components

The system is divided in five main components: smart contract, provider, verifier, whistleblower and recipient.

- **Smart contract:** Software that runs on a peer of the fabric network and offers an interface to interact with the blockchain. Among the functions available to Whistleblowers and Recipients are methods such as: *Subscribe*, allows Recipients to store their certificate in the blockchain; *GetPublications*, permits Recipients to view the list of anonymous disclosures; *PublishDisclosure*, enables Whistleblowers to publish anonymous disclosures; *GetSubscribers*, allows Whistleblowers to view the list of Recipients and their certificates. In addition, there are multiple methods to narrow down the information such as the date of disclosure, organization of the Recipient, etc. For an extended definition of the smart contract, see [4].
- **Provider (Issuer of group credentials):** Independent entity to fabric network and trustworthy. The Provider possesses the root certificates of the certification authorities in the fabric network and trust them. The Provider is responsible of offering group credentials to members of the fabric network, this requires that the Whistleblowers authenticate themselves against the Provider using their fabric identity in order to verify that the certificate has been issued by a trusted certification authority. It is worth mentioning that no personal information of the Whistleblowers is stored, only a digest of their certificates to avoid duplicated identities in the group. A thorough explanation of the registration process can be found in Sect. 3.2.
- **Verifier:** Permits the Whistleblowers interact with the fabric network (Interaction V) through the smart contract available in the peers, being possible to retrieve a list of Recipients and published disclosures. The Verifier must be managed by a trustworthy party belonging the fabric network. Disclosures are checked before publication: the signature of the disclosure must have been issued by a group member. In order to fulfill this attestation, the Verifier contacts the Provider to retrieve the public group key and ensures that the signature has been indeed issued by a group member and is valid.
- **Whistleblower:** Discloses confidential information in a totally anonymous way. In order for a disclosure to be published, Whistleblowers must have a valid group identity. Publication details are explained in Sect. 3.3.
- **Recipient:** Stores their certificate in the blockchain through the smart contract in the fabric network, allowing Whistleblowers to access them. In addition, Recipient can read disclosures directed towards them, reversing the process of the Whistleblower. A valid fabric identity is required to decrypt a disclosure.

## 3.2 Registration

**Recipient** Every member of the fabric network can register as Recipient (Stage 1 R), authorizing anonymous disclosures that can be read using the smart contract installed on the peers of the fabric network. The registration process consists in storing the Recipient certificate in the blockchain, granting Whistleblowers access to their certificate and be the target of disclosures.

**Whistleblower** The registration protocol (Stage 1 W) uses as essential basis the scheme PS16 implemented in `libgroupsig` [7] library. It consists of four steps, described in Fig. 2. The registration requires the exchange of three messages between Whistleblower and Provider—responsible of issuing group credentials. It is worth mentioning that the registration process is carried out only between Whistleblower and Provider, there is no interaction with third parties neither the fabric network—except at launch, when root certificates of certification authorities are collected.

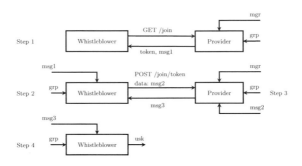

**Fig. 2.** Process to obtain a group credential

Registration begins with a TLS connection where both parties are mutually identified with a certificate: the Provider with a certificate issued by a trusted certification authority and the Whistleblower with his fabric certificate.

- **Step 1**: A Whistleblower interested in registering sends an HTTP GET request against the endpoint of the Provider, authenticating using their fabric certificate. The Provider generates a token, based on UUIDv4 in accordance with RFC4122, assigned to the digest of the certificate. In addition, the Provider generates a one-time random number as a challenge, using the group key ($grp$) to get the appropriate range. The Provider sends the token and the challenge ($msg1$) back to the Whistleblower.
- **Step 2**: The Whistleblower responds to the challenge, generating a private member key (if he have not done before) and using it to compute a zero-knowledge signature [2] of said private key on the challenge received. The Whistleblower sends an HTTP POST request against the endpoint and the token received in the previous step, including the response to the challenge ($msg2$).
- **Step 3**: The Provider verifies that the zero-knowledge signature received is a valid response to the challenge. If so, reuses the internal structure of the zero-knowledge signature to produce a member credential for the Whistleblower, which essentially consists of a PS16 signature blindly issued on the private key of the Whistleblower, using their group manager key ($mgr$). Said signature is sent back to the Whistleblower ($msg3$).

- **Step 4**: The Whistleblower verifies that the credential received is a valid PS16 signature, associated with the group key. From that point onwards, the Whistleblower will be able to use their member key ($usk$) to generate anonymous signatures on behalf of the group.

## 3.3  Publication

The disclosure publication process (Stage 2 R), begins with the gathering of Recipients identifiers. To that end, Whistleblower sends an HTTP GET request to the Verifier which responds with a list of identifiers belonging to Recipients interested in receiving anonymous disclosures. Immediately, Whistleblower sends an HTTP POST request, using the Recipient identifier as the body of the request, to which Verifier responds with the Recipient certificate. Finally, Whistleblower forges an ephemeral key pair ($eph\_ec$) using elliptic curve cryptography—more specifically SECP256R1—in order to derive the symmetric key to encrypt the disclosure.

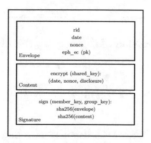

**Fig. 3.** Schema of the payload sent to the Verifier

Figure 3 shows the schema of the payload sent to the Verifier on disclosure publishing. The payload is composed of three parts: envelope, content and signature.

- **Envelope:** Includes Recipient identifier (rid), date of disclosure publication, a nonce and the public key of the ephemeral key pair created previously, all encoded in BASE64, as described in (1).

$$Base64(rid, date, nonce, eph\_ec_{pk}) \tag{1}$$

- **Content:** Includes the date, nonce and plain text of the disclosure. The content is encrypted using Fernet [11], which uses AES with a 128-bit key in CBC mode. The symmetric key to encrypt and decrypt the content is obtained using Diffie-Hellman applied to elliptic curves and a derivation function based on SHA256. In the encryption stage, described in (2), the key is calculated using the ephemeral secret key and the public key of the Recipient.

$$K_{enc} = kdf(SHA_{256}, eph\_ec_{sk} \times rec_{pk}) \tag{2}$$

In the decryption stage, detailed in (3), the key is calculated using the private key of the Recipient and the ephemeral public key found in the envelope.

$$K_{dec} = kdf(SHA_{256}, eph\_ec_{pk} \times rec_{sk}) \tag{3}$$

– **Signature:** Includes the group signature of the Envelope hash and Content hash, as shown in (4).

$$sign(usk, grp, (SHA_{256}(sobre), SHA_{256}(contenido))). \tag{4}$$

## 3.4   Blockchain

Figure 4 represents the format of Recipients and disclosures stored in the blockchain.

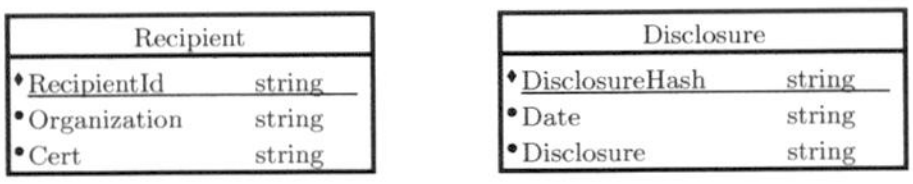

| Recipient | |
|---|---|
| ◆ RecipientId | string |
| ● Organization | string |
| ● Cert | string |

| Disclosure | |
|---|---|
| ◆ DisclosureHash | string |
| ● Date | string |
| ● Disclosure | string |

**Fig. 4.** Schema of the data in the blockchain

Recipients are defined by three elements: recipientId (rid), organization and cert.

– **RecipientId:** Unique identifier following the structure $x509{:}subject{:}issuer$, where *subject* and *issuer* match their homonymous in the Recipient certificate following RFC4514 format. Identifier is stored encoded in BASE64.
– **Organization:** Identifier of the MSP (*Membership Service Provider*) that manages the identity of the Recipient.
– **Cert:** Recipient certificate encoded in BASE64.

Same as Recipients, disclosures comprise three elements: disclosureHash, date and disclosure.

– **DisclosureHash:** Hash of the disclosure, using SHA256.
– **Date:** Date the disclosure publication, following the format $year{-}month{-}day$.
– **Disclosure:** Composed by envelope and content encoded in BASE64.

## 4   Conclusions

In this article we propose a system that provides anonymity to disclosures in an open and auditable system, based on Hyperledger Fabric, a permissioned blockchain. Our system provides confidentiality to the disclosures, so that only the Recipients are able to decrypt them. Although public disclosures could be easily supported.

A second goal is that only the Whistleblowers are able to revoke their anonymity, although this cannot be guaranteed cryptographically with PS16, so we take for granted that the Provider will not to store any member information on registration. In future works we will replace PS16 with DL21 [8], which is also available in `libgroupsig`. The scheme DL21, of type User-Controlled Linkability (UCL), offers total control to the user over which of their group signatures will be traceable due the lack of an Opener authority. At the same time, enables the Whistleblower to claim *a posteriori* credit of a disclosure. For instance, a Whistleblower creates a signature non-linkable to others they have previously generated and claims authorship one month later. Furthermore, our anonymity revocation process improves the complexity of similar works [13].

The use of elliptic curves over RSA provides a series of advantages such as: smaller key size with equivalent security to a larger one in RSA, requiring less processing power and thus allowing generation of robust ephemeral keys for each submission without notable drawbacks; or direct compatibility with fabric identities that use the same key algorithm.

The use of group signatures guarantee the anonymity of Whistleblowers in the anonymous set defined by the group they belong to. The Whistleblower only needs to prove their identity when they join the group. Ulterior actions are not linkable to the registration process (i.e., anonymity), however, it is guaranteed that only those who have followed the registration process can generate valid group signatures (i.e., traceability). Lastly, it is not possible to create a group signature to incriminate another group member (i.e., non-frameability), being the Whistleblower the only authorized entity to reveal the group signature authorship.

As improvements, we plan to suppress Recipients certificates and disclosures (envelope and content) from the blockchain due to unnecessary workload to the system as usage increases (i.e., scalability). We are replacing these items by a reference to an external storage system and a hash to the referenced element to ensure immutability. We also plan to support later communications between the Whistleblower and Recipient, taking advantage of the symmetric key they already share.

We have a prototype ready and our next steps aim to show the features of the system and make a quantitative comparison with other proposals in the literature. Then we will release the code of the prototype, as we made with AUTOAUDITOR [4].

# References

1. Camenisch, J., Lysyanskaya, A.: A signature scheme with efficient protocols. In: Cimato, S., Persiano, G., Galdi, C. (eds.) Security in Communication Networks, pp. 268–289. Springer, Berlin, Heidelberg (2003). https://doi.org/10.1007/3-540-36413-7_20
2. Chase, M., Lysyanskaya, A.: On signatures of knowledge. In: Dwork, C. (ed.) Advances in Cryptology—CRYPTO 2006, pp. 78–96. Springer, Berlin, Heidelberg (2006). https://doi.org/10.1007/11818175_5

3. Chaum, D., van Heyst, E.: Group signatures. In: Davies, D.W. (ed.) Advances in Cryptology—EUROCRYPT'91, pp. 257–265. Springer, Berlin, Heidelberg (1991)

4. Chica-Manjarrez, S.: Autoauditor: Semiautomatic vulnerabilities auditor using docker containers (2022). https://gitlab.gast.it.uc3m.es/schica/autoauditor

5. Chica-Manjarrez, S., Marín-López, A., Díaz-Sánchez, D., Almenares-Mendoza, F.: On the automation of auditing in power grid companies. In: Ambient Intelligence and Smart Environments, vol. 28, pp. 331–340 (2020). https://doi.org/10.3233/AISE200057

6. Council of European Union: Directive (EU) 2019/1937 of the European parliament and of the council of 23 October 2019 on the protection of persons who report breaches of union law (2019). https://eur-lex.europa.eu/eli/dir/2019/1937/oj

7. Diaz, J., Arroyo, D., Ortiz, F.: Libgroupsig: an extensible c library for group signatures. IACR Cryptol. ePrint Arch. **2015**, 1146 (2015)

8. Diaz, J., Lehmann, A.: Group signatures with user-controlled and sequential linkability. In: Garay, J.A. (ed.) Public-Key Cryptography—PKC 2021, pp. 360–388. Springer International Publishing, Cham (2021). https://doi.org/10.1007/978-3-030-75245-3_14

9. Habbabeh, A., Asprion, P., Schneider, B.: Mitigating the risks of whistleblowing; an approach using distributed system technologies. Ph.D. thesis (2020)

10. Marín-López, A., Chica-Manjarrez, S., Arroyo, D., Almenares-Mendoza, F., Díaz-Sánchez, D.: Security information sharing in smart grids: persisting security audits to the blockchain. Electronics **9**(11) (2020). https://doi.org/10.3390/electronics9111865. www.mdpi.com/2079-9292/9/11/1865

11. Parnell, K., Giménez, H., McDermott, K.: Fernet: delicious HMAC digest(if) authentication and AES-128-CBC encryption (2022). https://github.com/fernet/spec

12. Pointcheval, D., Sanders, O.: Short randomizable signatures. In: Sako, K. (ed.) Topics in Cryptology—CT-RSA 2016. Lecture Notes in Computer Science, vol. 9610, pp. 111–126. Springer, Cham (2016). https://doi.org/10.1007/978-3-319-29485-8_7

13. Tomaz, A.E.B., Nascimento, J.C., de Souza, J.N.: Blockchain-based whistleblowing service to solve the problem of journalistic conflict of interest. Ann. Telecommun., 1–18 (2021). https://doi.org/10.1007/s12243-021-00860-0

# Bottom-Up Trust Registry in Self Sovereign Identity

Kai Jun Eer[1]([✉]), Jesus Diaz[2], and Markulf Kohlweiss[1,2]

[1] University of Edinburgh, Edinburgh, UK
[2] Input Output Global, Madrid, Spain
{keeer,markulf.kohlweiss}@ed.ac.uk
jesus.diazvico@iohk.io

**Abstract.** Self sovereign identity is a form of decentralised credential management. During credential verification, data exchange only happens between the data owner and the verifier without passing through any third parties. While this approach offers a privacy-centric solution, it poses a challenge. How do verifiers trust that the credential is vouched by a trusted source? More specifically, how do verifiers know that the issuer has the reputation or is authorised to issue the credential? In this paper, we propose a trust registry design that handles the aspect of human trust in self sovereign identity. We also introduce an incentivisation mechanism for the trust registry in order to motivate each stakeholder to participate actively and honestly.

**Keywords:** Self sovereign identity · Trust registry

## 1 Introduction

In recent years, self-sovereign identity (SSI) as a decentralised identity system has been gaining traction, with the Sovrin blockchain being one of the most notable projects in this area. It aims to return identity data control to the identity owner through cryptographic proofs rather than relying on external identity providers [15]. However, as there is no central authority regulating the system, SSI faces challenges in key recoverability and trust in identity issuers, which needs to be addressed before SSI can realise its potential.

The foundation of SSI is verifiable credentials which contain attributes associated with an individual, signed by an issuer. Each credential issuer has an associated decentralised identifier (DID), which is recorded on the public blockchain [12]. A DID is resolvable into a public key and a service endpoint where interactions with the issuer is communicated. Therefore, while it is cryptographically verifiable that a credential is issued by a specific public-private key pair, there is still a layer of human trust needed. For a verifiable credential to be trusted by a verifier, the verifier needs to trust the issuer of the credential. In an ecosystem of SSI, there could be many valid identity issuers, however the

© The Author(s): under exclusive license to Springer Nature Switzerland AG 2023
J. Prieto et al. (Eds.): BLOCKCHAIN 2022, LNNS 595, pp. 423–433, 2023.
https://doi.org/10.1007/978-3-031-21229-1_39

verifier might not recognise all of them. A trust framework that governs the entities which are authorised to issue credentials is critical in managing the layer of human trust.

## 1.1 Contribution

The need for a secure and privacy-preserving digital identity system will only increase in the coming years. In this paper, we first analyse current trust frameworks in self sovereign identity, and propose an alternative framework that is more decentralised in nature.

The main research question of this paper is that in a decentralised setting without a central authority, how to design a trust framework that regulates and incentivises the participation of an ecosystem of identity issuers, which can be trusted by identity verifiers?

Our main contributions are in two areas of self sovereign identity:

– propose a trust registry framework that regulates a web of trust of identity issuers which can be trusted by identity verifiers
– propose an incentivisation mechanism in order for the trust registry to be adopted by identity issuers and identity verifiers, while minimising the risk of misbehaviour.

## 2    Related Work

During identity verification, the verifier receives a credential presentation proving that the credential is signed by the issuer holding the corresponding DID. However, there might be many identity issuers for the same credential schema, and the identity verifiers should not need to constantly update its local list of all issuers' DIDs. A verifier might not trust all the issuers too. A trust framework is necessary to govern the human trust aspect in self sovereign identity.

### 2.1 Credential Chaining

One of the current approaches to SSI trust framework is to use verifiable credentials [5]. Each valid issuer in the ecosystem holds a verifiable credential that is issued by an issuer one level above it. For example, education institutions hold a credential issued by the government that gives them the rights to issue degrees. During verification, the verifier obtains credential proof presentation by the identity owner. If the verifier does not recognise this level-two credential issuer, it requests another proof presentation from the credential issuer [14]. The level-two credential issuer then presents its own credential issued by a level-one issuer. This approach is similar to the certificates authority structure [3] widely adopted in web server authentication.

The foreseeable practicality challenge of the approach is availability and privacy. To verify credentials, verifiers need to create a communication channel with the issuer(s) in order to climb up the hierarchy and verify credentials. This will only work if the issuers' agents are always available to prove their own identities. Apart from that, privacy of identity owners decreases as issuers can now correlate the services that the owners access. Issuers gain more control, moving SSI as a user-centric identity model to an issuer-centric model.

## 2.2   Centralised Trust Registry

A more centralised approach to address the issue of trust in issuers is to have a trust registry that is managed by a trusted authority [5]. The trusted authority determines which issuers can be listed in the registry. During credential verification, the verifiers query the trust registry, where the trusted authority returns whether the credential issuer is listed in the registry. One drawback of such an approach is that all the verifiers need to trust the single authority to govern the registry, which is often not possible nor desirable in many use cases. If the single authority has biased interest or malicious intent, the trustworthiness of the entire registry will collapse. While there could be multiple trusted authorities that govern the trust registry instead of one, trust is still dependent on a small set of entities.

## 3   Bottom-Up Trust Framework Design

We designed a bottom-up trust framework. There are four direct stakeholders for the trust framework, namely the issuers, verifiers, credential holders and maintainers. The issuers issue credentials to credential holders, who are then able to present these credentials to the verifiers. The maintainers are specific to our trust framework design, where they are tasked to maintain the privacy of the trust registry and facilitate the incentivisation mechanism.

From a high-level view, we use a directed graph of issuers to build a web of trust. Verifiers can compute their level of trust to a particular issuer by traversing the directed graph. To compute the level of trust, verifiers need to pay a fee to a smart contract governing the graph, where the fee will be distributed to issuers. This is to incentivise issuers to issue credentials and form edges with other issuers in the web of trust. An overall protocol architecture is shown in Fig. 1.

### 3.1   Trust Registry Design

For the bottom-up trust registry, we designed a directed graph that connects identity issuers with other issuers that they trust. An example graph is shown in Fig. 2. The graph can be represented as $G = (V, E)$ where V is a set of identity issuers represented by their DIDs, while E is a set of directional edges connecting the issuers:

- $V = \{v_1, v_2, \ldots, v_n\}$ where n is the number of issuers listed in the registry.
- $E = \{e_{ij} \ldots, \}$ where $e_{ij}$ is the edge connecting from $v_i$ to $v_j$ and carries a weight of between 0 and 1.

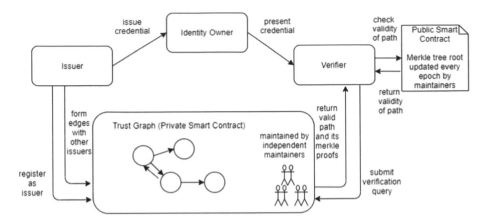

**Fig. 1.** System architecture that shows interaction of trust registry with different participants. The trust graph is kept private by utilising a private blockchain maintained by independent maintainers, in order to preserve privacy of issuers and set up an incentivisation mechanism.

An edge of $e_{ij}$ means that issuer $v_i$ trusts issuer $v_j$ by a weight, where higher weight indicates higher level of trust. The trust is strictly one-way, therefore $e_{ij}$ is not the same as $e_{ji}$. The value of the weight that is assigned is based on the risk that issuer $v_i$ is willing to take on behalf of issuer $v_j$, which is explained in more details in Sect. 4.2.

By implementing the graph as a smart contract, we promise the transparency and immutability of the graph as long as the majority of the blockchain validators are honest. Identity issuers first need to stake an amount of tokens to be listed in the public graph. This is to disincentivise malicious behaviours, where in the event of such behaviour, the involved issuer will lose its stake. Besides that, identity issuers also need to prove in control of the private key corresponding to their DIDs by creating digital signatures.

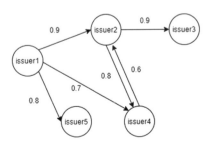

**Fig. 2.** An example trust graph that connects the identity issuers. Each edge has a weight which represents the trust score from the source issuer to the destination issuer.

## 3.2    Privacy Design

We suggest that the web of trust of identity issuers should be confidential. An identity issuer might want to keep the weight of an edge that it forms with other issuers private, for example, to avoid blackmailing. Therefore, the smart contract which forms the directed graph is deployed on a private and permissioned blockchain. For ease of reference, we will specify the private blockchain validators as **maintainers** going forward. The maintainers are independent third parties that help to maintain the privacy of the graph and transactions related to it.

As the graph is computed using a private blockchain run by a small set of maintainers, inevitably it becomes less decentralised. To improve on the trustworthiness, the maintainers have to publish a commitment on the state of the directed graph on a public smart contract at constant timestep. During verification, verifiers are able to compare the partial state returned by the maintainers with the commitment to verify the validity.

The commitment is in the form of a merkle tree root. The commitment is updated at a constant time interval, in which we name the time interval as epoch in this paper. At each epoch, the maintainers collect all the edges E formed in the graph and put them into a merkle tree data structure. Each edge is in the form of 'source-destination:score'. For example, 'issuer1-issuer2:0.9' means that issuer1 trusts issuer2 by a score of 0.9. Once the merkle tree is formed, the maintainers create a multi-signature on the merkle tree root and publish it in a public smart contract.

## 3.3    Private Blockchain Relay Server

The state of a private blockchain is only accessible by its validators, therefore anyone outside the set of validators does not have read and write permission to the blockchain. We introduce the use of relay servers as the bridge between external actors such as issuers and verifiers with the private blockchain. The relay servers are whitelisted as validators in the private blockchain. When external actors call a function in the private smart contract, they create a digital signature using their DID's private key on the function calls and send it to the relay servers. The relay servers then make the function call on their behalf to the private smart contract. This architecture is inspired by the Ethereum gas station network by OpenGSN [1].

# 4    Credential Verification

## 4.1    Traversing the Graph During Verification

Each verifier has its own list of trusted issuers. During verification, if the verifier receives a credential proof issued by an issuer it doesn't recognise, the verifier queries the private blockchain to derive the level of trust. The verifier specifies three arguments, which are its trusted issuers, the credential issuer and a threshold for the trust score.

Only maintainers have access to the directed graph. When verifiers make a verifcation request (through the relay server), the maintainers return the shortest path in the directed graph that starts from the trusted issuers specified by the verifiers and ends at the credential issuer that is above the threshold score. The path is calculated with a breadth first search algorithm. The trust score is calculated by multiplying the weights of the edges through the path. The reason for using mutiplication is to calculate the transitive trust score, as described in [10]. If the score is higher than the threshold score specified by the verifier, there is a valid path. If there are two or more paths that have the same path length, the path with the higher trust score is returned. If these paths have the same trust score, a path is randomly picked out of these paths to ensure fairness.

Recall that at every epoch, maintainers have to publish a merkle tree root commitment of the state of the graph. Once the maintainers have found a valid path, the maintainers generate a proof for each of the edges that forms the path. The maintainers return all the edges that form the path and their corresponding merkle tree proofs. The verifiers are able to check that these edges are accumulated in the merkle tree by verifying the proofs with the merkle tree root stored in the public smart contract. However, there could be privacy concern depending on use cases by simply revealing the edges that forms the path. We propose a alternative described in Sect. 6.

### 4.2 Economic Incentive Design for Trust Registry Participants

A common challenge for SSI is the lack of incentivisation for issuers. Unlike federated identity providers like Google who collect users behavourial data in exchange for providing free identity service, the privacy by design nature of SSI excludes issuers' involvement during the verification process, removing their source of benefits.

In our work, we propose that verifiers make a payment $P$ to the trust registry when making a verification query. This is logical as verifiers benefit from the SSI ecosystem, where they receive convenience and security when accepting credentials issued by trusted issuers. In traditional identity verification, verifiers usually have to pay the identity issuers directly too. Example of identity issuers are Onfido[1] and Jumio.[2] There are two criterias that we aim to achieve for a trustworthy trust registry:

- Issuers are incentivised to issue credentials
- Issuers are incentivised to form edges with other issuers in the trust graph, but only with issuers that they truly trust.

An issuer receives $0.5P$ for every verification query where it is the credential issuer, therefore they are incentivized to issue more credentials. When the maintainers return a shortest path in the graph that connects the verifiers' trusted issuers to the credential issuer, all issuers along the path receive an equal share

---

[1] https://onfido.com/.
[2] https://www.jumio.com/.

of the remaining $0.5P$ paid by the verifier. Thus, issuers are incentivised to form edges with other issuers, as it increases the chances that they are on a valid path during a verification query to the trust registry. Issuers are also incentivised to form an edge with a higher trust score to increase the probability that the path is a valid path with overall trust score higher than the threshold score submitted by the verifier.

For the trust registry to be trustworthy, we propose a staking mechanism for the issuers as the protocol governance, inspired by [2]. Issuers have to stake a minimum of $T$ tokens in order to be listed in the trust graph. A verifier that makes a verification query to the trust registry is able to challenge the credential issuer if it deems that the credential is issued maliciously. The maintainers then take a vote. If the challenge is successful, the credential issuer loses $T$ tokens, while other issuers that are connected directly adjacent to the issuer lose $T * s$ tokens based on the trust score $s$ that they form with the credential issuer. The stake tokens lost by the issuers are distributed to the verifier as compensation. With this mechanism, issuers are able to quantify the amount of risk they are willing to take on other issuers based on the trust score $s$ they assign [11].

When the issuers lose some stake tokens, its staking token balance in the trust registry might go below the minimum $T$ tokens required. These issuers need to top up the number of staking tokens to meet the minimum requirement, otherwise the maintainers remove the issuers from being a node in the trust graph. In such a scenario, the issuers will not earn any more rewards from future verification queries.

## 5   Implementation

We use a proof-of-authority (PoA) blockchain for the private blockchain, where each blockchain validator needs to have its identity verified before being allowed to participate. We set up an EVM private blockchain using the Hyperledger Besu framework to prototype our proposed design. The reason for using Hyperledger Besu is its capabilities to form a privacy group, where smart contracts in the privacy group and their corresponding transactions are protected from the public [8]. This private blockchain is used to maintain the state of the directed graph. In our work, we use four independent maintainers as the validators.

We use the Kovan blockchain (an Ethereum testnet) as the public blockchain. The smart contract Registry.sol which stores the directed graph is deployed on the private blockchain. The smart contract MerkleProof.sol which records the merkle tree root is deployed on Kovan and can be viewed on a block explorer such as Kovan Etherscan.[3]

## 6   Proposed Improvements

We explore the use of zero knowledge proofs [4] to reduce the trust needed on maintainers and limit the revealed information to the verifiers. Our current

---

[3] https://kovan.etherscan.io/address/0xc4516FD2317210a278da78F06f7F1CeBC495A4a8.

protocol relies on the maintainers to reveal a valid path and its corresponding merkle proofs when a verifier makes a verification query. The protocol also relies on the maintainers to distribute the verification fees to relevant issuers.

With the use of non-interactive zero knowledge (NIZK) proofs, maintainers are able to prove that they distribute the verification fee honestly. The proposed protocol is as follows, although we have not analysed this in details and is left as future work:

1. Maintainers keep a record of the balances of each issuer. The balances are stored in a merkle tree data structure, and the merkle tree root is recorded on a public smart contract.
2. The verifier pays the verification fee to a public smart contract, and subsequently makes a verification query to the maintainers, stating the trusted issuer, the credential issuer and the threshold trust score in order for the verification to be accepted.
3. The maintainers first check that the verifier has made the payment to the public smart contract. Then, the maintainers generate a NIZK proof to prove the existence of a valid path between the trusted issuer and credential issuer. The maintainers update the balances of issuers that are part of the valid path, giving rise to a new merkle tree. The NIZK proof also encapsulates that the balances of issuers are updated correctly. The maintainers update the new merkle tree root on the public smart contract and also publish the zero knowledge proof.
4. Time is divided into epochs, for example an epoch could be one month. During that epoch, steps 2 and 3 are repeated whenever a verifier makes a verification query.
5. At the end of the epoch, the maintainers reveal the balances of all the issuers. The correctness of these balances can be verified through the merkle tree root and the NIZK proofs generated previously.
6. Once the balances are revealed, issuers are able to claim their corresponding payments from the public smart contract.

The program used to generate the zero knowledge proof is described in Algorithm 1, where it first verifies that these edges are accumulated in the graph merkle tree, and lastly verifies that the balances of the issuers are updated correctly based on the edges.

With this proposed use of zero knowledge, we are able to reduce the trust on maintainers. If the maintainers coerce and update the issuers' balance maliciously, it is detectable as the correctness of the zero knowledge proof will be false. We maintain the privacy of the verification queries as the balances of the issuers are aggregated and only revealed at the end of an epoch, therefore it is difficult to correlate verifications to issuers in the directed graph, provided there is a large enough number of issuers in the graph [9]. The tradeoff to this approach is the reduction in scalability, as maintainers need to update the merkle tree root of the balances in the public smart contract for each verification query. Generating non-interactive zero knowledge proofs are also costly.

---

**Algorithm 1** Generate zero knowledge proof on existence of valid path and correct update of issuers' balance

---

**Require:** *edges, balances, proof$_{edges}$*                          ▷ private inputs
**Require:** *source, destination, root$_{graph}$, root$_{balance(previous)}$, root$_{balance(updated)}$,
threshold*                                                               ▷ public inputs
   $P_{cred} \leftarrow 0.5P$                          ▷ credential issuer gets half of the verification fee
   $P_{others} \leftarrow 0.5P/edges.length$  ▷ other issuers along the path gets the remaining half
   verify that *edges* is a valid path from *source* to *destination* and above *threshold*
   verify *balances* corresponds to *root$_{balance(previous)}$*
   **for** *edge* in *edges* **do**
       verify *edge* accumulated in *root$_{graph}$* by comparing against *proof$_{edges}$*
       *issuer* ← *edge.issuer*                          ▷ get the destination issuer in the edge
       **if** *issuer* is credential issuer **then**
           add $P_{cred}$ to its balance
       **else**
           add $P_{others}$ to its balance
       **end if**
   **end for**
   verify that updated *balances* corresponds to *root$_{balance(updated)}$*

---

# 7 Conclusion

Self sovereign identity (SSI) is an emerging field in this time where the world is undergoing rapid digital transformation. It offers a privacy-centric solution where users are always in control of their data, in terms of what data they consent to sharing and with whom. However, it is important to understand that human trust is still the underlying factor under the technology. Anyone is able to vouch for the correctness of a particular piece of data, however the trustworthiness of the vouched data depends on the reputation of the issuer. In a multi-party system where multiple issuers and verifiers exist, there is a need for a trust registry to manage the human trust layer of these different entities.

Our work introduces a bottom-up approach for a self sovereign identity trust registry using the web of trust model. For the trust registry to be adopted, we also introduce an incentivisation mechanism such that all direct stakeholders of the trust registry are incentivised to perform proactively and honestly. In comparison to the current common design and protocols for SSI trust registries, we are able to offer an alternative method, which can be useful in certain use cases where there is no single source of truth to verify the validity of the credential issuers.

## 7.1 Future Work

As mentioned in Sect. 6, the zero knowledge extension to our protocol is not fully experimented and evaluated yet. Besides that, our protocol currently relies on the maintainers to preserve the privacy of the directed graph. It is possible to use the technique of homomorphic encryption [13] in cryptography such that the directed graph is stored in an encrypted decentralised storage, and changes to the state of the graph can be made without revealing the graph itself. This makes it harder for a malicious maintainer to reveal the directed graph outside of the protocol, however the tradeoff is scalability. Other possible approaches worth exploring further include using secure multi party computation [6] to compute the trust score in the directed graph, removing the need for maintainers to answer verification queries. Related work is [7].

Our incentivisation mechanism relies on the verifiers paying a small fee to make a verification query. It is worth exploring different economic interactions such as the credential holder being the fee-paying entity. A game theory approach to validate the incentivisation mechanism will define clearer parameters too.

# References

1. Opengsn gas station network. https://docs.opengsn.org/#architecture. Accessed: 25 Feb 2022
2. Asgaonkar, A., Krishnamachari, B.: Token curated registries-a game theoretic approach. arXiv preprint arXiv:1809.01756 (2018)
3. Becker, G.: Merkle signature schemes, merkle trees and their cryptanalysis. Ruhr-University Bochum. Tech. Rep. **12**, 19 (2008)
4. Blum, M., Feldman, P., Micali, S.: Non-interactive zero-knowledge and its applications. In: Providing Sound Foundations for Cryptography: On the Work of Shafi Goldwasser and Silvio Micali, pp. 329–349 (2019)
5. Davie, M., Gisolfi, D., Hardman, D., Jordan, J., O'Donnell, D., Reed, D.: The trust over IP stack. IEEE Commun. Stand. Mag. **3**(4), 46–51 (2019)
6. Du, W., Zhan, Z.: A practical approach to solve secure multi-party computation problems. In: Proceedings of the 2002 Workshop on New Security Paradigms, pp. 127–135 (2002)
7. Dumas, J.G., Lafourcade, P., Orfila, J.B., Puys, M.: Dual protocols for private multi-party matrix multiplication and trust computations. Comput. Secur. **71**, 51–70 (2017)
8. Hasan, H.R., Salah, K., Jayaraman, R., Yaqoob, I., Omar, M.: Blockchain architectures for physical internet: a vision, features, requirements, and applications. IEEE Netw. **35**(2), 174–181 (2020)
9. Impagliazzo, R., Naor, M.: Efficient cryptographic schemes provably as secure as subset sum. J. Cryptol. **9**(4), 199–216 (1996). https://doi.org/10.1007/BF00189260
10. Jøsang, A., Gray, E., Kinateder, M.: Simplification and analysis of transitive trust networks. Web Intell. Agent Syst. Int. J. **4**(2), 139–161 (2006)
11. Litos, O.S.T., Zindros, D.: Trust is risk: a decentralized financial trust platform. In: International Conference on Financial Cryptography and Data Security, pp. 340–356. Springer (2017)
12. Mühle, A., Grüner, A., Gayvoronskaya, T., Meinel, C.: A survey on essential components of a self-sovereign identity. Comput. Sci. Rev. **30**, 80–86 (2018)
13. Naehrig, M., Lauter, K., Vaikuntanathan, V.: Can homomorphic encryption be practical? In: Proceedings of the 3rd ACM Workshop on Cloud Computing Security Workshop, pp. 113–124 (2011)

14. Preukschat, A., Reed, D.: Self-Sovereign Identity. Manning Publications (2021)
15. Tobin, A., Reed, D.: The inevitable rise of self-sovereign identity. The Sovrin Foundation **29** (2016)

# Integrating Web3 Features into Moodle

Urban Vidovič[(✉)], Vid Keršič, and Muhamed Turkanović

Faculty of Electrical Engineering and Computer Science,
University of Maribor, 2000 Maribor, Slovenia
{urban.vidovic2,vid.kersic,muhamed.turkanovic}@um.si

**Abstract.** While many innovations and possibilities of the world wide web were created, there is still a lot of unrealized potential due to the limitations of the current structure of the web, which is called web2. To solve current problems of the internet and bootstrap the new wave of innovation, the web3 started to evolve. Based on the distributed systems, most often on blockchain technology, the new use cases exploit advantages of the technology, such as data interoperability, decentralization, and self-sovereignty. In this paper, we present the approach of integrating new features and benefits of web3 to the existing web2 educational platform Moodle, which removes the need to recreate the whole platforms that take advantage of web3 from scratch. Furthermore, we present the evaluation of our approach and solution on the real world use-case.

**Keywords:** Web3 · Web2 · Blockchain · Smart contracts · Moodle · Plugin

## 1 Introduction

With the advancement of blockchain and distributed ledger technologies, new layers are added to the internet stack, such as identity, and store of value. Nevertheless, all these come with a cost. Because of the fast development and new paradigms, existing platforms cannot keep pace. Therefore, new platforms that introduce new technologies are built from scratch. However, this often results in double work. Thus the question arises: can new technological features be integrated into current platforms, and if yes, in what form?

Most of today's platforms belong to the web2 [16]. This includes websites and services where the user participates in creating and storing data - therefore, the synonym participative social web. However, a new paradigm is evolving, known as web3 [11]. It bases on blockchain technology - decentralized distributed ledgers [14]. web3 solves many problems of web2, one of the most significant being data silos. Data is locked in the existing platform and is most of the time not transferable to other platforms. This happens because the data is stored on platforms' servers and databases. web3 enables users to bring data to different places. It is stored on a shared distributed ledger, which makes platforms' data interoperable.

© The Author(s): under exclusive license to Springer Nature Switzerland AG 2023
J. Prieto et al. (Eds.): BLOCKCHAIN 2022, LNNS 595, pp. 434–443, 2023.
https://doi.org/10.1007/978-3-031-21229-1_40

With the main question being how to bring web3 functionalities to the existing web2 platforms, we propose the approach of integrating web3 functionalities into the web2 educational platform Moodle. The problems with web2 are central servers and databases and the platforms not being interoperable. They store user data in a centralized and use-case-specific fashion. Our goal was to solve the aforementioned challenge by proposing a web3 integration and designing a plugin that can be integrated into a Moodle. By defining, prototyping, deploying, and testing the plugin, we showed that web3 functionalities can be successfully integrated, elevating the level of interoperability.

The structure of the paper is as follows: Sect. 2 provides an overview of related work, Sect. 3 describes the technologies, Sect. 4 introduces the proposed solution, Sect. 5 evaluates the solution, and Sect. 6 provides the discussion. Section 7 concludes the paper.

## 2 Related Work

Over the last years, with the fast advance of web3 technologies and broader public awareness of their benefits, several solutions were proposed to utilize their advantages in existing web2 platforms. Platforms were enhanced for different purposes: standard and multi-factor authentication, increasing data privacy, self-sovereignty of users' data, payment methods.

Hyouk Lee proposes a BIDaaS (Blockchain-based Identity as a Service) system consisting of three parties: BIDaaS provider, partner, and user. The core contribution of the paper is the use of distributed ledgers to store users' credentials which ensures that the user data stays in the user's possession. This presents an excellent way to use the proposed service as SSO (Single Sign-On) [13]. João Antonio Aparecido Cardoso et al. proposed an integration of web3 Multi-Factor Authentication named Hydro Raindrop into websites created with WordPress. They focused on MFA (Multi-Factor Authentication), which provides an extra layer of security to an existing authentication system [8].

Makalesi and Karatas integrated a blockchain-based digital certificate platform into Moodle [15]. The certificate is registered and issued on the blockchain. The issuer is always a "hardcoded" certification authority, and the teacher has no control over when, how, and what certificate is issued [15].

Economides and Perifanou propose establishing a blockchain network that allows the use of so-called Smart CV (Curriculum Vitae), which store courses a user enrolls in, the projects a user is working on and final certificates [9]. The network would also enable validation of those immutable certificates, which a certification authority must issue. The proposal of this study in the current state presents an architecture, which could be integrated into, e.g., Moodle or Canvas, to bring additional functionalities to the existing platforms and cites the study which our web3 integration bases on [9].

This paper builds on the platform defined in the study EduCTX: A Blockchain-Based Higher Education Credit Platform [17]. EduCTX presents a blockchain and smart contract-based digital platform for managing micro-credentials issued as non-fungible tokens (NFTs) to students. Everything is stored and encrypted on-chain. The architecture consists of smart contracts implementing functionalities such as issuing, receiving, revoking, and validating certificates while keeping a hierarchy of certification authorities (University), authorized personnel (teachers), and receivers (students). The underlying network of the architecture is a consortium or permissioned Hyperledger Besu network. The research paper also presents a decentralized web3 application, enabling users with a suitable web2-like graphical interface. The communication with the dApp is handled using crypto wallets like MetaMask. With the extensible nature of the architecture, we built on top of it to make all functionalities available without leaving Moodle [3, 17].

## 3    Preliminaries

### 3.1    Web2

The web we know today is referred to as web2 or participating social web. It is the second iteration of the world wide web. Its predecessor, web1, consisted of static, non-interactive pages, data was stored in web files, and the websites were read-only. Web2 upgraded those aspects with dynamic interactive websites and separate databases. It allows users to participate in creating, storing, and retrieving the data. Websites are focused on the end-user, with the most popular categories being social networks, podcasts, RSS, and blogs. Examples of social networks are Facebook, Instagram, and Twitter. Some systems allow upgrades by allowing the implementation of third-party plugins, with the best examples being Learning Management Systems (LMS), such as Moodle, and Content Management Systems (CMS), such as WordPress [11].

**LMS and Moodle** Learning Management System (LMS) is a web-based application/service to elevate and enhance the learning process. It usually features functionalities such as giving a teacher the ability to upload and organize course material, create tasks with submission fields where the students upload their documents and files, create quizzes for students, monitor their attendance, and grade students' performance. It can also provide video conferences, forums, and newsletters [6].

Moodle is a highly customizable open-source LMS. It is the most popular and easy-to-use open-source self-hosted LMS. Moodle specifically has all of the above features with the ability to install custom plugins developed by a third party [5, 6].

### 3.2    Web3

web3 presents a new generation of the internet we know today. It originated from the introduction of blockchain technology and promotes principles such as decentralization, self-sovereignty, immutability, privacy, transparency, and at the same time, anonymity. The current internet is mostly populated with a group of a few big tech companies which provide us services in exchange for our data. Those companies have the power and represent a central authority, thus have the ability to exclude participants on their will. The main idea of web3 is to reduce the influence of such large corporations [11].

Currently, numerous blockchain networks exist, with the first one being Bitcoin [2], but due to its lack of support for smart contracts, it does not bring many capabilities of developing additional applications on top of it and is nowadays seen more as a store of value. At the time of writing this article, Ethereum [7] is the most significant blockchain in terms of ecosystem and was the first one to bring smart contracts to its network. Blockchains aspire to be Turing-complete and permissionless, allowing anyone to connect to the network and use or even develop applications on top of it [14]. For example, Khoury et al. proposed and implemented a voting mechanism where Ethereum acts as a source of truth [12]. Today, there are many decentralized apps (dApps) on different blockchain networks. They store their data on-chain and can be developed by writing smart contracts in supported programming languages such as Solidity [7].

## 4   Web3 Integration

This section presents our approach of integrating web3 components into a web2 educational platform. The demonstration and validation of the integration were done with a Moodle (web2) platform and the EduCTX platform with its permissioned blockchain network (web3). In this section, we present our proposed solution for the integration, its architecture and flow diagrams, and showcase the plugin in action.

The integration was done through a Moodle plugin functionality. The plugin serves as a user interface for teachers and students. It is a bridge between the client, Moodle server, and, in our case, the EduCTX blockchain network. For the plugin to be used, users need to have a MetaMask browser extension installed. The plugin utilizes MetaMask to connect to the EduCTX network and allow teachers to issue certificates and students to view their received certificates. EduCTX plugin can be added in the same way as any other existing plugin. In Moodle's terms, such plugins are referred to as Activity Modules.

### 4.1   Architecture

The plugin builds on top of the EduCTX network architecture. It uses Moodle's database to store the user's EduCTX ID, which is essential to correctly link EduCTX EVM accounts to Moodle users' accounts. It also enables Moodle specific features such as filtering the received certificates based on the currently selected course and enabling teachers to store certificate templates to speed up the process of bulk issuing the certificates. The architecture of the whole integration, which consists of two main components, can be seen in Fig. 1. The architecture is presented in with the C4 model for visualising software architecture, specifically the system context diagram.

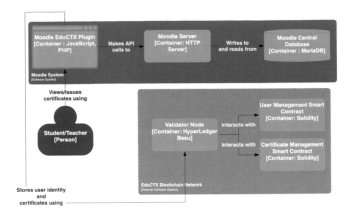

**Fig. 1.** The C4 based architecture system context diagram.

The first component is the EduCTX network - a permissioned blockchain where the smart contract-based business logic of the EduCTX platform is stored and executed. Smart contracts store the certificates and users' identities and present the "back-end" of the system.

The second component is the Moodle system. It consists of an Apache HTTP server, MariaDB database, and our proposed EduCTX plugin. Most actions taken by the user in the plugin are executed on the client-side. The Apache HTTP server hosts the Moodle platform and communicates with MariaDB. The figure also shows no communication between the server and the EduCTX network to emphasize the importance of correctly integrating web3 features.

### 4.2   Workflows

The use case consists of three main workflows - teacher's, student's and verifier's. We present only the first two in detail since these are connected to the Moodle plugin. A prerequisite for the usage of the system is that the teachers and students have a registered Moodle account (with the teacher registered as an authorized person) and a MetaMask wallet installed and set up (with the accounts created). The verifier's part is validation of all exported certificates and can be carried out on the EduCTX official platform [3].

**Teacher's Workflow** When a teacher lands on the plugin's page, the plugin fetches enrolled users in the current course. Everything is executed either on the client-side or the EduCTX network (smart contracts) from this point on. The teacher connects their MetaMask wallet to the plugin and issues the certificate. At this point, the plugin fetches the student's public key from EduCTX, encrypts and calculates the certificate's hash, and starts a transaction to store the encrypted certificate and its hash on the EduCTX network. By doing that, we achieve the availability of the certificate at any time while preserving privacy and authenticity. It is essential that the certificate encryption is carried out entirely on the client-side and that no critical data leaves the teacher's device exposed. The certificate can only be decrypted by the receiver using their private key. The entire workflow is shown in Fig. 2.

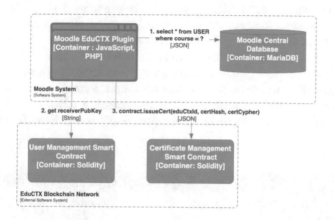

**Fig. 2.** Teacher's workflow when issuing a certificate.

**Student's Workflow** Students can only receive the certificates, meaning their only functionality is viewing and examining their digital credentials. When a student lands on the plugin's page, the initial part of the workflow proceeds in the same way as with the teacher. The difference comes when the MetaMask successfully establishes the connection with the plugin. Instead of the issue form, a field expecting a private key shows here. The certificates are stored and encrypted on-chain. To decrypt the certificates, the user needs to export their private key. The decryption is carried out entirely on the client-side, so the private key does not leave the browser and is deleted after attempted decryption. During this step, the certificates are fetched from EduCTX. The student then exports the private key from the MetaMask wallet and fills it into the input field. All certificates are decrypted and presented in a table upon clicking on the Decrypt button. If a student wants to export any of the certificates, he can do so on the main EduCTX platform [3].

**Verifier's Workflow** In contrast to the workflows mentioned beforehand, this one does not require the usage of our plugin since EduCTX aims to enable anyone to verify certificates, even if they do not have an EduCTX account, blockchain address, or are part of the Moodle platform. The advantage here is making validation of certificates available to everyone without additional requirements. Anyone can visit the platform [3] and verify any EduCTX-based credential without the need of having a wallet set up.

### 4.3  Prototype

As mentioned before, the plugin is written in PHP and JavaScript. The plugin's file structure is as suggested by Moodle documentation. It can be archived into a ZIP file and distributed to anyone to install it into their own Moodle server, which can be done by uploading the ZIP via the site administration page or by extracting it into correct folders. Moodle has predefined styles and templates which are to be used to ensure the same design and style throughout the plugin for the better user experience. Moodle developers prepared an API that simplifies the use of basic logic and database queries to conform to their desired standard. The better part of the plugin is written in JavaScript because of the client-side execution that allows us to omit the communication between the client and Moodle's server. Figure 3 shows us the form presented when a teacher wants to issue a certificate. Our plugin is open-sourced on Blockchain Lab:UM official GitHub.[1]

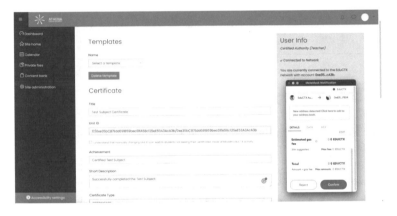

**Fig. 3.** EduCTX Moodle plugin issue certificate form.

## 5  Evaluation

The evaluation of the proposed web3 plugin was performed within the ecosystem of the ATHENA European university project, whereby Moodle is ATHENA's LMS of choice. ATHENA [10] is an Erasmus+ project connecting seven universities from seven European countries with the primary goal to bring the universities into a collaborative ecosystem and provide a higher quality of education with a positive impact on research, employability, and social advancement, as well as to promote interoperability between universities. ATHENA already hosts its own multi-university Moodle [1], currently in a test phase. Our plugin has

---

[1] Accessible on https://github.com/blockchain-lab-um/moodle-eductx-plugin.

been deployed and is being tested on ATHENA's LMS. The aim of the plugin in combination with EduCTX is to provide the ATHENA stakeholders an easier way of sharing and proving students' course completion, with the emphasis on the cross-border interactions.

As an example scenario, we introduce the student Bob from the University A (Slovenia), who goes on an ERASMUS+ exchange to the University B (Germany) for one semester. He finishes the subjects in University B and now needs to prove and present this information to the University A. Traditionally, the student would need signatures and written confirmations from the University B, confirming that he completed the subjects. With EduCTX, the communication between the University A and the University B and all the additional proofs are redundant. Upon the student completing the subject, the teacher, an authorized person on EduCTX, issues him a certificate that can be exported in PDF and JSON format. The student presents the exported certificate to his home university in an electronic fashion, which verifies the authenticity of the certificate via the EduCTX platform [3] by verifying the digital signatures and gets proof of the student's subject completion. As such, this approach follows the logic of micro-credentials [4].

By integrating the web3 functionalities into an existing LMS, the involved stakeholders (i.e., universities, students, academic staff) had an easier verification curve. On the contrary, the stakeholders would need to use two different platforms additionally to the required blockchain ecosystem components (e.g., Metamask wallet) to fulfill the tasks. The plugin is currently being incorporated into ATHENA's Moodle in a pre-production-ready environment.

## 6   Discussion

The main goal of our work was to analyze the possibilities of providing entities in the educational sphere, who use a Moodle-like LMS (web2) to use blockchain-related or web3 features seamlessly. Our approach focused on a case with such a requirement - i.e., being a multi-university case. Our contribution was the design, development and evaluation of a web3 plugin that can be integrated into a Moodle platform, thus easing the usage of new web3 features. By this we managed to prove the possibility to use web3 features seamlessly in legacy systems. The plugin development has turned out to be more complex than first expected. Since the blockchain architecture and necessary smart contracts have already been established and deployed, there have been various other issues, including third-party libraries needed for proper functionalities.

The main advantage is the potentially enhanced interoperability between multiple web2 education systems and the ability to export and show the certificates that can be verified without a third party. There is no need to build web2 platforms from scratch to support web3 functionalities. Web3 plugins can be used and integrated to enhance existing platforms.

There are also disadvantages to our proposal, which are present due to the requirements and limitations of the web3 ecosystem. One of them is the need for all the users to use a blockchain wallet, as it was the MetaMask in our case, which can be overwhelming for the new users of web3. With this in mind, each state-changing transaction with the web3 has to be manually signed by the user using the wallet, which could be hard to understand for many users. Other disadvantages come from the fact that blockchains can be either public or private, whereas public ones make them even harder to use due to transaction fees the users need to pay. However, many existing business ecosystems use their private, consortium-like blockchains, which omit the need to pay for transactions and the need to have a costly consensus mechanism in place (as it was with our evaluation case), but the users must fully trust the consortium running the blockchain network. In the future we plan to generalize our approach and the plugin in order for it to support more LMS systems or web2 platforms. We also plan to design and develop so-called mobile-based digital wallets, which will enable users to store their data either in the NFT format (EduCTX) or in the format of Verifiable Credentials (VC).

## 7   Conclusion

This paper proposes a modular approach for integrating web3-specific modules and features to existing web2 educational platform. We evaluated the method on an existing and widely used LMS Moodle. With the plugin integrated, teachers can issue platform-independent certificates to students, who can use these certificates anywhere without being locked to the Moodle ecosystem. The solution is based on blockchain technology, which serves as a public key infrastructure, coordination platform and the single source of truth.

While the solution proved beneficial and advantageous, many open questions could be considered in future work. The first is how to generalize the process of creating the plugin for many different platforms and therefore implementing a completely platform-independent plugin, whereas MetaMask Snaps could be the right direction. The second thing is how to streamline the process of issuing certificates and remove web3-specific interactions, e.g., signing transactions with MetaMask, as much as possible to come closer to web2 platform interactions. The last thing that could be improved is data sovereignty and giving data directly to users' wallets, which goes toward self-sovereign identity (SSI) principles.

**Acknowledgments.** This work was funded by the Slovenian Research Agency (research core funding No. P2-0057) and by the Erasmus+ KA2 European Universities Project ATHENA G. A. No. 101004096.

## References

1. Athena project. https://www.eu-athena.eu/moodle/. Last accessed 26 Apr 2022
2. Bitcoin: a peer-to-peer electronic cash system. https://bitcoin.org/bitcoin.pdf. Last accessed 26 Apr 2022
3. Eductx platform. https://platform.eductx.org. Last accessed, 26 Apr 2022

4. A European approach to micro-credentials — European education area. https://education.ec.europa.eu/education-levels/higher-education/micro-credentials. Last accessed 17 Jun 2022

5. Online learning with the world's most popular LMS—Moodle. https://moodle.com/. Last accessed 26 Apr 2022

6. Alias, A., Zainuddin, A.M.: Innovation for better teaching and learning: adopting the learning management system. Malays. Online J. Instr. Technol. **2**, 27–40 (2005)

7. Buterin, V.: Ethereum: a next-generation smart contract and decentralized application platform (2014)

8. Cardoso, J.A.A., Ishizu, F.T., de Lima, J.T., de Souza Pinto, J.: Blockchain based MFA solution: the use of hydro raindrop MFA for information security on wordpress websites. Braz. J. Oper. Product. Manag. **16**, 281–293 (2019). https://doi.org/10.14488/BJOPM.2019.v16.n2.a9. https://bjopm.emnuvens.com.br/bjopm/article/view/791

9. Economides, A.A., Perifanou, M.: Smart cv for lifelong qualifications certification based on blockchain. In: Proceedings of the International Conference on Education and New Developments (END Conference) 2021. https://doi.org/10.36315/2021end122

10. Escudeiro, N., Escudeiro, P., Almeida, R., Matos, P.: The ATHENA European university model for sustainable education: mainstreaming good practices for all-inclusive life-long sustainable learning in the digital era. In: ACM International Conference Proceeding Series (2020). https://doi.org/10.1145/3434780.3436682

11. Khan, A.G., Zahid, A.H., Hussain, M., Farooq, M., Riaz, U., Alam, T.M.: A journey of web and blockchain towards the industry 4.0: an overview. In: 3rd International Conference on Innovative Computing, ICIC 2019 (2019). https://doi.org/10.1109/ICIC48496.2019.8966700

12. Khoury, D., Kfoury, E.F., Kassem, A., Harb, H.: Decentralized voting platform based on Ethereum blockchain. In: 2018 IEEE International Multidisciplinary Conference on Engineering Technology, IMCET 2018 (2019). https://doi.org/10.1109/IMCET.2018.8603050

13. Lee, J.H.: Bidaas: Blockchain based ID as a service. IEEE Access **6**, 2274–2278 (2017). https://doi.org/10.1109/ACCESS.2017.2782733

14. Liu, Z., Xiang, Y., Shi, J., Gao, P., Wang, H., Xiao, X., Wen, B., Li, Q., Hu, Y.C.: Make web3.0 connected. IEEE Trans. Dependable Secure Comput. (2021). https://doi.org/10.1109/TDSC.2021.3079315

15. Makalesi, A., Karataş, E.: Developing Ethereum blockchain-based document verification smart contract for Moodle learning management system. J. Inf. Technol. **11**, 399–406 (2018). https://doi.org/10.17671/GAZIBTD.452686. https://dergipark.org.tr/en/pub/gazibtd/issue/40059/452686

16. Nath, K., Dhar, S., Basishtha, S.: Web 1.0 to web 3.0—evolution of the web and its various challenges. ICROIT 2014—Proceedings of the 2014 International Conference on Reliability, Optimization and Information Technology, pp. 86–89 (2014). https://doi.org/10.1109/ICROIT.2014.6798297

17. Turkanović, M., Hölbl, M., Košič, K., Heričko, M., Kamišalić, A.: Eductx: a blockchain-based higher education credit platform. IEEE Access **6**, 5112–5127 (2018). https://doi.org/10.1109/ACCESS.2018.2789929

# Doctoral Consortium

# Overview of Multiple User Encryption for Exchange of Private Data via Blockchains

Vanessa Carls[✉], Lambert Schmidt, and Marc Jansen

Institute of Computer Science, University of Applied Sciences Ruhr West,
46236 Bottrop, Germany
vanessa.carls@hs-ruhrwest.de

**Abstract.** In order to integrate the use of blockchain technologies into the daily routines of companies and institutions, it is essential to securely transmit the data stored and exchanged in these systems. Not only the security and reliability of the underlying system is important here, but also the encryption of the transferred information. Since there are usually more than two parties involved in transferring data over the blockchain in everyday applications, multi-party encryption is required. This paper introduces various encryption methods which can be used in the context of blockchain usage.

**Keywords:** Blockchain · Multiple user encryption · Information security

## 1 Introduction

The question of the security of data on the blockchain arises in particular when different parties want to exchange information using the distributed ledger technology, which is to be protected from access by third parties. Not only is this relevant when using public blockchains, it is also possible that information is to be kept secret from certain groups of the network when using private blockchains. There are various methods for encrypting this data, which can be used in different contexts and use cases. Some of these use cases can be found in the supply chains of companies, for example. In the case of order or delivery processes, it is sometimes necessary for companies, customers and suppliers to be able to view data, especially with regard to traceability, and multiple parties may be involved. Despite the transparency for all parties involved, the data should be protected from non-involved parties. This paper tries to answer the following question: Which types of multiple user encryption can be used in the context of storing and exchanging data via blockchains?

© The Author(s): under exclusive license to Springer Nature Switzerland AG 2023
J. Prieto et al. (Eds.): BLOCKCHAIN 2022, LNNS 595, pp. 447–453, 2023.
https://doi.org/10.1007/978-3-031-21229-1_41

## 2     State of the Art

Currently, hash methods and asymmetric encryption are frequently used to exchange private data via a blockchain. But the limitations of these methods are quickly reached, which makes it necessary to use other methods.

*Hashing:* Hashing is often used to store data compactly on blockchains and protect it from unauthorized access. Hashing generates a representation that maps a large input set to a smaller set. This smaller set is called a hash value. A frequently used procedure for hashing is the SHA256 function. The disadvantage of this method is that when using a hash function, the output value for a given input value is always the same and can not be decrypted again, since hashing is not an encryption method. So, when exchanging hash values, different parties can only find out the information sent if the original values are already known. This makes it nearly impossible to send unknown information [1].

*Asymmetric Encryption:* Another popular method for exchanging secret information is asymmetric encryption. Each participant has a private key and a public key, in other words, a pair of keys. The public key is used by the sender for encrypting messages and is publicly available. The private key of the receiver in contrast is used for decrypting the received message [9]. The classic public key encryption usually only works with one sender and one receiver if you consider sending single messages. In blockchain applications, however, it is often the case that individual messages are intended to be sent to several recipients, so that alternatives or extensions to known procedures are required here.

## 3     Multiple User Encryption and Key Exchange

This section explains various procedures for key exchange and encryption.

### 3.1     Diffie Hellman

The Diffie Hellman Key Exchange is known for sharing keys over unsecure channels. The classic approach (proposed in [2]) is used for exchanging keys between two parties, which are then used for symmetric encryption. For use with multiple users, this approach can be expanded as follows: Assuming that Alice, Bob, and Carol want to negotiate a key, a prime $p$ and a generator $g$ from the cyclic group $G = \{1, \ldots, p-1\}$ are selected. Every participant generates a private key $x_a$, $x_b$, $x_c$ and calculates their public key: $y_A = g^{x_A} \bmod p$, $y_B = g^{x_B} \bmod p$, $y_C = g^{x_C} \bmod p$. In the next step Alice sends $y_A$ to Bob, Bob sends $y_B$ to Carol and Carol sends $y_C$ to Alice. Now Alice, Bob and Carol correspondingly compute

$$z_{CA} = y_C^{x_A} \bmod p, \; z_{AB} = y_A^{x_B} \bmod p, \; z_{BC} = y_B^{x_C} \bmod p \qquad (1)$$

The results are again send to the "right neighbor" and the participants calculate the result to the power of their private key, that is

$$S_{BCA} = z_{BC}^{x_A} \bmod p, \ S_{CAB} = z_{CA}^{x_B} \bmod p, \ S_{ABC} = z_{AB}^{x_C} \bmod p \qquad (2)$$

In the end all participants share the common secret $S_{BCA} = S_{CAB} = S_{ABC}$. With this method, secrets can be exchanged between not only three but multiple parties by extending the principle to additional participants. The security of the Diffie Hellman method for key exchange is based on the Discrete Logarithm Problem. A problem that is already known in the classical approach is the man-in-the-middle-attack, which can also become a threat to this alternative.

### 3.2   El Gamal

Usually the El Gamal Cryptosystem (proposed in [4]) is used for a single sender and a single receiver similar to the Diffie Hellman Key Exchange. By expanding the classical approach, the following cryptosystem for encrypting a message by multiple senders results (proposed in [8]). The message then is decrypted by a single receiver. In this case, also a prime number $p$ and a number $g$ from the cyclic group $G = \{1, \ldots, p-1\}$ are first selected. The private key $a$ is also taken from this group. The public key consists of $\{p, g, b\}$, where $b$ is calculated as follows: $b = g^a \bmod p$. With a number of $n$ participants, $i \in \{1, 2, \ldots, n\}$ and the individual private keys $R_1, R_2, \ldots, R_n$, the following values are computed: $c_{1i} = g^{R_i} \bmod p$. After that, the value $c_2$ is calculated, with $u = \{x | x \bmod 2 = 0 \ and \ x <= n\}$ and $v = \{x | x \bmod 2 = 1 \ and \ x <= n\}$:

$$c_2 = \frac{(m * b^{R_2} * b^{R_4} * \ldots * b^{R_u})}{(b^{R_1} * b^{R_3} * \ldots * b^{R_v})} \bmod p \qquad (3)$$

The encrypted message now consists of all values $c_{1i}$ and $c_2$ and is send to the receiver in this form: $\{c_{11}, c_{12}, \ldots, c_{1n}, c_2\}$. For decrypting, m is calculated as in Eq. (4).

$$m = \frac{(c_2 * c_{11}{}^a * c_{13}{}^a * \ldots * c_{1v}{}^a)}{(c_{12}{}^a * c_{14}{}^a * \ldots * c_{1u}{}^a)} \bmod p \qquad (4)$$

The security of this El Gamal cryptosystem also relies on the Discrete Logarithm Problem, which is dependent on the size of the chosen prime number. A problem that occurred in [8] is the large increase in processing time when increasing the size of the prime numbers in steps of 100,000.

### 3.3   Multi Key Homomorphic Encryption

With Homomorphic Encryption, participants can perform calculations on encrypted data. This is particularly important when used in the context of blockchains, because the calculations can be executed on-chain. The data does not have to be decrypted first. The result of these calculations is again encrypted, which means that it must first be decrypted in order to obtain the information. The person who performs the calculations therefore knows neither the underlying data nor the result. Let $E$ be an encryption function and $F$ an operation that should be performed on the encrypted data $x_1, x_2$. The homomorphic encryption works as in Eq. (5).

$$F(E(x_1), E(x_2)) = E(F(x_1, x_2)) \tag{5}$$

The Multi Key Somewhat Homomorphic Encryption variant as presented in [7] offers the additional feature that calculations can be performed on encrypted texts that have been encrypted with different keys. Somewhat Homomorphic means that the operations addition and multiplication can be performed several times on a single data set. Given is the security parameter $\kappa$, a prime number $q$, a polynomial $\phi(x)$ of degree $n$ and an error distribution $X = X(\kappa)$ over the ring $R = \mathbb{Z}[x]/\langle\phi(x)\rangle$. Calculations are performed in the ring $R$ and with the message space $M = \{0, 1\}$. The parameters $\phi, n, q$ and $X$ are public. $f$ and $g$ need to be ring elements, i.e. $B$-bounded polynomials sampled from the distribution $X$ with $B = \text{poly}(n)$. $f^{-1}$ represents the inverse of $f$ in $R_q$. To generate the public key $(pk)$ and the private key $(sk)$, the polynomials $f'$ and $g$ need to be sampled from $X$. Then set $f = 2f' + 1$ to achieve $f \equiv 1 \, (mod \, 2)$. In the case, that $f$ is not invertible in $R_q$, it needs to be resampled. Then $pk$ and $sk$ are defined as follows:

$$pk = h = [2gf^{-1}]_q \in R_q, \; sk = f \in R_q \tag{6}$$

For encryption, the samples $s, e$ from $X$ are used to calculate the ciphertext

$$c = hs + 2e + m \in R_q \tag{7}$$

To decrypt the message, the private keys $sk_i$ from all participants $N$ are taken into consideration. The private keys are multiplied with the ciphertext and result in the original message:

$$\mu = [f_1 * f_2 * \ldots * f_N * c]_q \in R_q, \; m = \mu \, mod \, 2 \tag{8}$$

With the evaluate function two elements in the message space can be added or multiplied homomorphically, where $c$ and $c'$ are ciphertexts encrypted under different public keys $K$ and $K'$.

$$c_{add} = [c + c']_q \in R_q \tag{9}$$

The result of $c_{add}$ represents the encrypted sum of the decrypted values from $c$ and $c'$. This method can be applied if calculations are to be performed on-chain, where the result can only be decrypted in the collective. The result can only be extracted if all those involved give their consent, so to speak. The security of the presented Multi Key Homomorphic Cryptosystem relies on the Decisional Small Polynomial Ratio (DSPR) Assumption and on the Ring Learning With Errors (RLWE) Assumption. The magnitude of the coefficients of the joint secret key is exponentially dependent on the number of keys, which leads to an a-priori upper bound on $n$ [7].

## 3.4   Multi Key Functional Encryption

Functional Encryption works similarly to Homomorphic Encryption, except that the result of the calculations performed is already decrypted. This is useful when a blockchain user wants to use encrypted data to perform calculations, should not know this data, but needs to know the result. Generally speaking, the field of Functional Encryption is very broad, as it includes, among other methods identity-based encryption, searchable encryption, broadcast encryption or attribute-based encryption [6]. Given an encryption function $E$ and an operation $F$, that is to be executed on the data $x_1, x_2$, functional encryption works as follows:

$$F(E(x_1), E(x_2)) = F(x_1, x_2) \tag{10}$$

The attribute-based encryption makes it possible to enable decryption only for participants that have certain attributes. In this case, these could be certain roles e.g. on different levels within a company. [3] presents a Multi-Authority Attribute Based Encryption scheme, that allows multiple authorities to manage attributes and to distribute the secret keys needed to decrypt messages. In [6] a scheme is provided, that allows a various number of participants to encrypt messages and an evaluator to perform operations on all of these ciphertexts to get a decrypted result. In the context of blockchain, it could be discussed whether the aforementioned authority is defined via a smart contract and the monitoring of the attributes takes place via this contract, in order to avoid corruption.

## 4   Future Work and Conclusion

Following this research, it would also be interesting to look at how the metrics like runtimes, key size, overhead etc. of the different algorithms behave for key generation, encryption and decryption. Accordingly, it would also be useful to classify the encryption schemes by how many exchanges need to be performed to execute the previously mentioned methods. It would also be useful to analyse different schemes in terms of applicability, as well as to highlight the fundamental differences regarding the application scenarios. In addition, a mapping of the different approaches to usual business scenarios should be provided. The different areas of Functional Encryption need to be looked at in more detail and analyzed in relation to their application in the blockchain.

Future work also needs to explore the practical application of the theoretical procedures presented here. Furthermore, a comparison of the different approaches with respect to the necessary power of the underlying Smart Contract language is of relevance, especially because some Smart Contract languages are non-Turing-complete [5].

Table 1 provides a summary of the presented methods for encryption and key exchange. The summary includes the underlying security mechanisms as well as the main features and drawbacks.

**Table 1.** Overview of the presented encryption methods.

| Method | Security | Main features | Drawback |
|---|---|---|---|
| Multiple user Diffie Hellman | Discrete Logarithm Problem (DLP) | Key exchange between multiple users | As well as the classical approach, the scheme is vulnerable to man-in-the-middle-attacks |
| Multiple user El Gamal | Discrete Logarithm Problem (DLP) | Encryption of a message by a group of senders, decryption by a single receiver | Large increase of processing times for encryption and decryption when using large prime numbers |
| Multi key homomorphic encryption | RLWE assumption, DSPR assumption | Calculations can be performed on encrypted texts, that are encrypted under different keys | An upper bound on the number of participating keys is probably needed |

This paper has listed various encryption methods that can be used in the context of exchanging data on a blockchain. To promote the use of blockchain and increase trust in the storage of data, it is beneficial to use trustworthy and known methods to encrypt the affected data. However, there are various use cases that require the use of different encryption methods which have been presented and discussed in this paper. Moreover, this paper raises several additional research questions regarding the multiple user encryption in context of blockchains, that will be subject of further research.

# References

1. Andreeva, E., Mennink, B., Preneel, B.: Open problems in hash function security. Des. Codes Cryptogr., 611–631 (2015). https://doi.org/10.1007/s10623-015-0096-0
2. Diffie, W., Hellman, M.E.: New directions in cryptography. IEEE Trans. Inf. Theory **22**(6), 644–654 (1976)
3. Chase, M.: Multi-authority attribute based encryption. In: Theory of Cryp-tography, pp. 515–534. Springer Berlin Heidelberg, https://doi.org/10.1007/978-3-540-70936-7_28
4. Elgamal, T.: A public key cryptosystem and a signature scheme based on discrete logarithms. IEEE Trans. Inf. Theory **31**(4), 469–472 (1985). https://doi.org/10.1109/TIT.1985.1057074
5. Jansen, M., Hdhili, F., Gouiaa, R., Qasem, Z.: Do smart contract languages need to be turing complete? In: Prieto, J., Das, A.K., Ferretti, S., Pinto, A., Corchado, J.M. (eds.) BLOCKCHAIN 2019. AISC, vol. 1010, pp. 19–26. Springer, Cham (2020). https://doi.org/10.1007/978-3-030-23813-1_3
6. Libert, B., Țițiu, R.: Multi-client functional encryption for linear functions in the standard model from LWE. In: Galbraith, S.D., Moriai, S. (eds.) ASIACRYPT 2019. LNCS, vol. 11923, pp. 520–551. Springer, Cham (2019). https://doi.org/10.1007/978-3-030-34618-8_18
7. López-Alt, A., Tromer, E., Vaikuntanathan, V.: On-the-fly multiparty computation on the cloud via multikey fully homomorphic encryption. In: Proceedings of the 44th Symposium on Theory of Computing—STOC'12. ACM Press (2012). 10.1145/2213977.2214086

8. Ordonez, A.J., Medina, R.P., Gerardo, B.D.: Modified el Gamal algorithm for multiple senders and single receiver encryption. In: 2018 IEEE Symposium on Computer Applications and Industrial Electronics (ISCAIE). IEEE, 2018. https://doi.org/10.1109/iscaie.2018.8405470
9. Rivest, R.L., Shamir, A., Adleman, L.: A method for obtaining digital signatures and public-key cryptosystems. Commun. ACM **21**(2), 120–126 (1978). https://doi.org/10.1145/359340.359342

# Blockchain Adoption in the Energy Sector: A Comprehensive Regulatory Readiness Assessment Framework to Assess the Regulatory Readiness Levels of Countries

Karisma Karisma$^{(\boxtimes)}$ (iD) and Pardis Moslemzadeh Tehrani (iD)

University of Malaya, Kuala Lumpur, Malaysia
karisjay@outlook.com

**Abstract.** As we leverage blockchain technology, there is potential for it to become ubiquitous by advancing the operability and functionality of various sectors. Blockchain technology has gained significant scholarly attention in recent years, particularly due to its ability to manoeuvre and disrupt existing value chains and business models. Regulations play a primary role as the building blocks in successfully adopting blockchain technology. This research considers the legal and regulatory challenges circumjacent to blockchain-enabled energy solutions. It develops a set of regulatory criteria to proffer a comprehensive regulatory readiness assessment framework (RRAF) to assess the regulatory readiness levels (RRLs) of countries. It offers a valuable tool for evaluating the maturity, comprehensiveness, and efficacy of the existing regulatory mechanisms. The readiness and maturity levels of blockchain adoption in the energy sector can be evaluated in various jurisdictions using the RRAF model proposed in this paper. The RRAF model comprises multiple regulatory readiness criteria that can contribute to the practical and theoretical advancement of blockchain in the energy sector.

**Keywords:** Blockchain · Energy sector · Regulatory readiness · Governance

## 1 Introduction

Blockchain technology is gaining traction in the energy sector, given its potential to improve the digitalization process and optimize the operations of energy systems. The emergence of blockchain has drawn the interest of businesses, technology developers, and start-ups. In a literature review of the state-of-the-art technology in the energy sector, certain gaps can be identified, such as lack of transparency, increased operational costs, and the complexity of managing energy systems centrally. While blockchain technology appears to be a natural solution to these gaps, the adoption of this technology in the energy sector is rife with legal and regulatory challenges. These challenges have not been identified comprehensively in existing academic literature, despite the potential of the technology gaining momentum in the energy sector and its dependence on removing relevant barriers of entry. While there are a plethora of technological, operational, and

© The Author(s), under exclusive license to Springer Nature Switzerland AG 2023
J. Prieto et al. (Eds.): BLOCKCHAIN 2022, LNNS 595, pp. 454–460, 2023.
https://doi.org/10.1007/978-3-031-21229-1_42

societal challenges circumjacent to blockchain-based energy systems, this dissertation focuses on exploring the legal and regulatory challenges through a qualitative study. Hence, even with the adoption of blockchain by a growing number of countries, its potential is not fully exploited due to the legal and regulatory framework constraints. At this juncture, no academic literature on blockchain-based energy applications has comprehensively assessed these countries' regulatory readiness levels (RRLs). Based on the careful syntheses of academic studies, there is also an inadequate emphasis on the sufficiency and effectiveness of regulatory regimes in facilitating the adoption of blockchain technology in the energy sector. It is essential to assess the readiness of existing regulations to facilitate the adoption of blockchain technology as it can manoeuvre policy interventions. On the one hand, a widely worded technology-neutral regulatory framework might be able to address the challenges raised by the entry and operation of new technology and further provide a conducive and enabling landscape for its development and growth, thus having a high readiness level. On the other hand, the lack of clarity of regulations and laws may constitute an antithesis in the liberalization of the energy sector, thus resulting in low readiness levels. Therefore, due to the inchoate and nascent nature of the blockchain regulatory readiness framework, this work explores the readiness criteria and proposes a detailed Regulatory Readiness Assessment Framework (RRAF) that is carefully designed based on the criteria identified. The RRAF can then be used to explore and evaluate the RRLs of the selected countries.

Therefore, the key question is: *What regulatory readiness criteria are applicable in developing a detailed RRAF for blockchain-enabled energy trading?*

## 2  Related Work

Legal challenges raised in academic literature insufficiently address regulatory issues. Drawing from the core aim and in line with the research objective, it is pivotal to address the related work on these issues, which is a catalyst to the RRAF. This proposal delves into the following subsections that weave together and explore the widespread challenges circumjacent to blockchain-based energy systems.

### 2.1  The Multitude of Legal Challenges Surrounding the Application of Smart Contracts to Blockchain-Based Energy Systems

As identified in the literature, the concern is the legal recognition of smart contracts. Scholars propound the use of conventional contract law principles to determine the validity of smart contracts. Besides that, such contracts are practically "immutable" and cannot be altered, amended, or cancelled in light of the change in circumstances, such as when the contract terms are "commercially or physically impossible to fulfil" [1]. Besides that, smart contracts are formulated with "if-then" statements and cannot provide for force majeure clauses which are usually broadly phrased [1]. Henly et al. [2] state that it would be difficult to "encode a performance excuse clause that excuses seller from delivering energy based on circumstances beyond seller's reasonable control". Lee and Marie [3] raise the challenge of integrating smart contracts within the ambit of contract law. The prominent issue includes the difficulty in understanding and executing smart contracts, primarily written in the form of a code. There is a necessity to rely on those that are well versed in programming to ascertain whether smart contracts are properly coded. The recognition of smart contracts as legal contracts may create a barrier for those who are not "technologically inclined" [3]. Due to the inherent complexity of smart contracts, there is a need to assess further the legal challenges associated with the utilization of blockchain-based smart contracts in the energy sector. There is a lack of consensus from academic scholars on the regulation of smart contracts. The aspects of the scope, functioning, and legal recognition of smart contracts in the energy sector are considered critical legal issues. However, these legal challenges have not been thoroughly explored and reflected in the academic literature.

## 2.2  The Legal Issues Concerning Privacy and Data Protection Due to the Multidimensional Data Usage in Decentralized and Digitalized Blockchain-Based Energy Systems and Energy Trading

Scholars Andoni et al. and Ahl et al. [4, 5] discuss the data privacy breaches of participants actively partaking in blockchain-enabled energy applications, mainly due to the digitalization of energy systems. Andoni et al. [4] stipulate that even though the deployment of blockchain technology for data management purposes may improve smart metering, as every participant of the public blockchain network has access to the blockchain ledger databases, it may raise an array of privacy challenges. As stated by Thukral [6], while the General Data Protection Regulation (GDPR) provides that data controllers are required to implement suitable "technical and organizational measures" to ensure data protection, blockchain technology eliminates the need for centralized authorities or third-party intermediaries. Conversely, scholar Finck [7] states that in a private blockchain, it is likely to identify a data controller. However, in a public blockchain, "either no node qualifies as the data controller […] or, more likely, every node qualifies as a data controller." It depends on the absence or presence of self-determination of "purposes and means of the processing" of the data [7]. Scholars Lee and Marie [3] stipulate that peer-to-peer energy trading platforms can amount to data controllers by ascertaining the "means and purpose of processing energy data," thus invoking the application of the GDPR. Scholars have varying views regarding the recognition of data controllers in blockchain-based energy applications and the qualification of parties as data controllers. The lack of legal certainty may induce various legal challenges and implications in coherently applying the GDPR. There is a lack of comprehensive discussion on the legal challenges emanating from the myriad of uncertainties. It is pivotal to bridge the gap by providing a more wholesome overview of this aspect.

## 2.3  The Legal Implications of Security Breaches in Blockchain-Based Energy Applications

There are seminal contributions made to the resilience of blockchain technology. Blockchain technology is a feasible solution to address security issues due to its decentralized nature, consensus mechanisms, and cryptographic hash functions [4]. Nevertheless, blockchain technology can be deemed as a double-edged sword. Previous studies have shown that blockchain technology is susceptible to various security risks, such as

51% attacks and denial of service attacks [5, 6]. Recent research by Andoni et al. [4] suggests that the security risks emanate primarily from blockchain design flaws, insufficient experience in the development and implementation of intricate forms of blockchain applications coupled with the naivety of technology resulting in system malfunctions, and the lack of legislation and regulations to address the security threats. There is a lack of discussion in academic literature regarding these vulnerabilities, leading to various legal implications. With the emphasis on security in the energy sector, exploring these security threats in greater detail is essential.

## 2.4 The Legal Challenges Arising from the Lack of Licensing Requirements for Blockchain-Enabled Energy Transactions

Prior research suggests that subjecting prosumers who engage in peer-to-peer energy trading to licensing requirements similar to those imposed on energy companies may be unfeasible and inappropriate [8]. Due to the licensing requirements, prosumers may be required to deliver a regular and uninterrupted supply of electricity to meet the energy loads of the other party [8]. In furtherance, "business-to-consumer (B2C)" requirements as prescribed by legislative provisions can similarly be imposed on prosumers, which may hinder the growth of peer-to-peer energy trading systems [9]. On the contrary, with the development of peer-to-peer trading platforms and increased energy supply to energy consumers, it has to be evaluated whether the waiver of licensing requirements may create an "uneven playing field" [9]. This dissertation systematically addresses various uncertainties emanating from prosumer licensing.

## 2.5 The Legal Issues in the Creation and Evolution of Business Models in a Heavily Regulated Energy Sector, and the Deregulation and Liberalization of Energy Markets

There is a shift in approach from merely incorporating prosumers in existing business structures to empowering them to construct new business models [10]. As conventional utility companies cease to exist, blockchain-based business models thrive [5, 11]. In the absence of intermediaries, the question arises of the role, legal status, and obligations of blockchain-based energy platforms that facilitate energy trade between peers [9]. However, as elucidated by Henly et al. [2], it may be impractical and unfeasible to "cut the utility out" due to the considerable costs to "develop, operate, and maintain any distribution infrastructure" for the transmission of energy. There may be occurrences where conventional utility companies innovate and adapt their business models [11]. Emanating from that point, regulators have to account for the evolution of utility business models, which play a vital function in the growth and implementation of distributed energy resources [2].

## 2.6 Lack of Legal and Regulatory Readiness Framework/Maturity Model to Evaluate and Access Blockchain Technology

At this juncture, there is a deficit of regulations concerning blockchain technology in the energy sector. Blockchain-based energy applications operate in the face of strongly regulated energy landscapes. These applications have the "potential to extract greater value from existing regulatory regimes" and can be introduced or implemented within the existing regulatory framework [2, 12]. In certain circumstances, the deployment of blockchain to energy applications merely requires minor amendments to the regulatory framework. In the event of disruptive technological innovations, it may require wholesale reforms to existing regulations that may prove incompatible with blockchain-based energy applications [12]. A comprehensive readiness assessment is essential to deploy new technology successfully [13]. Nevertheless, research in this area is still in its infancy. There is a deficit of academic studies on the blockchain readiness framework [13]. In what follows, this dissertation will develop a RRAF that offers a holistic qualitative assessment of domestic legal and regulatory frameworks of various jurisdictions and evaluate whether they are sufficient and efficacious in (a) providing an answer to the legal challenges posed by the adoption of blockchain in the energy sector, and (b) facilitating the adoption of blockchain-enabled energy applications. In the preceding subsections, we have unlocked the current state-of-the-art of legal and regulatory challenges that hinder the widespread implementation of blockchain-based energy systems to craft and integrate readiness trajectories.

## 3  Hypothesis

The development of regulatory readiness criteria is fundamental for RRAF in determining the state of regulatory readiness of countries. These criteria include examining the current state of blockchain technology in the energy sector and the legal and regulatory challenges that warrant determining the level of regulatory readiness to bridge these challenges. In what follows, the RRAF adopts the regulatory readiness criteria to guide policymakers and regulators in creating a more conducive environment for blockchain technology adoption, including determining the adequacy and effectiveness of regulatory regimes in their countries.

## 4  Proposal

This dissertation aims to develop a multi-dimensional, all-inclusive RRAF in mainstreaming blockchain-based energy solutions. The RRAF encourages increased adoption of blockchain-based energy applications and transparency, clarity, and accountability of multi-stakeholders by providing a systematic and methodical architecture to regulate conduct and energy activities. The author develops comprehensive assessment parameters and practical tools for analysing the RRLs of countries by embracing an integrated RRAF with a complementary blend of facilitative modalities comprising three themes. While the author aims to carve a path through the RRAF that shapes the notion of regulatory models for blockchain-based energy solutions, it is by no means an exhaustive framework. At the forefront of the RRAF and in devising an authentic, all-encompassing, and rigorous framework, the author uses correlated themes to facilitate a complex academic inquiry. Given the complexity of the assessment, it is valuable to carefully design and contextualize the constituent elements in the proposed RRAF through different lenses. Having identified the prevailing themes and trajectories, presented as salient forces in this dissertation within the bounds of RRAF, circumjacent to blockchain-based fully peer-to-peer and community models, the author explores the (a) objective and merit, (b) feasibility, tenability, and viability, and (c) proposed design desiderata, including the indicators of the individual themes that can bridge the legal and regulatory gaps, enrich the analysis, and ameliorate policy strategies.

## 5 Preliminary Results and/or Evaluation Plan

As a starting point for the writing of this dissertation, the author develops common denominators, which are the essential matters circumjacent to blockchain-driven energy solutions, such as (a) trust-related framework, (b) industry-specific framework (including but not limited to the role of market actors, fiscal and non-fiscal incentives, licensing requirements, market access and authorization requirements, energy market structures, public utility regulations, and policy processes, amongst others), and (c) digital platform and electronic transactions framework, which emerge as dominant factors for successfully deploying blockchain-based energy solutions. These facilitative modalities are key enablers of rapid blockchain diffusion. They address the prevailing limits, barriers, and challenges by drawing meaningful parameters on accountability, data integrity and protection, and institutional and structural insights. More to the point, in postulating a multitude of indicators and sub-indicators for the RRAF, we consider the importance of infallibility and precision achieved through a broadly balanced framework that is widely representative. The focal point is to advance precision in the RRAF formulation by addressing the demands of 'transparency,' 'accessibility,' and 'congruence.' To achieve the quality of transparency is to emphasize clearly defined, comprehensible, and widely acknowledged indicators.

## 6 Reflections

In encouraging countries to participate fully and efficiently in the blockchain fora, countries should employ effective governance strategies to ensure resilient hard law and soft law instruments to resolve any shortfalls in existing regimes and strengthen blockchain adoption. The RRAF, with various facilitative modalities, can advance a systematic and methodical architecture in regulating blockchain-enabled energy activities. In essence, the author proposes the development of the RRAF to guide policymakers and other stakeholders in examining the RRLs of countries. This guide would encourage the enactment of comprehensive policies and regulations to increase legal readiness and facilitate blockchain adoption. This research targets a problem that, once addressed, can contribute to and be relevant for most countries worldwide that are actively adopting blockchain technology in the energy industry. Hence, the rise in technology readiness levels necessitates the adoption of RRLs in the relevant countries.

**Acknowledgements.** The research for this publication is supported by a research grant from Quanta RegTech Capital PLC, International Funding of the University of Malaya under Grant Number IF057B-2018.

# References

1. Schneiders, A., Shipworth, D.: Community energy groups: can they shield consumers from the risks of using blockchain for peer-to-peer energy trading? Energies **14**, 3569 (2021)
2. Henly, C., Hartnett, S., Mardell, S., Endemann, B., Tejblum, B., Cohen, D.S.: Energizing the future with blockchain. Energy L. J. **39**, 197–232 (2018)
3. Lee, J., Khan, V.M.: Blockchain and smart contract for peer-to-peer energy trading platform: legal obstacles and regulatory solutions. UIC Rev. Intell. Prop. L. **19**, 285–308 (2019)
4. Andoni, M., et al.: Blockchain technology in the energy sector: a systematic review of challenges and opportunities. Renew. Sust. Energ. Rev. **100**, 143–174 (2019)
5. Ahl, A., et al.: Exploring blockchain for the energy transition: opportunities and challenges based on a case study in Japan. Renew. Sust. Energ. Rev. **117**, 109488 (2020)
6. Thukral, M.K.: Emergence of blockchain-technology application in peer-to-peer electrical-energy trading: a review. Clean Energy **5**, 104–123 (2021)
7. Finck, M.: Blockchains and data protection in the European Union. Eur. Data Prot. L. Rev. **4**, 17–35 (2018)
8. Ahl, A., Yarime, M., Tanaka, K., Sagawa, D.: Review of blockchain-based distributed energy: implications for institutional development. Renew. Sust. Energ. Rev. **107**, 200–211 (2019)
9. Lavrijssen, S., Parra, A.C.: Radical prosumer innovations in the electricity sector and the impact on prosumer regulation. Sustainability **9**, 1207 (2017)
10. Diestelmeier, L.: Changing power: shifting the role of electricity consumers with blockchain technology—policy implications for EU electricity law. Energy Policy **128**, 189–196 (2019)
11. Mengelkamp, E., Gärttner, J., Rock, K., Kessler, S., Orsini, L., Weinhardt, C.: Designing microgrid energy markets: a case study: the Brooklyn microgrid. Appl. Energy **210**, 870–880 (2018)
12. Amenta, C., Sanseverino, E.R., Stagnaro, C.: Regulating blockchain for sustainability? The critical relationship between digital innovation, regulation, and electricity governance. Energy Res. Soc. Sci. **76**, 102060 (2021)
13. Balasubramanian, S., Shukla, V., Sethi, J.S., Islam, N., Saloum, R.: A readiness assessment framework for blockchain adoption: a healthcare case study. Technol. Forecast. Soc. Change **165**, 120536 (2021)

# Author Index

Printed in the United States
by Baker & Taylor Publisher Services